FUNDAMENTALS OF
OCEANOGRAPHY
FOURTH EDITION

ALISON B. DUXBURY
Seattle Community College

ALYN C. DUXBURY
University of Washington

KEITH A. SVERDRUP
University of Wisconsin–Milwaukee

Mc
Graw
Hill

Boston   Burr Ridge, IL   Dubuque, IA   Madison, WI   New York   San Francisco   St. Louis
Bangkok   Bogotá   Caracas   Kuala Lumpur   Lisbon   London   Madrid   Mexico City
Milan   Montreal   New Delhi   Santiago   Seoul   Singapore   Sydney   Taipei   Toronto

McGraw-Hill Higher Education

A Division of The **McGraw-Hill** *Companies*

FUNDAMENTALS OF OCEANOGRAPHY, FOURTH EDITION

Published by McGraw-Hill, a business unit of The McGraw-Hill Companies, Inc. 1221 Avenue of the Americas, New York, NY 10020. Copyright © 2002, 1999, 1996, 1992 by The McGraw-Hill Companies, Inc. All rights reserved. No part of this publication may be reproduced or distributed in any form or by any means, or stored in a database or retrieval system, without the prior written consent of The McGraw-Hill Companies, Inc., including, but not limited to, in any network or other electronic storage or transmission, or broadcast for distance learning.

Some ancillaries, including electronic and print components, may not be available to customers outside the United States.

This book is printed on recycled, acid-free paper containing 10% postconsumer waste.

1 2 3 4 5 6 7 8 9 0 QPD/QPD 0 9 8 7 6 5 4 3 2 1

ISBN 0–07–242790–6

Executive editor: *Margaret J. Kemp*
Developmental editor: *Renee Russian*
Executive marketing manager: *Lisa L. Gottschalk*
Project manager: *Shelia M. Frank*
Production supervisor: *Kara Kudronowicz*
Coordinator of freelance design: *David W. Hash*
Cover designer: *Nathan Bahls*
Cover image: © *Rick Doyle/CORBIS*
Senior photo research coordinator: *Lori Hancock*
Photo research: *Alexandra Truitt & Jerry Marshall,*
www.pictureresearching.com<http://www.pictureresearching.com>
Supplement produce: *Sandra M. Schnee*
Media technology producer: *Judi David*
Compositor: *Precision Graphics*
Typeface: *10/12 Cochin*
Printer: *Quebecor World Dubuque, IA*

The credits section for this book begins on page 337 and is considered an extention of the copyright page.

Library of Congress Cataloging-in-Publication Data

Duxbury, Alison B.
   Fundamentals of oceanography—4th ed. / Alison B. Duxbury, Alyn C. Duxbury.
   Keith A. Sverdrup.
      p. cm.
   Includes index.
   ISBN 0–07–242790–6
   1. Oceanography.   I. Duxbury, Alyn C., 1932– .   II. Sverdrup, Keith A.   III. Title.

GC11.2.D88   2002
551.46–dc21   2001030921
                                                                CIP

www.mhhe.com

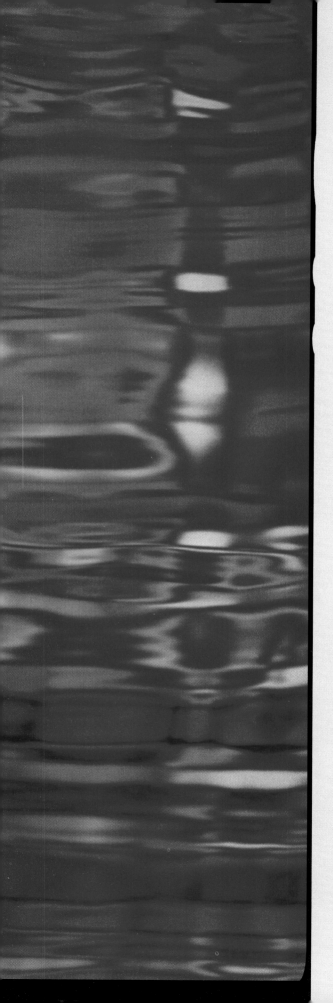

We shall not cease from
exploration
And the end of all our
exploring
Will be to arrive where we
started
And know the place for the
first time.
Through the unknown,
remembered gate
When the last of earth left
to discover
Is that which was the
beginning;
At the source of the longest
river
The voice of a hidden
waterfall . . .
Not known, because not
looked for,
But heard, half heard, in
the stillness
Between two waves of the sea.
T.S. Eliot
From "Little Gidding," The
Four Quartets

contents

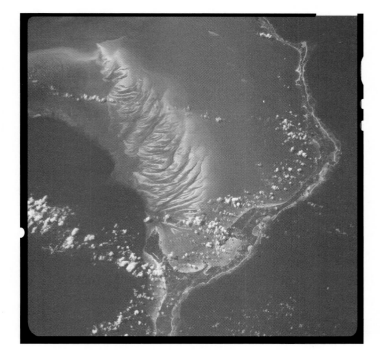

Waves and Tides

Coasts, Estuaries, and Environmental Issues

Oceanic Environment and Production

chapter eleven

Life in the Water

chapter twelve

Life on the Sea Floor

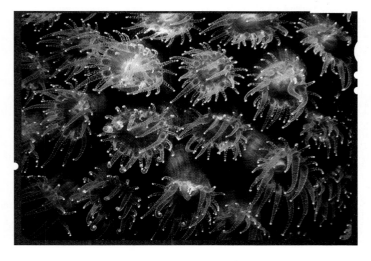

preface

undamentals of Oceanography is intended for professors and students who need a more basic oceanography text to better serve less intensive college oceanography courses, courses tailored for nonscience majors interested in learning more about the fascinating, and often mysterious, marine world that covers 71% of the Earth's surface and yet is more poorly mapped than the surface of the Moon. This fourth edition is an extensive revision in response to the suggestions and ideas of instructors as well as advances in the ocean sciences. Our goal in this edition is to provide students with up-to-date information, and to make each chapter as clear and readable as possible without sacrificing scientific accuracy. This fourth edition continues to emphasize principles, processes, and properties of the oceans.

Because oceanography embraces immense amounts of geological, physical, chemical, biological, and engineering information related to the marine environment, and because of the interdependency among these subject areas, the choice of topics to be included in a fundamentals text presents a complex challenge. We have endeavored to choose those topics that best illustrate basic processes and at the same time answer students' questions about the oceans while encouraging their interest. We invite instructors to change the sequence of material to best fit their own presentations and to elaborate on subjects as desired.

Six new *Items of Interest* have been added to this edition for a total of twelve topics designed to cover issues that are not addressed directly in the text. The new topics include the recovery of black smokers from the seafloor, the use of sound to measure ocean temperature, tsunami warning systems, national ma-

rine sanctuaries, whale falls, and biodiversity in the oceans. The discussion of plate tectonics and plate boundaries in chapter 3 has been extensively revised. In chapter 4 the material concerning marine sediments has been reorganized to emphasize sediment classification and characteristics. New measurement techniques and new instruments used by oceanographers are discussed and illustrated in chapter 7. The environmental problems presented by "dead zones" and the toxic *Pfiesteria* organism are new to chapter 9, and chapter 10 has a new section on extremophiles, microorganisms that live under conditions of extreme temperature. Chapter 12 provides new data on the world's coral reefs, and all fisheries catch data has been updated in chapters 11 and 12. Short essays on the responsibilities of a chief scientist and ship's Captain in planning and executing a successful oceanographic expedition have been included in the middle of the text.

Fundamentals of Oceanography continues to present students with numerous aids to facilitate their study of oceanography. Each chapter opens with learning objectives, and review questions are presented as self-checks for the student at the end of each section. Throughout this text, information is presented in table form to help the student organize, summarize, and compare. Chapters end with a concise summary to aid in review, critical thinking questions to encourage reflection about chapter topics, suggested readings to explore subjects in more detail, and internet references related to topics of discusssion. Internet references have also been added to figure captions in the text where appropriate. Three appendices are included: (1) methods of deriving latitude and longitude, (2) taxonomic

classifications of plankton, nekton, and benthos, and (3) scientific notation and units. All quantities in the text are given in both metric and traditional units.

We acknowledge that this book is a product of many experiences, in the field, at sea, and in the classroom. We extend our thanks to many friends and colleagues who have graciously answered our questions, helped us with current information, and provided access to their photo files. We particularly thank:

Marcia McNutt
Monterey Bay Aquarium Research Institute

Ian Young
Monterey Bay Aquarium Research Institute

Steve Nathan
University of Massachusetts

We also wish to extend a special thanks to the instructors who used and reviewed the first three editions of this text and to those who contributed to the development of this fourth edition.

Christopher T. Baldwin
Sam Houston State University

Heather Gallacher
Cleveland State University

Lawrence A. Krissek
Ohio State University

Donald W. Lovejoy
Palm Beach Atlantic College

Leslie A. Melim
Western Illinois University

Donald F. Palmer
Kent State University

Keith Simmons
Hartnell College

Finally, we express our sincere gratitude to McGraw-Hill and the outstanding staff members working with us to bring you this textbook.

History of Oceanography

Where are your monuments, your battles, martyrs?
Where is your tribal memory? Sirs,
in that grey vault. The sea. The sea
has locked them up. The sea is History.

Derek Walcott

from The Sea is History

By the seventeenth century significantly improved charts and navigation instruments were available to mariners.

Learning Objectives

After reading this chapter, you should be able to

- Understand the diversity of the sciences collected to form "oceanography."
- Understand the development of oceanography as a science.
- Follow the development of ocean knowledge from early voyages of exploration and discovery.
- Understand the significance of navigation in describing the oceans and making accurate maps of their extent and boundaries.
- Recognize the contributions of early U.S. oceanography to the development of marine commerce.
- Understand the role of early scientific voyages in the investigation of the world's oceans.
- Recognize the role of U.S. agencies in developing oceanography before and after World War II.
- Recognize the relationship between technology, international cooperation, and the development of recent large-scale oceanographic programs.
- Discuss the programs being planned for the future.

Marine Archaeology

Oceanography is a broad field in which many sciences focus on the common goal of understanding the oceans. Geology, geography, geophysics, physics, chemistry, geochemistry, mathematics, meteorology, botany, and zoology all play roles in expanding our knowledge of the oceans. Geological oceanography includes the study of the Earth at the sea's edge and below its surface and the history of the processes that formed the ocean basins. Physical oceanography investigates how and why the oceans move; marine meteorology, the study of heat transfer, water cycles, and air-sea interactions, is often included in this discipline. Chemical oceanography studies the composition and history of seawater, its processes, and its interactions. Biological oceanography concerns itself with the marine organisms and the relationship between these organisms and the environment of the oceans. Ocean engineering is the discipline of designing and planning equipment and installations for use at sea.

Our progress toward the goal of understanding the oceans has been uneven, and it has frequently changed direction. The interests and needs of nations as well as the intellectual curiosity of scientists have controlled the rate at which we study the oceans, the methods we use to study them, and the priority we give to certain areas of study. To gain some perspective on the current state of knowledge about the oceans, we need to know something of the events and incentives that guided previous investigations of the oceans.

The Early Times   1.1

People have been gathering information about the oceans for millennia, passing it on by word of mouth. Curious individuals must have acquired their first ideas of the oceans from wandering the seashore, wading in the shallows, and gathering food from the ocean's edges. As early humans moved slowly away from their inland centers of development, they took advantage of the sea's food sources when they first explored and later settled along the ocean shore. The remains of shells and other refuse found at the sites of ancient shore settlements show that our early ancestors gathered shellfish, and certain fish bones suggest that they also began to use rafts or some type of boat for offshore fishing.

Early information about the oceans was mainly collected by explorers and traders. These voyages left little in the way of recorded information. Using descriptions passed down from one voyager to another, early sailors piloted their way from one landmark to another, sailing close to shore and often bringing their boats up onto the beach each night.

Some historians believe that seagoing ships of all kinds are derived from early Egyptian vessels. The first recorded voyage by sea was led by Pharaoh Snefru about 3200 B.C. In 2750 B.C. Hannu led the earliest documented exploring expedition from Egypt to the southern edge of the Arabian Peninsula and the Red Sea.

The Phoenicians, who lived in present-day Lebanon from about 1200 B.C. to 146 B.C., were well known as excellent sailors and navigators. While their land was fertile it was densely populated so they were compelled to engage in trade with others to acquire many of the goods they needed. They accomplished this by establishing land routes to the East and marine routes to the West. The Phoenicians were the only nation in the region at that time who had a navy. They traded throughout the Mediterranean Sea with the inhabitants of North Africa, Italy, Greece, France, and Spain. They also ventured out of the Mediterranean Sea to travel north along the coast of Europe to the British Isles and south to circumnavigate Africa in about 590 B.C. In 1999 the wreckage of two Phoenician cargo vessels circa 750 B.C. was explored using remotely operated vehicles (ROVs) that could dive to the wreckage and send back live video images of the ships. The ships were discovered about 48 km (30 miles) off the coast of Israel at depths of 300 to 900 m (roughly 1000–3000 ft).

Extensive migration throughout the Southwestern Pacific may have begun by 2500 B.C. These early voyages were relatively easy because of the comparatively short distance between islands in the far Southwestern Pacific region. By 1500 B.C. the Polynesians had begun more extensive voyages to the east where the distance between islands grew from tens of miles at the edge of the western Pacific to thousands of miles in the case of voyages to the Hawaiian Islands. They successfully reached and colonized the Hawaiian Islands sometime between A.D. 450 and 600. By the eighth century A.D., they had colonized every habitable island in a triangular region roughly twice the size of the United States bound by

Hawaii on the north, New Zealand in the southwest, and Easter Island to the east.

A basic component of navigation throughout the Pacific was the careful observation and recording of where prominent stars rise and set along the horizon. Observed near the equator, the stars appear to rotate from east to west around the Earth on a north-south axis. Some rise and set farther to the north and some farther to the south, and they do so at different times. Navigators created a "star structure," dividing the horizon into 32 points where the stars for which the points are named rise and set. These points form a compass that provides a reference for recording information about the direction of winds, currents, waves, and the relative positions of islands, shoals, and reefs (fig. 1.1). The Polynesians also navigated by making close observations of waves and cloud formations. Observations of birds and the distinctive smells of land such as flowers and wood smoke alerted them to possible landfalls. Once islands had been discovered, their locations relative to one another and to the regular patterns of sea swell and waves bent around islands could be recorded with stick charts constructed of bamboo and shells (fig. 1.2).

As early as 1500 B.C. Arabs of many different ethnic groups and regions were exploring the Indian Ocean. In the seventh century A.D. they were unified under Islam and began to control the trade routes to India and China and consequently the commerce in silk, spices, and other valuable goods (this monopoly wasn't broken until Vasco da Gama defeated the Arab fleet in 1502).

These early sailors did not investigate the oceans; for them the sea was only a dangerous road, a pathway from here to there. This situation continued for hundreds of years. However, the information that they accumulated became a body of lore to which sailors and voyagers added from year to year.

While the Greeks traded and warred throughout the Mediterranean, they observed and also asked themselves questions about the sea. Aristotle (384–322 B.C.) believed that the ocean occupied the deepest parts of the Earth's surface; he knew that the Sun evaporated water from the sea surface, which condensed and returned as rain. He also began to catalog marine organisms. The brilliant Eratosthenes (c. 265–194 B.C.) of Alexandria, Egypt, mapped his known world and calculated the circumference of the Earth to be about 40,250 kilometers (km) or 25,000 miles (mi) (today's measurement is 40,067 km or 24,881 mi). Posidonius (c. 135–50 B.C.) reportedly measured an ocean depth to about 1800 meters (6000 ft) near the island of Sardinia, according to the Greek geographer Strabo (c. 63 B.C.–c. A.D. 21). Pliny the Elder (A.D. 23–79) related the phases of the Moon to the tides and reported on the currents moving through the Strait of Gibraltar. Ptolemy, in A.D. 127–151, produced the first world atlas and established world boundaries: to the north the British Isles, northern Europe, and the unknown lands of Asia; to the south an unknown land, "Terra Australis Incognita," including Ethiopia, Libya, and the Indian Sea; to

Figure 1.1

A traditional navigator from the Caroline Islands instructing others in the use of a "star structure" showing the relative points on the horizon where key stars rise and set.

More Information: www.museum.upenn.edu/Navigation/Misc/contents.html

Figure 1.2

A navigational chart *(rebillib)* of the Marshall Islands. Sticks represent a series of regular wave patterns (swells). Curved sticks show waves bent by the shorelines of individual islands. Islands are represented by shells.

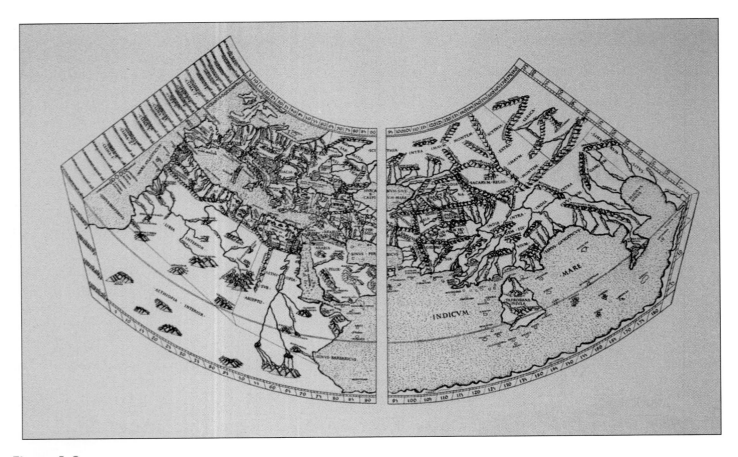

Figure 1.3

A chart from an Italian fifteenth-century edition of Ptolemy's *Geographia*.

the east China; and to the west the great Western Ocean reaching around the Earth to China on the other side (fig. 1.3). His atlas listed more than 8000 places by latitude and longitude, but his work contained a major error: he had accepted a value of 29,000 km (18,000 mi) for the Earth's circumference. This shortened Earth distances and allowed Columbus, more than a thousand years later, to believe that he had reached the eastern shore of Asia when he landed in the Americas.

Name the subfields of oceanography.

What did early sailors use for guidance during long ocean voyages?

What kind of "compass" did the Polynesians use for navigation?

How long ago was the circumference of the Earth first calculated?

How did Ptolemy's atlas contribute to a greater understanding of world geography, and how did it produce confusion?

The Middle Ages 1.2

After Ptolemy, intellectual activity and scientific thought declined in Europe for about one thousand years. However, shipbuilding improved during this period; vessels became more seaworthy and easier to sail allowing sailors to extend their voyages. The Vikings (Norse for *piracy*) were highly accomplished seamen who engaged in extensive exploration, trade, and colonization for nearly three centuries from about 793 to 1066 (fig. 1.4). They sailed to Iceland in 871 where it is believed that as many as 12,000 immigrants eventually settled. During this time they also journeyed inland on Russian rivers throughout central and eastern Europe and western Asia. Voyages into the Mediterranean Sea led them to Rome and Baghdad. Erik Thorvaldsson (known as Erik the Red) sailed west from Iceland in 982 and discovered Greenland. He lived there for three years before returning to Iceland to recruit more settlers. Icelander Bjarni Herjolfsson, on his way to Greenland to join the colonists in 985–6, was blown off course, sailed south of Greenland, and is believed to have come within sight of Newfoundland before turning back and reaching Greenland. Leif Eriksson, son of Erik the Red, sailed west from Greenland

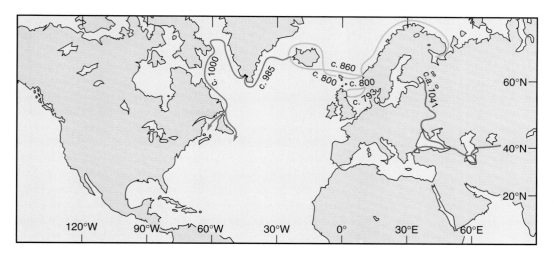

Viking Routes
Earliest   Leif Eriksson
Erik the Red   Ingvar

Figure 1.4

Major routes of the Vikings to the British Isles, Asia, and across the Atlantic to Iceland, Greenland, and North America.

More Information:
www.mariner.org/age/vikingexp.html

in 1002 and reached North America roughly 500 years before Columbus.

To the south, in the Mediterranean region after the fall of the Roman Empire, the Arabs preserved the knowledge of the Greeks and the Romans, on which they continued to build. The Arabic writer El-Mas'údé (d. 956) gave the first description of the reversal of the ocean currents due to the seasonal monsoon winds. Using this knowledge of winds and currents, the Arabs established regular trade routes across the Indian Ocean. In the 1200s large Chinese junks with crews of 200 to 300 sailed the same routes (between China and the Persian Gulf) as the Arab dhows.

During the Middle Ages, while scholarship about the sea remained static, the knowledge of navigation increased. Harbor-finding charts, or *portolanos*, appeared. These charts carried a distance scale and noted hazards to navigation, but they did not have latitude or longitude. With the introduction of the magnetic compass to Europe from Asia in the thirteenth century, compass directions were added. One example, a Dutch navigational chart from Johannes van Keulen's *Great New and Improved Sea-Atlas or Water-World* of 1682–84, is shown in figure 1.5.

As scholarship was reestablished in Europe, Arabic translations of early Greek studies were translated into Latin, which made them again available to northern European scholars. By the 1300s, Europeans had established successful trade routes, including some partial ocean crossings. An appreciation of the importance of navigational techniques grew as trade routes were extended.

> What advances occurred during the Middle Ages that allowed longer ocean voyages?
>
> During the tenth century, which oceans were explored and by what people?
>
> Where did the Vikings establish a large colony in the North Atlantic?

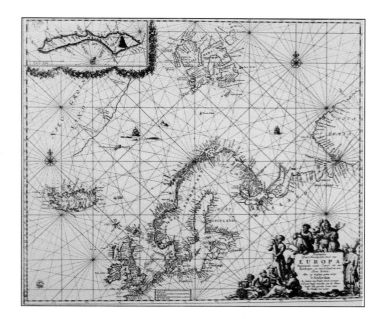

Figure 1.5

A navigational chart of northern Europe from Johannes van Keulen's *Sea-Atlas* of 1682–84.

Voyages of Discovery   1.3

Early in the fifteenth century the Chinese organized seven voyages to explore the Pacific and Indian Oceans. More than 300 ships, one more than 122 m (400 ft) long, participated in these ventures to extend Chinese influence and demonstrate the power of the Ming dynasty. The voyages ended in 1433 when their explorations led the Chinese to believe that other societies had little to offer, and the government of China withdrew within its borders beginning a 400-year period of isolation.

In Europe the desire for riches from new lands persuaded wealthy individuals, often representing their countries, to underwrite the costs of long voyages to all the oceans of the world. The individual most responsible for the great age of European discovery was Prince Henry the Navigator (1394–1460) of Portugal. He established a naval observatory for the teaching of navigation, astronomy, and cartography about 1450. Prince Henry sent expedition after expedition down the west coast of Africa to secure trade routes and to establish colonies. Bartholomeu Dias (1450?–1500) rounded the Cape of Good Hope in 1487 in the first of the great voyages of discovery (fig. 1.6). Dias sailed in search of new and faster routes to the spices and silks of the East.

Portugal's slow progress down the west coast of Africa in search for a route to the east finally came to fruition with Vasco da Gama (1469–1524). In 1498 he followed Bartholomeu Dias' route to the Cape of Good Hope and then continued beyond along the eastern coast of the African continent. He successfully mapped a route to India but was challenged along the way by Arab ships. In 1502, da Gama returned with a flotilla of 14 heavily armed ships and defeated the Arab fleet. By 1511, the Portuguese mastered the spice routes and had access to the Spice Islands. In 1513, Portuguese trade extended to China and Japan.

Christopher Columbus (1451–1506) made four voyages across the Atlantic Ocean in an effort to find a new route to the East Indies by traveling west rather than east. By relying on inaccurate estimates of the Earth's size he badly underestimated the distances involved and believed he had found islands off the coast of Asia when he reached the New World. The Italian navigator Amerigo Vespucci (1454–1512) made several voyages to the New World (1499–1504) for Spain and Portugal; he accepted South America as a new continent, not part of Asia. In 1507, the German cartographer Martin Waldseemüller applied the name "America" to the continent in Vespucci's honor. Vasco Nuñez de Balboa (1475–1519) crossed the Isthmus of Panama and found the Pacific Ocean in 1513. All claimed the new lands they found for their home countries. Although they had sailed for riches, not knowledge, they more accurately documented the extent and properties of the oceans, and the news of their travels stimulated others to follow.

Ferdinand Magellan (1480–1521) left Spain in September, 1519 with 270 men and five vessels in search of a westward passage to the Spice Islands. The expedition lost two ships before finally discovering and passing through the Strait of Magellan and rounding the tip of South America in November 1520. Magellan crossed the Pacific Ocean and arrived in the Philippines in March 1521 where he was killed in a battle with the natives on April 27, 1521. Two of his ships sailed on and reached the Spice Islands in November 1521 where they loaded valuable spices for a return home. In an attempt to guarantee that at least one ship made it back to Spain the two ships parted ways. The *Victoria* continued sailing west and successfully crossed the Indian Ocean, rounded Africa's Cape of Good Hope, and arrived back in Spain on September 6, 1522 with 18 of the original crew. This was the first circumnavigation of the Earth (fig. 1.7). Magellan's skill as a navigator makes his voyage probably the most outstanding single contribution to the early charting of the oceans. In addition, during the voyage he established the length of a degree of latitude and measured the circumference of the Earth.

By the latter half of the sixteenth century, adventure, curiosity, and hopes of finding a trading shortcut to China spurred efforts to find a sea passage around the north side of North America. Sir Martin Frobisher (1535?–94) made three voyages in the 1570s, and Henry Hudson (d. 1611) made four voyages between 1607 and 1610, dying with his son when set adrift in Hudson Bay by his mutinous crew. The Northwest Passage continued to beckon, and William Baffin (1584–1622) made two attempts in 1615 and 1616.

While European countries were setting up colonies and claiming new lands, Francis Drake (1540–96) set out in 1577 with 165 crewmen and five ships to show the English flag around the world (fig. 1.7). He was forced to abandon two of

Figure 1.6

The routes of Bartholomeu Dias and Vasco da Gama around the Cape of Good Hope and Christopher Columbus' first voyage.

More Information:
www.mariner.org/age/princehenry.html,
www.mariner.org/age/firstvoyage.html,
www1.minn.net/~keithp/, and
www.mariner.org/age/dagama.html

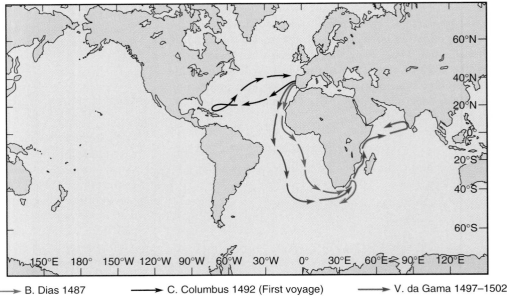

→ B. Dias 1487   → C. Columbus 1492 (First voyage)   → V. da Gama 1497–1502

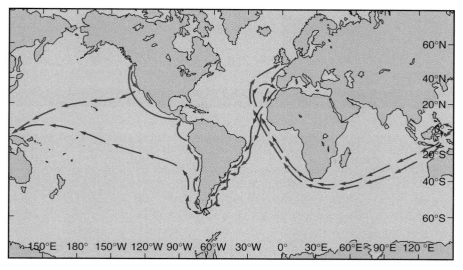

Figure 1.7

The sixteenth-century circumnavigation voyages by Magellan and Drake.

More Information: www.mariner.org/age/magellan.html, www.mariner.org/age/drake.html, and www.mcn.org/2/oseeler/drake.htm

→ F. Magellan 1519–22

→ F. Drake 1577–80

his ships off the coast of South America. He was separated from the other two ships while passing through the Straits of Magellan. During the voyage Drake plundered Spanish shipping in the Caribbean and in Central America and loaded his ship with treasure. In June 1579, Drake landed off the coast of present-day California and sailed north along the coast to the present United States–Canadian border. He then turned southwest and crossed the Pacific Ocean in two months time. In 1580 he completed his circumnavigation and returned home in the *Golden Hind* with a cargo of Spanish gold, to be knighted and treated as a national hero. Queen Elizabeth I encouraged her sea captains' exploits as explorers and raiders because, when needed, their ships and their knowledge of the sea brought military victories as well as economic gains.

techniques. To obtain the precise location of landfall or ship's position, it is necessary to know the location of the Sun or the stars related to time, and because early clocks did not work well on rolling ships, precise navigational measurements were not possible on early voyages. In 1714 the British Parliament offered 20,000 pounds sterling for a clock that could keep time with an error not greater than two minutes on a voyage to the West Indies from England. John Harrison, a clock maker, accepted the challenge and built his first sea-going clock in 1735. It was not until 1761 that his fourth model (fig. 1.8) met the test, losing only 51 seconds, on the 81-day voyage.

What stimulated the long voyages of the fifteenth and sixteenth centuries?

Who was Amerigo Vespucci, and how was he honored?

Why was Magellan's voyage of such great importance?

What is the Northwest Passage? Why was there an interest in finding it?

The Importance of Charts and Navigational Information   1.4

As colonies were established far away from their home countries, and as trade and travel expanded, there was renewed interest in developing better charts and more accurate navigation

Figure 1.8

John Harrison's fourth chronometer. A copy of this chronometer was used by Captain James Cook on his 1772 voyage to the southern oceans.

More Information: www.nmm.ac.uk/

Captain James Cook (1728–79) made his three great voyages to chart the Pacific Ocean between 1768 and 1779 (fig. 1.9). During his voyages, he explored and charted much of the South Pacific and the coasts of New Zealand, Australia, and northwest North America. He took a copy of Harrison's fourth chronometer on his second voyage of discovery to the south seas, and with this timepiece he was able to produce accurate charts of new areas and to correct previously charted positions. He searched for a way to the Atlantic from the Bering Sea and discovered the Hawaiian Islands, where he was killed. He made soundings to depths of 400 m (1300 ft) and logged accurate observations of winds, currents, and water temperatures. Cook takes his place as one of history's greatest navigators and sailors as well as a fine scientist. His careful and accurate observations produced much valuable information and made him one of the founders of oceanography.

In the United States, Benjamin Franklin (1706–90) became concerned about the amount of time required for news and cargo to travel between England and America. With Captain Timothy Folger, his cousin and a whaling captain from Nantucket, he constructed the 1769 Franklin-Folger chart of the Gulf Stream current (fig. 1.10), which encouraged captains to sail within the Gulf Stream enroute to Europe and to avoid it on the return passage. Since the Gulf Stream carries warm water from low latitudes to high latitudes it is possible to map its location with satellites that measure sea surface temperature. Compare the Franklin-Folger chart in figure 1.10 with a map of the Gulf Stream shown in figure 1.11 based on the average sea surface temperature during 1996.

The U.S. Naval Hydrographic Office, now the U.S. Naval Oceanographic Office, was set up in 1830. In 1842, Lieutenant Matthew F. Maury (1806–73) was assigned to the Hydrographic Office and founded the Naval Depot of Charts. He began a systematic collection of wind and current data from ships' logs. He produced his first wind and current charts of the North Atlantic in 1847; these became a part of the first published atlases of sea conditions and sailing directions. The British estimated that Maury's sailing directions took thirty days off the passage from the British Isles to California, twenty days off the voyage to Australia, and ten days off the sailing time to Rio de Janeiro. In 1855, he published *The Physical Geography of the Sea*. This work includes chapters on the Gulf Stream, the atmosphere, currents, depths, winds, climates, and storms, and the first contour chart of the North Atlantic sea floor. See figure 1.12 for the Gulf Stream chart from this book and compare the change in detail and style with the Franklin-Folger chart in figure 1.10. Many consider Maury's book the first textbook of what we now call oceanography and consider Maury the first true oceanographer. Again, national and commercial interests were the driving forces behind the study of the oceans.

Why was John Harrison's clock important to ocean exploration?

What did Captain James Cook's voyage contribute to the science of oceanography?

Why was Benjamin Franklin interested in the Gulf Stream?

Who was Matthew F. Maury, and what was his contribution to ocean science?

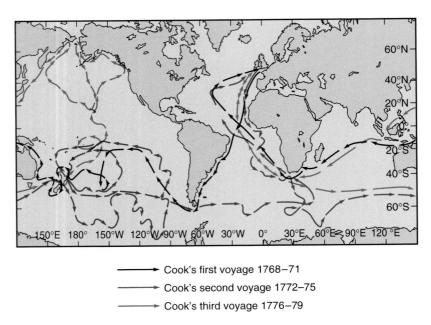

Cook's first voyage 1768–71
Cook's second voyage 1772–75
Cook's third voyage 1776–79

Figure 1.9

The three voyages of Captain James Cook.

More Information: www.mariner.org/age/cook.html, winthrop.webjump.com/jcook.html, and pacific.vita.org/pacific/cook/cook1.htm

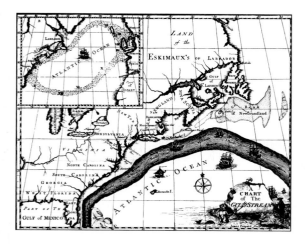

Figure 1.10

The Franklin-Folger map of the Gulf Stream, 1769.

More Information: podaac-www.jpl.nasa.gov/kids/history.html

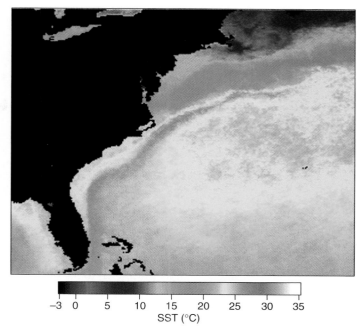

-3 0   5   10   15   20   25   30   35
SST (°C)

Figure 1.11

Average sea surface temperature for the year 1996. The red-orange streak of 28°C to 30°C water shows the Gulf Stream.

More Information: dcz.gso.uri.edu/avhrr-archive.html

Figure 1.12

Matthew F. Maury's chart of the Gulf Stream and North Atlantic Ocean surface currents from *Physical Geography of the Sea*, 1855.

Ocean Science Begins   1.5

As charts became more accurate and as information about the oceans increased, the oceans captured the interest of naturalists and biologists. Charles Darwin (1809–82) joined the survey ship *Beagle* and served as the ship's naturalist from 1831 to 1836. He described, collected, and classified organisms from the land and sea. His theory of atoll formation (described in chapter 4) is still the accepted explanation. At approximately the same time, another English naturalist, Edward Forbes (1815–54), began a systematic survey of marine life around the British Isles and the Mediterranean and Aegean Seas. He collected organisms in deep water and, based on his observations, proposed a system of ocean depth zones, each characterized by specific animal populations. However, he also mistakenly theorized that the environment below 550 m (1800 ft) was without life. His announcement is curious, since twenty years earlier the Arctic explorer Sir John Ross (1777–1856) had taken bottom samples at over 1800 m (6000 ft) depth in Baffin Bay and had found worms and other animals living in the mud. His nephew, Sir James Clark Ross (1800–62) took even deeper samples from Antarctic waters and noted their similarity to the Arctic species recovered by his uncle. Still, Forbes's systematic attempt to make orderly predictions about the oceans, his enthusiasm, and his influence make him another candidate as a founder of oceanography.

The investigation of the minute drifting plants and animals of the ocean was not seriously undertaken until the German scientist Johannes Müller (1801–58) began to examine these organisms microscopically. Victor Hensen (1835–1924) introduced the quantitative study of these minute drifting organisms and gave them the name *plankton* in 1887. A portion of a plate from an 1899 publication describing these organisms is seen in figure 1.13.

Although science blossomed in the seventeenth and eighteenth centuries, there was little scientific interest in the sea beyond the practical needs for navigation, tide prediction, and safety. By the early nineteenth century, ocean scientists were still few and usually only temporarily attracted to the sea.

How did Edward Forbes advance understanding of the oceans, and in what way was he mistaken?

What kinds of organisms were investigated by Müller and Hensen?

The *Challenger* Expedition   1.6

Charles Wyville Thomson (1830–82), a professor of natural history at Edinburgh University in Scotland, participated in a series of deep-sea expeditions during the years 1868–70. The deep-sea organisms recovered by these expeditions caught the public's attention, and with public interest running high, the British Royal Society was able to persuade the British Admiralty to organize the *Challenger* expedition, the most comprehensive single oceanographic expedition ever undertaken. The leadership of this expedition was offered to the Scottish professor Charles Wyville Thomson, who named as his assistant a young geologist, John Murray (1841–1914). At this time, Thomson also wrote *The Depths of the Sea*, based on his participation in the previous scientific expeditions. This very popular book was published in 1873 and is regarded by some as the first book on oceanography.

The naval corvette *Challenger* was refitted with laboratories, winches, and equipment; it sailed from England on December 21, 1872, for a voyage that was to last nearly three and a half years (fig. 1.14). During this voyage the vessel logged 110,840 km (68,890 mi), and returned to England on May 24, 1876. The *Challenger* expedition's purpose was scientific research; during the voyage the crew took soundings at 361 ocean stations (the deepest at 8180 m or 26,850 ft), collected

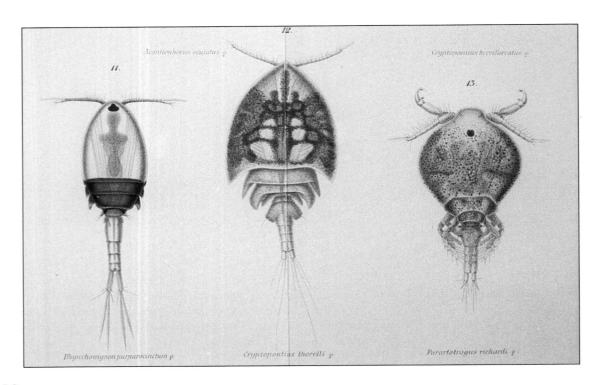

Figure 1.13

This illustration of microscopic drifting animals is from volume 25 of *Fauna und Flora des Golfes von Neapel (Fauna and Flora of the Gulf of Naples)*, published in 1899.

(a)

(b)

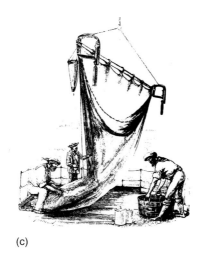

(c)

(d)

Figure 1.14

The *Challenger* expedition: December 21, 1872–May 24, 1876. Engravings from the *Challenger Reports*, volume 1, 1885, (a) "H.M.S. *Challenger*—Shortening Sail to Sound," decreasing speed to take a deep-sea depth measurement. (b) "Dredging and Sounding Arrangement on board *Challenger*." Rigging is hung from the ship's yards to allow the use of over-the-side sampling equipment. A biological dredge can be seen hanging outboard of the rail. The large cylinders in the rigging are shock absorbers. (c) A biological dredge used for sampling bottom organisms. Note the frame and skids that keep the mouth of the net open and allow it to slide over the sea floor. (d) "H.M.S. *Challenger* at St. Paul's Rocks," in the equatorial mid-Atlantic. (e)The cruise of the *Challenger*, 1872–76, the first major oceanographic research effort.

More Information: www.soc.soton.ac.uk/OTHERS/CSMS/hmschall.html

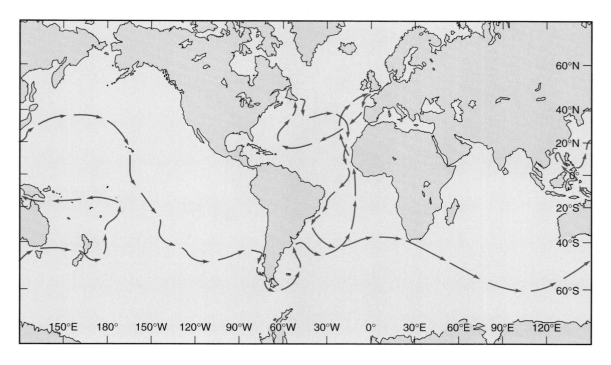

Cruise of the *Challenger*, 1872–76

(e)

deep-sea water samples, investigated deep-water motion, and made temperature measurements at all depths. Thousands of biological and sea-bottom samples were collected. *Challenger* brought back evidence of an ocean teeming with life at all depths and opened the way for the era of descriptive oceanography that followed.

Although the *Challenger* expedition ended in 1876, the work of organizing and compiling information continued for twenty years, until the last of the fifty-volume *Challenger Reports* was issued. John Murray edited the reports after Thomson's death and wrote many of them himself. He is considered the first geological oceanographer. William Dittmar (1833–92) prepared the information on seawater chemistry for the *Challenger Reports*, confirming the findings of earlier chemists that in seawater around the world, the proportion of the major dissolved elements to each other is constant. Oceanography as a modern science is usually dated from the *Challenger* expedition. The *Challenger Reports* laid the foundation for the science of oceanography.

> Explain the significance of the *Challenger* expedition and the *Challenger Reports*.
>
> The *Challenger* and its expedition are often called unique. Why is this term used?

Exploratory Science   1.7

During the late nineteenth century and the early twentieth century, intellectual interest in the oceans increased. Oceanography was changing from a descriptive science to a quantitative one. Oceanographic cruises now had the goal of testing hypotheses by gathering data. Theoretical models of ocean circulation and water movement were developed. The Scandinavian oceanographers were particularly active in the study of water movement. One of them, Fridtjof Nansen (1861–1930) (fig. 1.15), a well-known athlete, explorer, and zoologist, was interested in the current systems of the polar seas. He decided to test his ideas about the direction of ice drift in the Arctic by freezing a vessel into the polar ice pack and drifting with it; he expected in this way to reach the North Pole. To do so he had to design a special vessel that would be able to survive the great pressure from the ice; the 39-meter (128-ft) *Fram* ("to push forward"), shown in figure 1.15, was built with a smoothly rounded, wooden hull and planking over 60 cm (2 ft) thick. Nansen with thirteen men departed from Oslo in June 1893.

The ship was frozen into the ice nearly 1100 km (700 mi) from the North Pole and remained in the ice for thirty-five months. During this period, measurements made through holes in the ice showed that the Arctic Ocean was a deep ocean basin and not the shallow sea that had been expected. The crew recorded water and air temperatures, analyzed water chem-

(a)

(b)

Figure 1.15

(a) Fridtjof Nansen, Norwegian scientist, explorer, and statesman (1861–1930), using a sextant to determine his ship's position. (b) The *Fram*, frozen in ice. As the ice pressure increased, it lifted her specially designed and strengthened hull so that she would not be crushed.

More Information: www.nrsc.no/nansen/fritjof_nansen.html, www.nobel.se/laureates/peace-1922-bio.html, and www.uninett.no/ietf45/social/fram-museum.html

istry, and observed the great plankton blooms of the area. Nansen became impatient with the slow rate and the direction of drift, and with F.H. Johansen left the *Fram* locked in the ice some 500 km (300 mi) from the pole. They set off with a dogsled toward the pole, but after four weeks they were still more than 300 km (200 mi) from the pole, with provisions running low and the condition of their dogs deteriorating. The two men turned away from the pole and spent the winter of 1895–96 on the ice living on seals and walrus. They were found by a British polar expedition in June 1896 and returned to Norway. The crew of the *Fram* continued to drift with the ship until they freed the vessel from the ice in 1896 and also returned to Oslo. Nansen's expedition had laid the basis for future Arctic research. His name is familiar today from the Nansen bottle, which he designed to collect and isolate water samples at depth.

The *Fram* paid another visit to polar waters carrying the Norwegian explorer Roald Amundsen (1872–1928) to Antarctica on his successful 1911 expedition to the South Pole. It was also Amundsen who finally made a Northwest Passage in the

Gjoa (fig. 1.16), leaving Norway in 1903 and arriving in Nome, Alaska, three years later.

Fluctuations in the abundance of commercial fish in the North Atlantic and adjacent seas, and the effect of these changes on national fishing programs, stimulated oceanographic research and international cooperation. As early as 1870, researchers began to realize their need for knowledge of ocean chemistry and physics in order to understand ocean biology. Advances in theoretical oceanography often could not be verified with practical knowledge until new instruments and equipment were developed. Lord Kelvin (1824–1907) invented a tide-predicting machine in 1872 that made it possible to combine tidal theory with astronomical predictions to produce predicted tide tables. Deep-sea circulation could not be systematically explored until approximately 1910, when Nansen's water-sampling bottles were combined with thermometers designed for deep-sea temperature measurements, and an accurate method for determining the water's salt content was devised by the chemist Martin Knudsen (1871–1949). The reliable and accurate measurement of ocean depths had to wait until the development of

Figure 1.16

(b)

(a) Roald Amundsen, Norwegian explorer of the Arctic and Antarctic (1872–1928). (b) The *Gjoa*, preparing for her journey through the Northwest Passage (1903).

More Information: www.south-pole.com/p0000101.htm and collections.ic.gc.ca/arctic/explore/amundsen.htm

the echo sounder, which was given its first scientific use on the 1925–27 German cruise of the *Meteor* in the central and southern Atlantic Ocean.

Who was Fridtjof Nansen, and why was his work important?

Give examples of nineteenth- and early twentieth-century advances in technology that helped ocean studies.

U.S. Oceanography in the Twentieth Century 1.8

In the United States, government agencies related to the oceans proliferated during the nineteenth century. These agencies were concerned with gathering information to further commerce, fisheries, and the Navy. After the Civil War, the replacement of sail by steam lessened government interest in studying winds and currents and in surveying the ocean floor. Private institutions and wealthy individuals took over the support of oceanography in the United States. Alexander Agassiz (1835–1910), marine scientist and Harvard Professor, financed a series of expeditions that greatly expanded knowledge of deep-sea biology. One of Agassiz's students, William E. Ritter, a professor at the

University of California, Berkeley, established a permanent marine biological field station in San Diego in 1903 with financial support from members of the Scripps family, who had made a fortune in newspaper publishing. This was the beginning of the University of California's Scripps Institution of Oceanography (fig. 1.17a). In 1927, a National Academy of Sciences committee recommended that ocean science research be expanded by creating a permanent marine science laboratory on the East Coast. This led to the establishment of the Woods Hole Oceanographic Institution in 1930 (fig. 1.17b). It was funded largely by a grant from the Rockefeller Foundation. The Rockefeller Foundation allocated funds to stimulate other programs in marine research and construct additional laboratories at this same time and oceanography began to move onto university campuses.

Oceanography mushroomed during World War II, when practical problems of military significance had to be solved quickly. The United States and its allies needed to move men and materials by sea to remote locations, to predict ocean and shore conditions for amphibious landings, to know how explosives behave in seawater, to chart beaches and harbors from aerial reconnaissance, and to find and destroy submarines. Academic studies ceased as oceanographers pooled their knowledge in the national effort.

After the war, oceanographers returned to their classrooms and laboratories with an array of new sophisticated instruments, including radar, improved sonar devices, automated wave detectors, and temperature-depth recorders. They also returned with large-scale government funding for research and education. The Earth sciences in general and oceanography in

(a)

(b)

Figure 1.17

(a) The Scripps Institution of Oceanography in La Jolla, California. Established in 1903 by William Ritter, a zoologist at the University of California, Berkeley with financial support from E. W. Scripps and his daughter Ellen Browning Scripps. The first permanent building was erected in 1910. (b) The Woods Hole Oceanographic Institution in Woods Hole, Massachusetts. In a rare moment all three WHOI research vessels are in port. *Knorr* is in the foreground with *Oceanus*, bow forward, and the new *Atlantis*, stern forward, on the opposite side of the pier.

More Information: www.sio.ucsd.edu/, and www.whoi.edu/

particular blossomed during the 1950s. The numbers of scientists, students, educational programs, research institutes, and professional journals all increased.

Major funding for applied and basic ocean research was supplied by both the Office of Naval Research (ONR) and the National Science Foundation (NSF). The Atomic Energy Commission (AEC) financed oceanographic work at the South Pacific atoll sites of atomic tests. During the 1950s the Coast and Geodetic Survey expanded its operations and began its seismic sea wave (tsunami) warning system. International cooperation brought about the 1957–58 International Geophysical Year (IGY) program, in which sixty-seven nations cooperated to explore the sea floor and made discoveries that completely revolutionized geology and geophysics. As a direct result of the IGY program, special research vessels and submersibles were built to be used by both federal agencies and university research programs.

The decade of the 1960s brought giant strides in programs and equipment. In 1963–64 another multinational endeavor, the Indian Ocean Expedition took place. In 1965 a major reorganization of governmental agencies occurred. The Environmental Science Services Administration (ESSA) was formed by consolidating the Coast and Geodetic Survey and the Weather Bureau among others. Under ESSA, federal environmental research institutes and laboratories were established, and the use of satellites to obtain data became a major focus of ocean research. In 1968 the Deep Sea Drilling Program (DSDP), a cooperative venture between research institutions and universities, began to sample the Earth's crust beneath the sea (fig. 1.18a, discussed in more detial in chapter 3) using the specially built drill ship *Glomar Challenger*. It was finally retired in 1983 after 15 years of extraordinary service. The *Glomar Challenger* was named after the ship used during the *Challenger* expedition of 1872–76. Electronics developed for the space program were applied to ocean research. Computers went aboard research vessels, and for the first time data could be sorted, analyzed, and interpreted at sea. This made it possible for scientists to adjust experiments while they were in progress. Government funding allowed large-scale ocean experiments. Fleets of oceanographic vessels from many institutions and nations carried scientists studying all aspects of the oceans.

In 1970 the U.S. government reorganized its Earth science agencies once more. The National Oceanic and Atmospheric Administration (NOAA) was formed under the Department of Commerce. NOAA combined several formerly independent agencies including the National Ocean Survey, National Weather Service, National Marine Fisheries Service, Environmental Data Service, National Environmental Satellite Service, and Environmental Research Laboratories. NOAA also administers the National Sea Grant College Program. This program consists of a network of 29 individual programs located in each of the coastal and Great Lakes states. Sea Grant encourages cooperation in marine science and education among government, academia, and industry. The ten-year International Decade of Ocean Exploration (IDOE) occurred in the 1970s. IDOE was a multi-national effort to survey seabed mineral resources, improve environmental forecasting, investigate coastal

ecosystems, and modernize and standardize the collection, analysis, and use of marine data.

Oceanography in the 1970s faced a reduction in funding for ships and basic research; nonetheless, the discovery of deep-sea hot water vents and their associated animal life and mineral deposits renewed the excitement over deep-sea biology, chemistry, geology, and ocean exploration in general. Instrumentation grew ever more sophisticated and expensive as deep-sea moorings, deep-diving submersibles, and the remote sensing of the ocean by satellites were used more and more often.

It is not possible to equip enough research vessels to study more than a small area of one ocean at one time, but with satellites oceanographers can study the oceans as a global system. Each satellite makes millions of observations every day as it follows changing conditions across the world's oceans. The huge amounts of information satellites provide are processed and manipulated by sophisticated computers to increase knowledge of currents, waves, plant life, sea ice, storms, and

(a)

(b)

Figure 1.18

(a) The *Glomar Challenger*, the Deep Sea Drilling Program drill ship used from 1968 to 1983. (b) The *JOIDES Resolution*, the Ocean Drilling Program drill ship in use since 1985.

More Information: www-odp.tamu.edu/

even the sea floor. Oceanographic vessels are linked to satellites via computers allowing scientists to use immediate data to plan their sampling programs while at sea.

SEASAT, a specialized oceanographic satellite, was launched in June 1978 but remained operational only until October. Its radar could measure the distance between the satellite and the sea surface with an accuracy of about 5 cm (2 in), allowing the measurement of wave heights. Sea surface temperatures, wind speeds, sea ice cover, currents, and plant production were also monitored.

During the 1970s and 1980s Earth scientists began to recognize the signs of global degradation and the need for management of living and nonliving resources. As more and more nations turned to the sea for food and as technology increased their abilities to harvest the sea, problems of resource ownership, dwindling fish stocks, and the need for fishery management became more and more evident.

In 1983 the Deep Sea Drilling Project became the Ocean Drilling Program (ODP). The objectives of the ODP included drilling into the thick sediments near continental margins. ODP uses a larger drilling ship, the *JOIDES Resolution* (fig. 1.18b), to replace the retired *Glomar Challenger* (fig. 1.18a). The *JOIDES Resolution* is named after the *HMS Resolution*, used by Captain James Cook to explore the Pacific Ocean basin over 200 years ago.

NASA's *NIMBUS-7* satellite, launched in 1978, carried a sensor package called the Coastal Zone Color Scanner (CZCS) that detected multiband radiant energy from chlorophyll in sea and land plants. The sensor operated from 1978 to 1986 when it finally failed. CZCS images can be used to determine levels of biological productivity in the oceans. They can also indicate how, and where, physical processes in the oceans influence the distribution and health of marine biological communities, particularly the small marine plants called phytoplankton, which are discussed in detail in chapter 11 (fig. 1.19).

In 1985 the U.S. Navy Geodynamic Experimental Ocean Satellite (GEOSAT) was launched. It was designed to collect data for military purposes but its orbit was changed to replace the failed SEASAT. From 1986 to 1990 it monitored sea-level topography, surface winds and waves, local gravity changes, and abrupt "boundaries" in the ocean caused by changes in salinity and temperature.

How did the U.S. oceanographic research focus change after the Civil War?

How has each of the following affected twentieth-century oceanography? (a) economics, (b) commerce and transportation, (c) military needs.

Compare the U.S. government's institutional support of oceanography before and after World War II.

How did the emphasis on ocean research change in the 1960s, 1970s, and 1980s?

Why are oceanographers interested in data collected by satellite as well as data from research vessels?

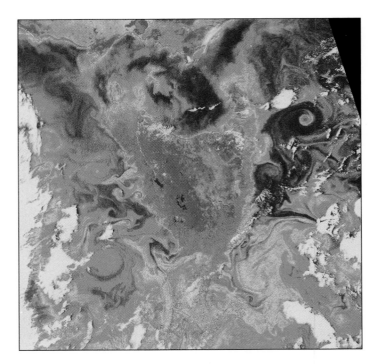

Figure 1.19

False color image of the oceans centered on the island of Tasmania. Tasmania is located south of the eastern coast of Australia. Yellows and reds indicate high concentrations of phytoplankton, greens and blues low concentrations, and dark blue and purple very low concentrations. The complex current interactions around the island have significant influence on the distribution of phytoplankton.

More Information: daac.gsfc.nasa.gov/CAMPAIGN_DOCS/OCDST/classic_scenes/00_classics_index.html

Oceanography of the Recent Past, Present, and the Future 1.9

Today's scientists see the Earth not as a single system but as a complex of systems and subsystems acting as a whole. Projects that emerged in the 1990s and continue in the 2000s require that scientists cross from one discipline to another and share information for common goals. Satellites are used for global observation. Earth and ocean scientists are able to manipulate online and archival data quickly by computer at sea or on land, and they share it rapidly by the Internet. In addition successful integrated approaches to Earth studies require that governments, agencies, universities, and national and international programs agree to set common priorities and to share program results.

Several large-scale oceanographic programs have been developed to better understand the role of the oceans in processes of the atmosphere-ocean-land system. These programs provide data for models that scientists use to predict the evolution of the Earth's environment as well as the consequences of human-influenced changes. The World Ocean Circulation Experiment (WOCE) studies the world oceans using computer models and chemical tracers to model the present state of the oceans and then predict ocean evolution in relation to long-term changes in the atmosphere. This effort combines sampling by ship, satellites, and floating independent buoys with sensors. The U.S. Joint Global Ocean Flux Study (JGOFS) studies the relationship between ocean plant production and solar radiation. Scientists are monitoring the worldwide abundance of plant life by ship and satellite to understand how carbon and other biologically active elements move between the ocean, atmosphere, and land. The Global Ocean Atmosphere-Land System (GOALS) studies the energy transfer between the atmosphere and the tropical oceans to better understand El Niño and its effects and to provide improved large-scale climate prediction.

Two large international programs, the Deep Ocean Drilling Program (DODP) and the Ridge Interdisciplinary Global Experiment (RIDGE) explore the Earth's ocean floors and the margins of the continents. They investigate the structure and history of the Earth and probe the ocean's great mountain range systems and their relationships to the chemistry of the oceans.

In August 1992, the satellite *TOPEX/Poseidon* was launched in a joint U.S.-French mission to explore ocean circulation and its interaction with the atmosphere. *TOPEX/Poseidon* measures sea level along the same path every 10 days. This information is used to relate changes in ocean currents with atmospheric and climate patterns. The measurements obtained allow scientists to chart the height of the seas across ocean basins with an accuracy of less than 10 centimeters (4 inches). *TOPEX/Poseidon*'s three-year prime mission ended in fall 1995 and is now in its extended observational phase. A major follow-on mission to continue these studies is scheduled to begin early in 2001 with the launch of *Jason-1*. *Jason-1*'s mission will be the same as *TOPEX/Poseidon*'s but it is designed to acquire continuous data over longer periods of time in order to measure long-term circulation processes more accurately.

A meeting of representatives from twenty-four nations recommended the development of a Global Ocean Observing System (GOOS) to include satellites, buoy networks, and research vessels. The goal of this program is to enhance our understanding of ocean phenomena so that events such as El Niño (see section 6.14 in chapter 6) and its impact on climate can be predicted more accurately and with greater lead time. The successful prediction of the 1997–98 El Niño six months in advance of its peak had a dramatic effect on the public.

An integral part of the GOOS is a project called Argo, named after the mythical vessel used by the ancient Greek seagoing hero Jason. Argo is an international project that will deploy an array of 3000 independent instruments, or floats, throughout the oceans by the year 2005 (fig. 1.20a). The first of these floats are being deployed and tested now. Each float will be programmed to descend to a depth of 2000 m (6560 ft, or about 1.25 miles) where it will remain for about 10 days (fig. 1.20b). It will then ascend to the surface measuring temperature and salinity as it rises. When it reaches the surface it will relay the data to shore via satellite and then descend once again and wait for its next cycle. In this fashion the entire ar-

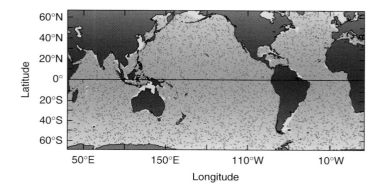

(a)

(b)

Figure 1.20

(a) The Argo array will consist of 3000 autonomous floats spread throughout the oceans. (b) Every ten days the float will ascend to the surface, measuring temperature and salinity as it rises. When it reaches the surface it will transmit the data to shore via satellite and then descend to 2000 m (6560 ft) once again.

More Information: www.argo.ucsd.edu/

ray will provide detailed temperature and salinity data of the upper 2000 m (6560 ft) of the oceans every ten days.

In 1998 the National Research Council's Ocean Studies Board released a report highlighting three areas that are likely to be the focus of future research. These three areas are (1) improving the health and productivity of the coastal oceans, (2) sustaining ocean ecosystems for the future, and (3) predicting ocean-related climate variations.

Although large-scale, federally funded studies are presently at the forefront of ocean studies, it is important to remember that studies driven by the specific research interests of individual scientists are essential to point out new directions for oceanography and other Earth sciences. In the following chapters, you will follow the development of ideas that have enabled us to build an understanding of the dynamic and complex systems that are the Earth's oceans.

> List some of oceanography's current research targets.
>
> In what ways do satellites create ocean study opportunities that are not available to research vessels?
>
> Why are oceanographers interested in a global approach to ocean science?
>
> What role does the individual scientist play in advancing our knowledge?

Summary

Oceanography is a multiscience field in which geology, geophysics, chemistry, physics, meteorology, and biology are all used to understand the oceans. Early information about the oceans was collected by explorers and traders such as the Phoenicians, the Polynesians, the Arabs, and the Greeks. Eratosthenes calculated the circumference of the Earth, and Ptolemy produced the first world atlas.

During the Middle Ages, the Vikings crossed the North Atlantic, while shipbuilding and chartmaking improved. In the fifteenth and sixteenth centuries the Chinese, Dias, Columbus, da Gama, Vespucci, and Balboa made voyages of discovery. Magellan's expedition became the first to circumnavigate the Earth. In the sixteenth and seventeenth centuries, some explorers searched for the Northwest Passage while others set up trading routes to serve developing colonies.

By the eighteenth century, national and commercial interests required better charts and more accurate navigation techniques. Cook's voyages of discovery to the Pacific produced much valuable information, and Franklin sponsored a chart of the Atlantic's Gulf Stream. A hundred years later, the U.S. Navy's Maury collected wind and current data to produce current charts and sailing directions and then wrote the first book on "oceanography."

Ocean science began with the nineteenth-century expeditions and research of Darwin, Forbes, Müller, and others. The three-and-a-half-year *Challenger* expedition laid the foundation for modern oceanography with its voyage, which gathered large quantities of data on all phases of oceanography. Exploration of the Arctic and Antarctic oceans was pursued by Nansen and Amundsen into the beginning of the twentieth century.

In the twentieth century, private institutions played an important role in developing U.S. oceanographic research, but the largest single push came from the needs of the military during World War II. After the war, large-scale government funding and international cooperation allowed oceanographic projects that made revolutionary discoveries about the ocean basins. Development of electronic equipment, deep-sea drilling programs, research submersibles, and use of satellites continued to produce new and more detailed information of all kinds. At present, oceanographers are focusing their research on global studies and the management of resources as well as continuing to explore the interrelationships of the chemistry, physics, geology, and biology of the sea.

Critical Thinking

1. Why have Captain James Cook, Edward Forbes, Charles Wyville Thomson, and Matthew F. Maury all been called the "founder of oceanography?" Can one individual be considered as oceanography's founder? Why or why not? If one individual were to be selected, which would you select and why?
2. How is oceanography affecting global trade, resource use, and military operations at the present time? Search for examples in print media, television, and the Internet as you continue your study of the oceans.
3. What are some advantages of using satellites for oceanographic research? Are there any disadvantages?
4. The ability to establish a precise ocean location in both space and time has occupied humans for thousands of years. Beginning with the stick charts of the Polynesians and ending with today's satellites, discuss the importance of this endeavor.
5. What do you think should be the goals of oceanographic science for the next ten to twenty years? Try answering this question again when you have finished your class in oceanography and compare your answers.

Suggested Readings

Bailey, H. S. 1953. The Voyage of the *Challenger.* In *Ocean Science, Readings from Scientific American* 188(5):8–12.

Bascom, W. 1988. *The Crest of the Wave: Adventures in Oceanography.* Harper & Row, New York.

Brosse, J. 1983. *Great Voyagers of Discovery, Circumnavigators and Scientists, 1764–1843.* Facts on File publications, New York.

Deacon, M. 1997. *Scientists and the Sea 1650–1900. A Study of Marine Science,* 2d ed. Ashgate, Brookfield, VT.

Glick, D. 1994. Windows on the World. *National Wildlife* 32(2): 6–13.

International Congress on the History of Oceanography. 1980. *Oceanography, the Past,* ed. M. Sears and D. Merriman. Springer-Verlag, New York.

McGovern, T. H., and S. Perdikaris. 2000. The Viking's Silent Saga. *Natural History* 109(8):50–57.

Orange, D. L. 1996. Mysteries of the Deep. *Earth* 5(6):42–45.

Schuessler, R. 1984. Ferdinand Magellan: The Greatest Voyager of them all. *Sea Frontiers* 30(5):299–307.

Sobel, D. 1995. *Longitude: The True Story of a Lone Genius Who Solved the Greatest Scientific Problem of His Time.* Walker Publishing Company, Inc., New York, 184 pp.

Zaburunov, S. A. 1992. Monitoring Our Global Environment. *Earth* 1(4):46–53.

internet references

worldwide websites

The Internet makes available oceanographic information and data to researchers, and it also provides images and information in many forms to instructors and students. Public agencies and museums, universities and research laboratories, satellite and oceanographic projects, interest groups and individuals all over the planet provide information. These information sites may be accessed most easily through a World Wide Web (WWW) browser using a Uniform Research Locator (URL) or address. However URLs are not permanent and some searching by title or subject may be both necessary and desirable. The listings at the end of each chapter have been selected, in part, because they are less likely to be removed, and they receive frequent updating. Using the Internet is the best way to learn what kinds of oceanographic information are available from this resource.

General References

National Oceanic and Atmospheric Administration (NOAA)—all with information and images

Home Page
http://www.noaa.gov/

Central Library
http://www.lib.noaa.gov/

Environmental Information Services
http://www.esdim.noaa.gov

Environmental Research Laboratories
http://www.erl.noaa.gov/

National Geophysical Data Center
http://www.ngdc.noaa.gov/

Oceanography Resources on the Internet
http://wwwesdim.noaa.gov/ocean_page.html

Ocean Resources, Conservation & Assessment (ORCA)
Information Service
http://www-orca.nos.noaa.gov/

NOAA Ship *Discoverer*
http://www.pmc.noaa.gov/di/index.html

Oceanographer of the Navy
http://oceanographer.navy.mil/

International Council for Exploration of the Sea (ICES)
http://www.ices.dk/

NASA Goddard Space Flight Center
http://pao.gsfc.nasa.gov/gsfc.html

United States Geological Survey (USGS)
http://www.usgs.gov/

The World-Wide Web Virtual Library-Oceanography
http://www.mth.ues.ac.uk/ocean/vl/

Oceanography and Meteorology Servers
http://www.coaps.fsu.edu/Ocean-Met.html

Global Positioning System (GPS)
http://www.colorado.Edu/geography/gcraft/notes/gps/gps_f.html

Magazines

Scientific American
http://www.sciam.com/

Discover
http://discover.com/

National Geographic
http://www.nationalgeographic.com/

Chapter 1

Time Line of Ocean Exploration
http://www.mariner.org/age/menu.html

Polynesian Voyages, Navigation, and Wayfinding
http://www.pbs.org/wayfinders
http://leahi.kcc.hawaii.edu/org/pvs/
http://www.pbs.org/wayfinders/wayfinding2.html
http://www.museum.upenn.edu/Navigation/Misc/contents.html

The Middle Ages

The Vikings
http://www.mariner.org/age/vikingexp.html

Voyages of Discovery

Prince Henry the Navigator
http://www.mariner.org/age/princehenry.html

Christopher Columbus
http://www.mariner.org/age/firstvoyage.html
http://www1.minn.net/~keithp/

Vasco da Gama
http://www.mariner.org/age/dagama.html

Ferdinand Magellan
http://www.mariner.org/age/magellan.html

Sir Francis Drake
http://www.mariner.org/age/drake.html
http://www.mcn.org/2/oseeler/drake.htm

James Cook
http://www.mariner.org/age/cook.html
http://pacific.vita.org/pacific/cook/cook1.htm
http://winthrop.webjump.com/jcook.html

Charts and Navigational Information

Benjamin Franklin
http://podaac-www.jpl.nasa.gov/kids/history.html

Matthew Fontaine Maury
http://xroads.virginia.edu/~UG97/monument/maurybio.html

http://www.vmi.edu/~arcmaury/mfmpaprs.html
http://oceanographer.navy.mil/maury.html

The *Challenger* Expedition
http://www.hmschallenger.org/
http://www.soc.soton.ac.uk/OTHERS/CSMS/
hmschall.html
http://www.bartleby.com/65/ch/ChallengEx.html

Exploratory Science

Fridtjof Nansen
http://www.nrsc.no/nansen/fritjof_nansen.html
http://www.nobel.se/laureates/peace-1922-
bio.html
http://www.uninett.no/ietf45/social/fram-
museum.html

Roald Amundsen
http://www.south-pole.com/p0000101.htm
http://collections.ic.gc.ca/arctic/explore/
amundsen.htm

Twentieth Century

ODP and DSDP
http://www-odp.tamu.edu/

Joint Oceanographic Institutions
http://www.joi-odp.org/

Recent Past, Present, and the Future

SeaWiFS Project—satellite images and information
http://seawifs.gsfc.nasa.gov/SEAWIFS.html

TOPEX/Poseidon Satellite
http://topex-www.jpl.nasa.gov/

Coastal Zone Color Scanner (CZCS) images
http://daac.gsfc.nasa.gov/CAMPAIGN_DOCS/
OCDST/classic_scenes/00_classics_index.html

Global Ocean Observing System's Argo Project
http://www.argo.ucsd.edu/

Oceanographic Institutions, Museums, and Aquaria

Scripps Institution of Oceanography
http://www.sio.ucsd.edu/

Woods Hole Oceanographic Institution
http://www.whoi.edu/

Mystic Aquarium: Institute for Exploration
http://www.ife.org/index1.cfm

Monterey Bay Aquarium Research Institute
http://www.mbari.org/

Monterey Bay Aquarium
http://www.mbayaq.org/

National Aquarium in Baltimore
http://www.aqua.org/

National Maritime Museum, Greenwich, England
http://www.nmm.ac.uk/

More than two thousand years of war and trading have left thousands of wrecks scattered across the ocean floor. These wrecks are great storehouses of information for archaeologists and historians about ships, trade, warfare, and the details of personal lives.

Wrecks that lie in deep water are initially much better preserved than those in shallow water, because deep-water wrecks lie

plates, bowls, serving vessels, and decorative pieces; a bronze astrolabe for determining latitude; and a bronze and glass compass. Most of the hull had been destroyed by shipworms and currents; surviving pieces were measured and then covered with sand for protection.

The Swedish man-of-war, *Vasa* (fig. 2), sank in 1628 in Stockholm Harbor at the beginning of its maiden voyage. It was located in

Marine Archaeology

below the depths of the strong currents and waves that break up most shallow-water wrecks. Shallow-water wrecks are often the prey of treasure hunters who destroy the history of the site while they search for adventure and items of market value. Over long periods the cold temperatures and low oxygen content of deep water favor preservation of wooden vessels by slowing decomposition and excluding the marine organisms that bore into wood in shallow areas. Also the rate at which muds and sands falling from above cover objects on the deep sea floor is much slower than the same process in coastal water.

Marine archaeologists are using techniques developed for oceanographic research to find, explore, recover, and preserve wrecks and other artifacts lying under the sea. Sound beams that sweep the sea floor produce images that are viewed on board ship to locate an object of interest, and magnetometers are used to detect iron from sunken vessels. Once the initial contact has been made, the find is verified by divers in shallow water and by research submersibles carrying observers in deeper water. Unmanned, towed camera and instrument sleds or remotely operated vehicles (ROVs) equipped with underwater video cameras explore in deep water or in areas that are difficult or unsafe for divers and submersibles. ROVs and submersibles, as well as divers, collect samples to help identify wrecks.

Figure 1

The hull of the *Mary Rose* in its display hall at Portsmouth Dockyard, England. To preserve and stabilize the remaining wood, it is constantly sprayed with a preservative solution.

The Ships

A Bronze Age merchant vessel was discovered in 1983 more than 33 m (100 ft) down in the Mediterranean Sea off the Turkish coast. Divers have recovered thousands of artifacts from its cargo, including copper and tin ingots, pottery, ivory, and amber. Using these items archaeologists have been able to learn about the life and culture of the period, trace the ship's trade route, and understand more about Bronze Age people's shipbuilding skills.

In the summer of 1545 the English warship, *Mary Rose*, sank as it sailed out to engage the French fleet. It was first studied in place and then raised in 1982. More than 17,000 objects were salvaged, ranging from a surgeon's walnut medicine chest to archers' longbows and arrows. These objects give archaeologists and naval historians insights into the personal as well as the working lives of the officers and crew of a naval vessel at that time. The portion of its hull that was buried in the mud was preserved and is on display at Portsmouth, England (fig. 1).

The Spanish galleon, *San Diego*, sank December 14, 1600, as it engaged two Dutch vessels off the Philippine coast. The European Institute of Underwater Archaeology and the National Museum of the Philippines began an intensive two-year archaeological excavation in 1992. Relics recovered include 570 stoneware storage jars for water, wine, and oil; 800 pieces of intact Chinese porcelain, including

1956 and raised to the surface in 1961. Because the water of the Baltic Sea is much less salty than the open ocean, there were no shipworms to destroy its wooden hull, and it is one of the few complete ships ever recovered from the sea floor. The *Vasa* was a great ship over 200 feet long, excluding its bowsprit. Like others of its period, it was fantastically decorated with carvings and statues. Divers searched the seabed every summer between 1963 and 1967 and recovered 700 sculptures and carved details that had adorned it. Several chests and barrels containing crewmen's possessions, wooden and earthenware utensils used by the seamen, and pewter dishes used by the officers have also been recovered. The *Vasa* is now on exhibit in Stockholm.

Two expeditions have found more modern vessels sunk in the very deep sea. Robert Ballard of the Institute for Exploration in Mystic, Connecticut found and surveyed both the *Titanic* in 1985 and the *Bismarck* in 1989. The passenger liner *Titanic* sank in 1912 after striking an iceberg on its maiden voyage; the wreck was located 4000 m (13,000 ft) down in the North Atlantic. The contact was made by a towed underwater camera sled, computer controlled from the vessel above. A manned submersible was used for direct observation, and further inspection was made using ROVs that could be maneuvered to take pictures inside and outside the vessel. The World War II German battleship *Bismarck* (fig. 3) was located 4750 m (15,600 ft) below the

Figure 2

The *Vasa*, a seventeenth-century man-of-war, is on permanent display in Stockholm, Sweden. Most of the *Vasa* is original, including two of the three masts and parts of the rigging.

(a)

(b)

Figure 3

The World War II German battleship *Bismarck* was sunk in 1941 and lies 4750 m (15,600 ft) below the surface in the North Atlantic. In 1989, the towed camera sled *Argo* photographed sections of the wreck: (a) a 4.1 inch antiaircraft gun, (b) part of the ship's super structure.

surface of the North Atlantic. A photographic survey documented both battle damage and destruction due to sinking.

Ancient Mediterranean Shipwrecks

In 1997, a team of oceanographers, engineers, and archaeologists led by Robert Ballard discovered a cluster of five Roman ships, along with thousands of artifacts, on the bottom of the Mediterranean Sea. The ships lie in 762 m (2500 ft) of water beneath an ancient trade route where they were lost sometime between about 100 B.C. and A.D. 400. Prior to this discovery, no major ancient shipwreck sites had been discovered and explored by archaeologists in water deeper than 61 m (200 ft). The ships were remarkably well-preserved, having been protected from looting by divers and encrustation with coral by the deep water in which they sank. The vessels were spread over an area of about 32 km^2 (20 mi^2) and probably sank when they were caught in sudden, violent storms. The ships were initially located using the U.S. Navy's nuclear submarine *NR-1*. The *NR-1* is capable of searching over large areas for extended periods of time and its long-range sonar can detect objects at a much greater distance than conventional sonar

systems typically used by oceanographers. Once the site had been identified, the small ROV *Jason* was used for detailed mapping, observation, and the recovery of over 100 artifacts selected to help the archaeologists date the ancient wrecks.

Readings

Ballard, R. D. 1985. How We Found *Titanic*. *National Geographic* 168(6):698–722.

———. 1986. A Long Last Look at *Titanic*. *National Geographic* 170 (6):698–727.

———. 1989. Finding the *Bismark*. *National Geographic* 176 (5):622–37.

———. 1998. Roman Shipwrecks. *National Geographic* 193 (4):32–41.

Bass, G. F. 1987. Oldest Known Shipwreck Reveals Splendors of the Bronze Age. *National Geographic* 172 (6):693–734.

Gerard, S. 1992. The Caribbean Treasure Hunt. *Sea Frontiers* 38 (3):48–53.

Goddio, F. 1994. The Tale of the *San Diego*. *National Geographic* 186 (1):35–56.

Marx, R. F. 1990. In Search of the Perfect Wreck. *Sea Frontiers* 36 (5):46–51.

Roberts, D. 1997. Sieur de la Salle's Fateful Landfall. *Smithsonian* 28 (1):40–52. (Recovering the 300-year-old *Belle*.)

Internet Resources

Maritime Archaeology Links
http://www.mm.wa.gov.au/Museum/toc/links.html

Institute of Nautical Archaeology
http://ina.tamu.edu/

Department of the Navy: Naval Historical Center
http://www.history.navy.mil/

National Park Service

Submerged Cultural Resources Unit
http://www.nps.gov/scru/home.htm

National Maritime Initiative
http://www.cr.nps.gov/history/maritime/nmi.html

Southeast Archaeological Center
http://www.cr.nps.gov/seac/seac.htm

The *Mary Rose*
http://www.maryrose.org

Vasa Museum
http://www.vasamuseet.se/indexeng.html

Titanic Web Sites
http://seawifs.gsfc.nasa.gov/OCEAN_PLANET/HTML/
titanic.html
http://www.discovery.com/stories/science/titanic/reports/
live08161.html

Ancient Roman Shipwrecks
http://www.nationalgeographic.com/society/ngo/events/97/
ballardngt/index.html
http://www.cnn.com/TECH/9707/30/ancient.shipwrecks/

chapter two

Introduction to Earth

Sunset over the Pacific Ocean and the Scripps Institution
of Oceanography pier in La Jolla, California.

Learning Objectives

After reading this chapter, you should be able to

- Describe the development of the Earth, its atmosphere, and its oceans.

- Relate the size, shape, and scale of the Earth and its crustal features.

- Know the age of the Earth and understand radiometric dating and the geologic time scale.

- Describe the location system of latitude and longitude.

- Differentiate and describe a map projection, a bathymetric chart, and a physiographic map.

- Understand natural time cycles including day, year, and seasons.

- Explain the hydrologic cycle.

- Relate the distribution of oceans and land with elevation.

- Compare the dimensions of the four oceans.

- Know the kinds of electronic navigation aids used in oceanography.

billions of shimmering masses of stars, known as galaxies, move through the space we call the universe. About one-third of the way toward the center of one whirling mass of some 100 billion stars called the Milky Way galaxy, there is a fairly ordinary star, the Sun. Around the Sun move nine planets in their predictable and nearly constant orbits. The Sun and its nine planets are called the solar system, and the third planet from the Sun is called Earth (fig. 2.1).

Although we call our planet Earth, it is more accurate to consider it as the water planet. Water did not exist on the Earth in the beginning, but its formation on a planet that was not too far from the Sun nor too close, not too hot nor too cold, changed the Earth and allowed the development of life. In this chapter, we begin to investigate this planet Earth on which the largest bodies of water are known as the oceans.

The Beginnings   2.1

In recent years, observational data have provided increasing evidence that the universe originated in an event known as the **Big Bang.** The Big Bang model envisions all energy and matter in the universe as initially having been concentrated in an extremely hot, dense singularity much smaller than an atom. Roughly 13 billion years ago this singularity experienced a cataclysmic explosion that caused the universe to rapidly expand and cool as it grew larger. One second after the Big Bang, the temperature of the universe was about 10 billion K (roughly 1000 times the temperature of the Sun's interior). On the Kelvin (K) temperature scale, 0 K is absolute zero, the coldest possible temperature. At 0 K all atoms and molecules would stop moving. Room temperature is about 300 K (see Appendix C for conversions from K to °C and °F). At this time the universe consisted mostly of elementary particles, light, and other forms of radiation. The elementary particles, such as protons and electrons, were too energetic to combine into atoms. One hundred seconds after the Big Bang the temperature had cooled to about 1 billion K (roughly the temperature in the centers of the hottest stars at the present time). At this temperature the universe would still have been dominated by radiation but matter, in the form of nuclei of simple atoms such as hydrogen, deuterium, helium, and lithium, began to form. Later, as the universe cooled, matter took over. Eventually, when the temperature had dropped to a few thousand degrees, electrons and nuclei would have started to combine to form atoms. It was then possible for small variations in the distribution of matter to begin to grow by gravitational attraction, increasing their mass and density even further. A billion years or so after the Big Bang, gravity began to pull matter into the structures we see in the universe today. The first stars probably formed billions of years before the Sun.

Present theories attribute the beginning of our solar system to the collapse of a single, rotating cloud of gas and dust that included material that was produced within older stars and liberated into space when the older stars exploded. This rotating cloud, or **nebula,** formed about 5 billion years ago. The shock wave from a nearby exploding star, or supernova, is thought to have imparted spin to the cloud, pushing it together and causing it to shrink even further from its own gravitational pull. As the collapsing cloud's rotation speed increased, its temperature rose, and the gas and dust became disk shaped. At the center of this disk a star, the Sun, was formed. Self-sustaining nuclear reactions kept the Sun hot, but the outer regions of the disk began to cool and the gases began to interact chemically. The interactions produced particles of matter that grew through further chemical reactions and collisions with other particles. In this manner the planets of our solar system began to form.

The early Earth was bombarded by particles of all sizes, and a portion of their energy was converted into heat on impact. Each new layer of accumulated material buried the previous one, trapping the heat and raising the temperature of the Earth's interior. At the same time, the growing weight of the accumulating layers compressed the interior, raising the Earth's internal temperature. Atoms of radioactive elements, such as uranium and thorium, disintegrated by emitting subatomic particles that were absorbed by the surrounding matter, further raising the temperature.

Some time during the first few hundred million years after the Earth formed,

Figure 2.1

Earth as seen from space is the water planet.

More Information: www.nssdc.gsfc.nasa.gov/planetary/planets/earthpage.html

its interior reached the melting point of iron and nickel. When the iron and nickel melted, they migrated toward the center of the plant, displacing less dense material that moved upward and spread over the surface, cooling and solidifying. The process of melting and solidifying probably happened repeatedly, separating the lighter, less dense compounds from the heavier, denser substances in the interior of the planet. In this way the Earth became completely reorganized and differentiated into a layered system that will be discussed further in chapter 3.

The Earth's oceans are probably by-products of both this heating and differentiation. As the Earth warmed and partially melted, the components of water locked in the minerals as hy-

drogen and oxygen were released and carried to the surface in volcanic eruptions to form water vapor mixed with other gases. As the Earth's surface cooled, the water vapor condensed to form the oceans.

It is also believed that during the process of differentiation, gases released from the Earth's interior formed the first atmosphere, which was primarily made up of water vapor, hydrogen gas, hydrogen chloride, carbon monoxide, carbon dioxide, and nitrogen. Any free oxygen present would have quickly combined with the metals of the crust. Oxygen gas could not accumulate in the atmosphere until it was produced in amounts large enough to exceed its loss by chemical reactions with the crust. This did not occur until life evolved to a level of complexity in which green plants could convert carbon dioxide and water with the energy of sunlight into organic matter and free oxygen. This process and its significance to life are discussed in chapter 10.

> How and when is our solar system thought to have been formed?
>
> What processes added heat to the early Earth, and how was the Earth changed by this heat?
>
> What was the source of the early Earth's water and atmosphere?

The Age of the Earth 2.2

The Earth is an active planet, and its original surface rocks no longer exist. The oldest materials on the Earth's surface have been dated at slightly more than 4 billion years old. Moon samples returned by the Apollo mission are dated at a little over 4.4 billion years old. Meteorites that have struck the Earth have been dated between 4.5 billion and 4.6 billion years old. The substances found in many meteorites appear to represent materials that condensed out of the hot gases present at the beginning of the solar system. These ages agree with theoretical calculations made for the age of the Sun. This information sets the accepted age of the Earth at about 4.6 billion years.

The method of age-dating rock samples is known as **radiometric dating,** which is based on the decay of radioactive **isotopes** of elements such as uranium and potassium found in the rocks. An atom of a radioactive isotope has an unstable nucleus that changes, or decays. The time at which any single nucleus decays is unpredictable, but with large numbers of atoms, it is possible to predict that a certain fraction of the isotope will decay over a certain period of time. The time over which one-half of the atoms of an isotope decay is known as its **half-life.** The half-life of each isotope is characteristic and constant; for example, uranium-235 has a half-life of 704 million years. If a substance starts out containing only atoms of uranium-235, in 704 million years, the substance will be one-half uranium-235; the other half will be one of several decay products, primarily lead-207. Because each isotope system behaves differently in nature, data must be carefully tested, compared, and evaluated. The best analyses are those in which different isotope systems give the same date.

> What is the generally accepted age of the Earth?
>
> How are rocks dated?
>
> How much of a radioactive isotope would be left after two half-lives had passed?

Geologic Time 2.3

To refer to events in the history and formation of the Earth, scientists use geologic time (fig. 2.2 and table 2.1). The principal divisions of geologic time are the four eons: the Hadean (4.6 to 4.0 billion years ago), the Archean (4.0 to 2.5 billion years ago), the Proterozoic (2.5 billion to 570 million years ago), and the Phanerozoic (since 570 million years). Fossils are known from other eons, but they are common only in the Phanerozoic.

This eon is divided into three eras: the Paleozoic era of ancient life, the Mesozoic era of middle life (popularly called the Age of Reptiles), and the Cenozoic era of recent life (the Age of Mammals). Each of these eras is subdivided into periods and epochs; today, for instance, we live in the Holocene epoch of the Quaternary period of the Cenozoic era. The appearance or disappearance of fossil types was first used to set the boundaries of the time units. Radiometric dating has allowed scientists to assign numbers to these time-scale boundaries.

Very long periods of time are incomprehensible to most of us. We often have difficulty coping with time spans of more than ten years—what were you doing exactly ten years ago today? We have nothing with which to compare the 4.6-billion-year age of the Earth. In order to place geologic time in a framework we can understand, let us divide the Earth's age by 100 million. Then we can think of the Earth as being just forty-six years old. What has happened over that forty-six years?

There is no remaining record of events during the first three years. The earliest history preserved can be found in some rocks of Canada, Africa, and Greenland that formed forty-three years ago. Sometime between thirty-five and thirty-eight years ago the first primitive living cells of bacteria-type organisms appeared. Oxygen production by living cells began about twenty-three years ago, half the age of the planet. Most of this oxygen combined with iron in the early oceans and did not accumulate in the atmosphere. It took about eight years, or until roughly fifteen years ago, for enough oxygen to accumulate in the atmosphere to support significant numbers of complex oxygen-requiring cells. Oxygen reached its present concentration in the atmosphere approximately eleven years later. The first invertebrates (animals without backbones) developed seven years ago and two years later the first vertebrates (animals with backbones) appeared. Primitive fish first swam in the oceans and corals appeared just five years ago. Three years, eight and one-half months ago the first sharks could be found in the oceans and roughly five months later reptiles could be found on land. A massive extinction struck the planet just two and one-half years ago, killing ninety-six percent of all life. Following this catastrophic event the dinosaurs appeared just twenty-seven months and ten days ago. Three and one-half weeks later the first mammals developed. Just eighteen and one-half months ago the first birds flew in the air and they would have seen the first flowering plants a little less than five months later. A second major extinction struck the Earth 237 days ago, killing off the dinosaurs as well as many other species. Two hundred eleven days ago the mammals,

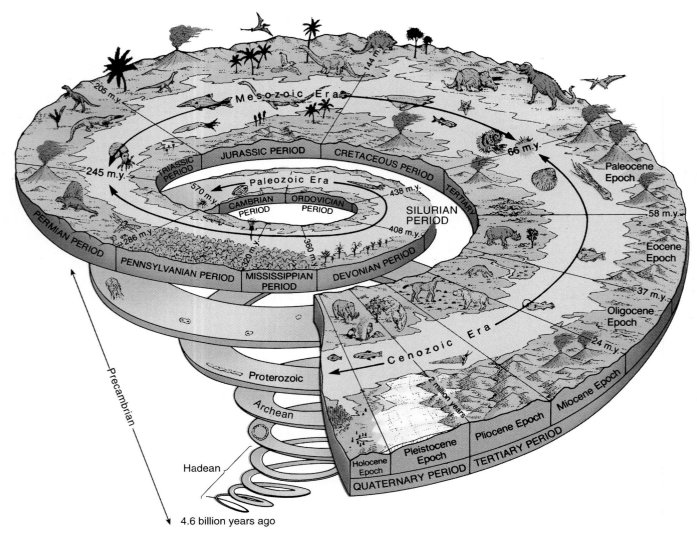

Figure 2.2

The history of the Earth showing evolution of life-forms.

More Information: www.vulcan.wr.usgs.gov/Glossary/geo_time_scale.html

Source: After Geologic Time, *U.S. Geological Survey publication.*

birds, and insects became the dominant land animals. Our first human ancestors (the first identifiable member of the genus *Homo*) appeared just a little less than six days ago. About half an hour ago, modern humans began the long process we know as recorded civilization, and only one minute ago, the Industrial Revolution began changing the Earth and our relationship with it for all time.

> How were the divisions of geologic time established before age-dating of rocks was possible?
>
> What is the significance of the Industrial Revolution occurring "one minute ago" in the history of the Earth?

The Shape of the Earth 2.4

As the Earth cooled and turned on its axis, gravity and the forces of rotation produced its nearly spherical shape. The Earth sphere has a **mean,** or average, radius of 6371 km (3959 mi). It has a shorter polar radius (6356.9 km; 3950 mi) and a longer equatorial radius (6378.4 km; 3963 mi), a difference of 21.5 kilometers or 13 miles (fig. 2.3). The difference between the polar and equatorial radii is due primarily to the Earth's spin, which tends to flatten its shape, producing a slight bulge at the equator due to centrifugal force.

The top of Mount Everest, the Earth's highest mountain, is 8846 m (29,022 ft) above sea level, while the deepest ocean depth, the bottom of the Challenger Deep in the Mariana

table 2.1 The Geologic Time Scale

Eon	Era	Period	Epoch	Began (millions of years ago)	Life forms/Events
Phanerozoic	Cenozoic "Age of Mammals"	Quaternary	Holocene	0.01	Modern humans
			Pleistocene	1.6	Earliest humans
		Tertiary	Pliocene	5.3	
			Miocene	23.7	Earliest hominids
			Oligocene	36.6	Flowering plants
			Eocene	57.8	Earliest grasses
					Mammals, birds, and insects dominant
			Paleocene	65	

Cretaceous-Teritiary boundary: The extinction of dinosaurs and many other species at the end of the Mesozoic Era (65 million years ago).

Eon	Era	Period	Epoch	Began (millions of years ago)	Life forms/Events
	Mesozoic "Age of Reptiles"	Cretaceous		144	Earliest flowering plants (115) Dinosaurs in ascendence
		Jurassic		208	First birds (155) Dinosaurs abundant
		Triassic		245	First turtles (210) First mammals (221) First dinosaurs (228) First crocodiles (240)

Permian-Triassic boundary: The greatest mass extinction of all time. Ninety-six percent of all life on Earth perishes at the end of the Paleozoic Era (245 million years ago).

Eon	Era	Period	Epoch	Began (millions of years ago)	Life forms/Events
	Paleozoic	Permian	"Age of Amphibians"	286	Extinction of trilobites and many other marine animals
		Carboniferous		360	First reptiles (330) Large coal swamps Amphibians abundant
		Devonian	"Age of Fishes"	408	First seed plants (365) First sharks (370) First insect fossils (385) Fishes dominant
		Silurian		438	First vascular land plants (430)
		Ordovician	"Age of Invertebrates"	505	First land plants similar to lichen (470) First fishes (505) Earliest corals Marine algae
		Cambrian		570	Abundant shelled invertebrates Trilobites dominant
Proterozoic Archean Hadean	collectively these are popularly known as the Precambrian			2500	First invertebrates (700) Earliest shelled organisms (~750) First fossil evidence of single-celled life with a cell nucleus: eukaryotes (1500)
				4000	First evidence of by-products of eukaryotes (2700) Earliest primitive life, bacteria and algae: prokaryotes (3500–3800). These will dominate the world for the next 3 billion years. Oldest surface rocks (4030)
				4600	Oldest single mineral (4300) Oldest Moon rocks (4440) Oldest meteorites (4560)

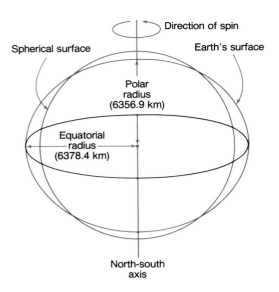

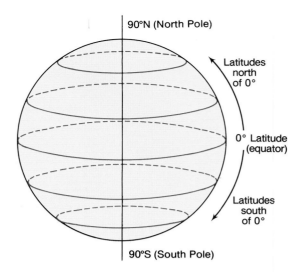

Figure 2.3

The rotation of the Earth on its axis stretches the equator. Notice that the equatorial radius is longer than the polar radius.

More Information: 164.214.2.59/GandG/geolay/TR80003A.HTM#ZZ5

Figure 2.4

Latitude lines are drawn parallel to the equatorial plane.

More Information: www-spof.gsfc.nasa.gov/stargaze/Slatlong.htm_A.html

Trench of the western Pacific Ocean, is 11,022 m (36,163 ft) deep. On a scale model of the Earth the size of a basketball, the topography of the surface of the Earth would be similar to the texture of a basketball, or on a scale model the size of a grapefruit the topography would be similar to the texture of grapefruit peel. The topography of the surface of the Earth, its high mountains and deep oceans, are minor compared to the size of the entire planet.

> What is the shape of the planet Earth?
>
> Relate the Earth's surface heights and depths to its size.

Latitude and Longitude   2.5

Whether one is an oceanographer, a geographer, an airplane pilot, or a ship's captain, it is necessary to fix one's location on the Earth's surface with precision and accuracy. To do so, a grid of reference lines that cross at right angles over the Earth's surface is used. These grid lines are determined by internal Earth angle measurements and are called lines of **latitude** and **longitude** (see Appendix A). Lines of latitude, also called parallels, begin at the equator, 0° latitude. Other latitude lines encircle the Earth parallel to the equator, north to 90°N or the North Pole, and south to 90°S, the South Pole (fig. 2.4). Notice that parallels of latitude describe increasingly

smaller circles as the poles are approached. All parallels of latitude north of the equator must be designated as either north or positive latitude and those south of the equator must be designated as either south or negative latitude. Since the distance expressed in whole degrees is large (1 degree of latitude equals 60 nautical miles (nm) (about 111 km at the Earth's surface), each degree is divided into 60 minutes of arc and each minute into 60 seconds of arc. Thus 1 minute of latitude equals 1 nm (1.852 km or 1852 m) and 1 second of latitude equals about 101 ft (roughly 31 m).

Lines of longitude, also called meridians, are formed at right angles to the latitude grid, extending from the North Pole to the South Pole. The longitude grid begins at an arbitrarily chosen meridian: 0° longitude is the meridian that passes directly through the Royal Naval Observatory in Greenwich, England, just outside of London (fig. 2.5). The 0° longitude meridian is called the **prime meridian.** On the opposite side of the Earth from the Prime meridian is the 180° longitude line. The 180° meridian approximates the **international date line.** Longitude may be recorded in two different ways, either halfway around the world east and west of the prime meridian or all the way around the world moving continuously east of the prime meridian (fig. 2.6). Thus longitude may be reported as either 0°–180° east (also known as positive longitude) and 0°–180° west (also known as negative longitude) or 0°–360° east (always positive). In this manner −90°, 90°W, and 270° longitude all mark the same meridian. Because longitude lines rotate with the turning Earth, it is necessary to know the position of the Sun or the stars with time to determine longitude. The reference time used is clock time, set to 12 noon when the Sun is directly above 0° longitude. This is **Greenwich Mean Time (GMT),** now called **Universal Time** or Zulu time. Since Sun time changes by one hour for each 15° of longitude, the Earth has been divided into 24 time zones that are 15° of longitude wide (fig. 2.7). Longitude is determined by comparing local Sun time and Universal Time. If the Sun appears directly overhead at 1000 UT, the observation location is 30°E of the prime meridian.

> Sketch parallels of latitude and meridians of longitude on a sphere; include the equator and the prime meridian.
>
> Define the equator and the prime meridian.
>
> Why is time important to the measurement of longitude?
>
> Why have time zones been established?

East ◄————— 0° —————► West

Figure 2.5

The Royal Naval Observatory at Greenwich, England. The brass strip set into the courtyard marks the prime meridian, the division between east and west longitudes.

More Information: www.rog.nmm.ac.uk/

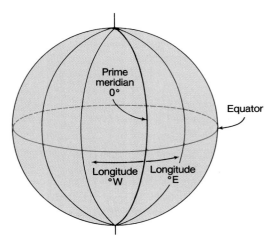

Figure 2.6

Longitude lines are drawn east and west of the prime meridian.

More Information: www-spof.gsfc.nasa.gov/stargaze/Slatlong.htm

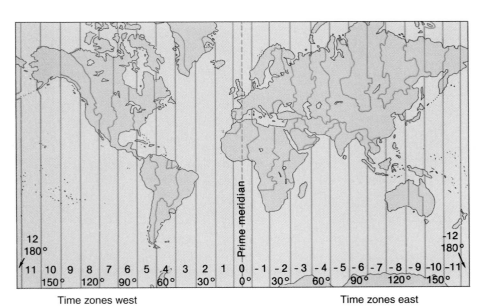

Time zones west   Time zones east

Figure 2.7

Distribution of the world's time zones. Degrees of longitude are marked along the bottom of the diagram. Time zones are positive (west zones) or negative (east zones). Time at Greenwich is determined by adding the number of hours represented by the zone number to the local time.

More Information: www.worldtimezone.com/ and www-spof.gsfc.nasa.gov/stargaze/Slatlong.htm

Chart Projections   2.6

Maps and charts show the Earth's three-dimensional surface on a flat or two-dimensional surface. Maps usually show the Earth's land features, while charts depict the sea and sky. Making flat maps or charts from a curved Earth surface produces distortion, and the user must select the most accurate, most convenient, and least distorted type for his or her purpose.

In order to understand this problem of distortion, imagine a transparent globe with the continents and the latitude and longitude lines painted on its surface. Place a light in the center of the globe and let the light rays shine through it. Hold a piece of paper up to the outside of the globe. The light will project the shadows of the continents and the latitude and longitude lines onto the paper forming a map or chart **projection.** Different projections are obtained by varying the position of the light and the shape of the surface on which the projection is made. Most projections are modifications of three basic types: cylindrical, conic, and tangent plane, all shown in figure 2.8.

Compare the three parts of figure 2.8 and notice that distortion on a map or chart increases as the distance from its place of contact to the globe increases. Consider Greenland. In figure 2.8c its size and shape are very close to its true form on Earth. In figure 2.8b the island has grown larger, and in figure 2.8a both its size and shape are greatly distorted. Notice that each type of chart or map has its own characteristics.

Charts of the ocean that show lines connecting points of similar depth below the sea surface are **bathymetric** charts (fig. 2.9). The connecting lines are known as **contours** of depth or elevation. Color, shading, and perspective drawing may be added to indicate elevation changes and produce a **physiographic** map. Compare the physiographic map in figure 2.10

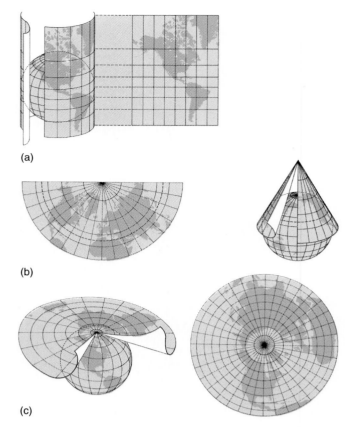

(a)

(b)

(c)

Figure 2.8

The three basic types of map projections: (a) An equatorial cylindrical projection. (b) A simple polar conic projection. (c) A polar tangent plane projection.

More Information: www.aquarius.geomar.de/omc/ and hammondmap.com/proani.html

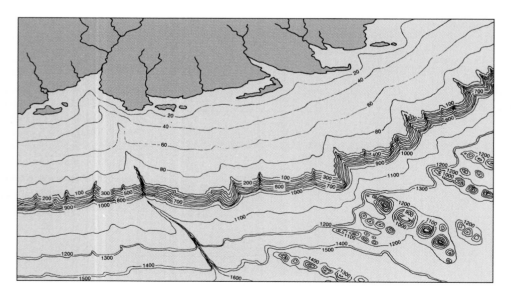

Figure 2.9

A bathymetric, or contour, chart of the sea floor along a section of generalized coast. Water depth in discrete increments is indicated by contour lines. Changes in the pattern and spacing between contour lines reflect variations in the slope and orientation of the bottom. Closely spaced contours indicate steep slopes while widely spaced contours indicate gentler slopes.

Figure 2.10

A physiographic map of the same area shown in figure 2.9.

with the bathymetric chart in figure 2.9. Today computers use electronic ocean depth measurements to form detailed bathymetric charts that are converted into three-dimensional, color images of the sea floor from any angle (fig. 2.11).

> Which of the projections shown in figure 2.9 would provide the most accurate map of a. Alaska? b. Central America? c. Northern Europe? d. Cape Horn?
>
> How are bathymetric charts and physiographic maps similar?
>
> How do they differ?

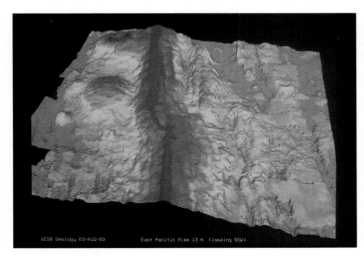

Figure 2.11

Three-dimensional computer processed image of a section of the East Pacific Rise. Data for this image were obtained by using the Sea Beam echo sounder.

Natural Cycles   2.7

Natural cycles affect the Earth's atmosphere, its water, and its land. An understanding of these cycles helps us understand processes that occur at the ocean surface and are discussed in later chapters: climate zones, winds, currents, vertical water motion, plant life, and animal migration. The average time for the Earth to make one rotation relative to the Sun is one day or twenty-four hours. The Earth completes one orbit about the Sun in about $365^{1}/_{4}$ days or one year. For convenience, the year is set at 365 days with an extra day added every four years, except years ending in hundreds and not divisible by 400.

As the Earth follows its nearly circular orbit around the Sun, those who live in temperate zones and polar zones are very conscious of the seasons and of the differences in the lengths of the periods of daylight and darkness. The reason for these seasonal changes is seen in figure 2.12. The Earth moves with its axis tilted $23^{1}/_{2}°$ from the vertical to its orbit. Therefore, during the year the Earth's North Pole is sometimes tilting toward the Sun and sometimes tilting away from it. The Northern Hemisphere receives its maximum hours of sunlight when the North Pole is tilted toward the Sun; this is the Northern Hemisphere's summer. During the same period the South Pole is tilted away from the Sun, so that the Southern Hemisphere receives the least sunlight; this period is winter in the Southern Hemisphere. At the other side of the Earth's orbit, the North Pole is tilted away from the Sun, creating the Northern Hemisphere's winter, and the South Pole is inclined toward the Sun, creating the Southern Hemisphere's summer. Note also that during summer in the Northern Hemisphere it is always light around the North Pole and always dark around the South Pole; the opposite is true during the Northern Hemisphere's winter.

The periods of daylight in the Northern Hemisphere increase as the Sun moves north to stand above $23^{1}/_{2}°$N latitude,

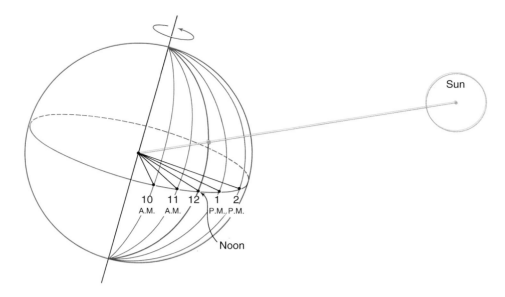

Figure 2.12
The time on a meridian relative to the Sun changes by one hour for each 15° change in longitude.

the **Tropic of Cancer.** This occurs at the **summer solstice** on or about June 22, the day of the year with the longest period of daylight and the beginning of summer in the Northern Hemisphere. On this day the Sun does not sink below the horizon above $66\frac{1}{2}$°N latitude, the **Arctic Circle,** nor does it rise above $66\frac{1}{2}$°S latitude, the **Antarctic Circle.** Following the summer solstice the Sun appears to move southward until on or about September 23, the **autumnal equinox,** it stands directly above the equator. On this day the periods of daylight and darkness are equal all over the world. The Sun continues its southward movement until about December 21, when it stands over $23\frac{1}{2}$°S latitude, the **Tropic of Capricorn,** at the **winter solstice.** On this day the daylight period is the shortest in the Northern Hemisphere, and above the Arctic Circle the Sun does not rise. In the Southern Hemisphere the daylight is longest, and above the Antarctic Circle the Sun does not set. The Sun then begins to move northward, and about March 21, the **vernal equinox,** it stands again above the equator, and the periods of daylight and darkness are once more equal around the world.

The greatest annual variation in intensity of direct solar illumination occurs in the **temperate** zones. In the polar regions, the seasons are dominated by the long periods of light and dark, but the intensity of solar heating is small, as the Sun is always low on the horizon. Between the Tropics of Cancer and Capricorn, there is little seasonal change in solar radiation levels, as the Sun never moves beyond these boundaries.

Notice on figure 2.12 that the Sun is not in the center of the path traced by the Earth in its orbit. During the Northern Hemisphere's summer, the Earth is farther away from the Sun (152.2×10^6 km or 94.5×10^6 mi) than it is in winter (148.5×10^6 km or 92.2×10^6 mi).

Diagram the Earth's orbit of the Sun, and explain why the seasons change during the orbit.

What is the latitude of the Tropic of Cancer, Tropic of Capricorn, Arctic Circle, and Antarctic Circle?

Why are the Arctic and Antarctic Circles displaced from the poles by $23\frac{1}{2}$°?

How will the seasons change over a calendar year at each of these latitudes?

The Hydrologic Cycle 2.8

The Earth's water is found as a liquid in the oceans, rivers, lakes, and below the ground surface; it occurs as a solid in glaciers, snow packs, and sea ice; it takes the form of droplets and gaseous water vapor in the atmosphere. The places in which water resides are called **reservoirs,** and each type of reservoir, when averaged over the entire Earth, contains a certain amount of water at any one instant. But water is constantly moving into and out of reservoirs. This movement of water through the reservoirs, diagrammed in figure 2.13, is called the **hydrologic cycle.**

Water is taken out of the oceans and enters the atmosphere by evaporation where it may condense to form clouds. Most of this water returns directly to the sea by precipitation, but air currents carry some water vapor over the continents. Precipitation in the form of rain and snow transfers this water from the atmosphere to the land surface, where it percolates

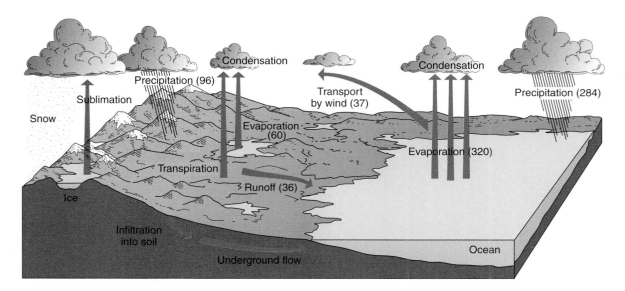

Figure 2.13

The hydrologic cycle and annual transfer rates for the whole Earth. Precipitation transfer rate includes both snow and rain. Evaporation transfer rate from the continents includes evaporation of surface water, transpiration, (the release of water to the atmosphere by plants), and sublimation (the direct change in state from ice to water vapor). Runoff from the continents includes both surface flow and underground flow. Annual transfer rates are in thousands of cubic kilometers (10^3 km^3).

More Information: www.epa.gov/seahome/groundwater/src/cycle.htm#cycle and ga.water.usgs.gov/edu/

into the soil, is taken up by plants, fills rivers, streams, and lakes, or remains for longer periods as snow and ice in some areas. Some of this water will return to the atmosphere by evaporation of surface water, **transpiration,** the release of water by plants, and **sublimation,** the conversion of ice directly to water vapor. Melting snow and ice, rivers, groundwater, and land runoff move water from the continents back to the oceans. For a comparison of the water volume stored in different reservoirs, see table 2.2 and figure 2.14.

The properties of climate zones are principally determined by their surface temperature (mean Earth surface temperature is 16°C) and their evaporation-precipitation patterns: the moist, hot equatorial regions, the dry, hot subtropic deserts, the cool, moist temperate areas, and the cold, dry polar zones. Differences in these properties, coupled with the movement of air between the climate zones, moves water through the hydrologic cycle from one reservoir to another at different rates. The transfer of water between the atmosphere and the oceans alters the salt content of the oceans' surface water and, with the seasonal latitude changes in surface temperature, determines many of the characteristics of the world's oceans, which will be explored in chapters 6 and 7.

Table 2.2   The Earth's Water Supply

Reservoir	Approximate Water Volume (km³)	Approximate Water Volume (mi³)	Approximate Percent of Total Water
Oceans & Sea Ice	1,326,370,000	317,000,000	97.24
Ice caps & Glaciers	29,289,000	7,000,000	2.14
Groundwater	8,368,000	2,000,000	0.61
Freshwater lakes	125,500	30,000	0.009
Saline lakes & Inland seas	105,000	25,000	0.008
Soil moisture	67,000	16,000	0.005
Atmosphere	13,000	3,100	0.001
Rivers	1,250	300	0.0001
Total water volume	1,364,339,000	326,000,000	100

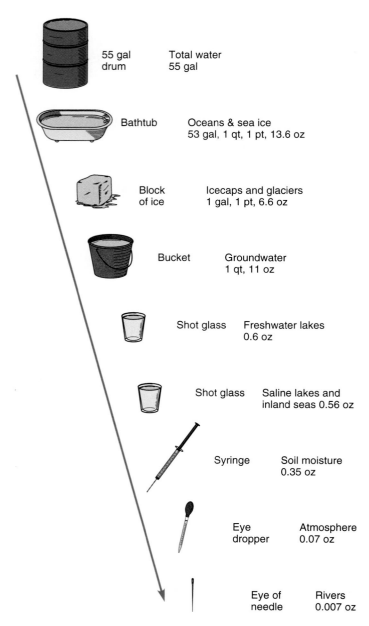

Figure 2.14

Comparison of the amount of the world's water supply held in each of the major water reservoirs. For purposes of illustration the Earth's total water supply has been scaled down to the volume of a 55-gallon drum.

Use the annual transfer amounts in figure 2.13 to show that ocean volume does not change during the year.

Why does the whole Earth cycle shown in figure 2.13 not apply to a specific region of the Earth?

What is the relationship between the hydrologic cycle and climate zones?

Distribution of Land and Water   2.9

The oceans cover 361 million square kilometers (361×10^6 km^2 or 139×10^6 mi^2) of the Earth's surface. (If you are unfamiliar with scientific notation to express very large numbers, see Appendix C.) Because these numbers are so large, they do not convey a clear idea of size; therefore, an easier concept to remember is that 71% of the Earth's surface is covered by the oceans, and only 29% is land above sea level.

The volume of water in the oceans is enormous: 1.37 billion cubic kilometers (1.37×10^9 km^3). A cubic kilometer of seawater is very large indeed. Consider this: the largest building in the world in cubic capacity is the main assembly plant of the Boeing Company in Everett, Washington. This building is used for the manufacture of Boeing airplanes and has a cubic capacity of 0.0847 km^3; in other words, 11.8 of these buildings would fit into one cubic kilometer. Another way to express the oceanic volume is to think of a smooth sphere with exactly the same surface area as the Earth uniformly covered with the water from the oceans. Such a sphere would be covered by a layer of water 2680 m (8800 ft) deep. If the water from all other sources in the world were added, the depth would rise to 2743 m (9000 ft).

To understand the distribution of land and water over the Earth consider the Earth when it is viewed from the north (fig. 2.15a) and from the south (fig. 2.15b). About 70% of the Earth's landmasses are in the Northern Hemisphere, and most of the land lies in the middle latitudes. The Southern Hemisphere is the water hemisphere, with its land located mostly in the tropic latitudes and in the polar region.

Another method used by oceanographers to depict land-water relationships is shown in figure 2.16. This graph of depth or elevation versus area is called a **hypsographic curve.** Since volume is the product of height and area, the hypsographic curve can also be used to show both the volume of land above sea level and the volume of the oceans below sea level.

How much of the Earth's surface is land, and how much is covered by the oceans?

How do the land-water distributions of the Northern and Southern Hemispheres differ?

What does the hydrographic curve show us about the relationship of the Earth's land and water?

The Oceans   2.10

The distribution and shape of the continents divides the world ocean into at least three, and some would argue as many as five, individual oceans. Oceanographers all recognize the three major

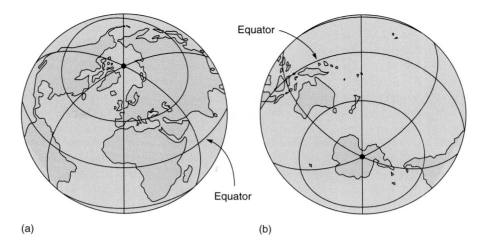

(a)   **(b)**

Figure 2.15

Continents and oceans are not distributed uniformly over the Earth. (a) The Northern Hemisphere contains most of the land; (b) the Southern Hemisphere is mainly water.

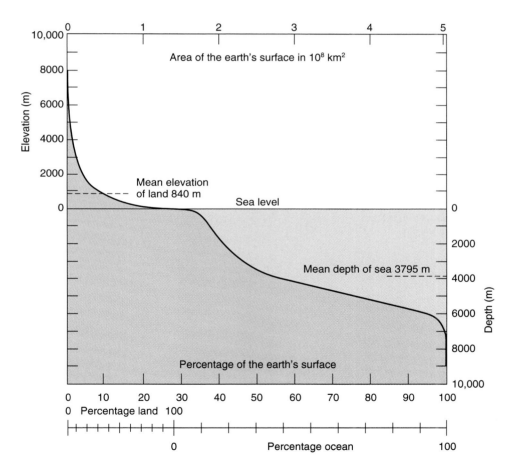

Figure 2.16

The hypsographic curve displays the area of the Earth's surface at elevations above and below sea level.

oceans—the Pacific, Atlantic, and Indian Oceans—that extend northward as three fingers from a common source of undivided ocean surrounding Antarctica. This circumpolar body of water can either be thought of as an additional ocean at latitudes above 50°S, commonly called the Southern Ocean or the Antarctic Ocean, or it can be divided along lines of longitude to become the southernmost extensions of the Pacific, Atlantic, and Indian Oceans. The Arctic Ocean is often considered an extension of the North Atlantic but because of its size and relative isolation many people consider it to be an independent ocean basin. The four most commonly recognized oceans, the Pacific, Atlantic, Indian, and Arctic, are shown in figure 2.17. Each of these oceans has its own characteristic surface area, volume, and mean depth.

The Pacific Ocean has greater surface area, volume, and mean depth than any of the other oceans. The Pacific was named by Ferdinand Magellan in 1520 for the calm weather they enjoyed while crossing it (*paci* = peace). It covers a little over a third of Earth's surface and just over half of the world ocean's surface. At its maximum width near 5°N it stretches 19,800 km (12,300 mi) from Indonesia to Columbia. There are roughly 25,000 islands in the Pacific, the majority of which are south of the equator. This is more than the total number of islands in the rest of the oceans combined. There are a number of marginal seas along the edges of the Pacific including the Celebes Sea, Coral Sea, East China Sea, Sea of Japan, Sulu Sea, and Yellow Sea.

The Atlantic Ocean is the second largest. Its name is derived from Greek mythology and means "Sea of Atlas" (Atlas was a Titan who supported the heavens by means of a pillar on his shoulders). The land area that drains into the Atlantic is four times larger than that of either the Pacific or Indian Oceans. There are relatively few islands given the size of the basin. The irregular coastline of the Atlantic includes a number of bays, gulfs, and seas. Some of the larger ones include the Caribbean Sea, Gulf of Mexico, Mediterranean Sea, Black Sea, and Baltic Sea.

The Indian Ocean is primarily a Southern Hemisphere ocean; it is the third largest but is quite deep. The northernmost extent of the Indian Ocean is in the Persian Gulf at about 30°N. The Indian Ocean is separated from the Atlantic Ocean to the west by the 20°E meridian and from the Pacific Ocean to the east by the 147°E meridian. It is nearly 10,000 km (6,200 mi) wide between the southern tip of Africa and Australia. For centuries it has had tremendous strategic importance as a trade route between Africa and Asia.

The Arctic Ocean is the smallest of the four, occupying a roughly circular basin over the North Pole region. It is connected to the Pacific Ocean through the Bering Strait and to the Atlantic Ocean through the Greenland Sea. Its floor is divided into two deep basins by an underwater mountain range.

Figure 2.17

The world's four major oceans in order of decreasing size. (a) Pacific Ocean (the deepest spot in the oceans is located in the Mariana Trench); (b) Atlantic Ocean; (c) Indian Ocean; and (d) Arctic Ocean.

More Information: oceanographer.navy.mil/pacific.html (Specific information about the Pacific Ocean. To obtain similar information for the other oceans just replace *pacific* in the address with the name of the appropriate ocean.)

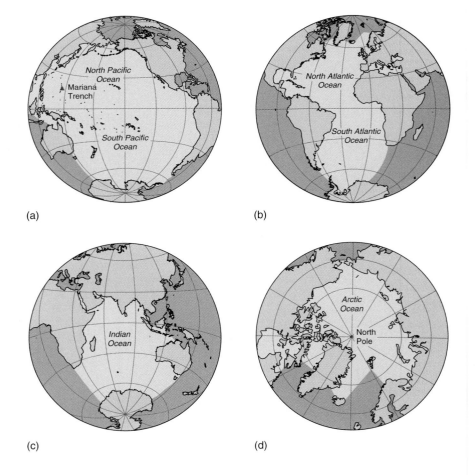

(a)

(b)

(c)

(d)

Table 2.3   Ocean Depths, Areas, and Volumes

Ocean	Average Depth	Area	Volume	Percent of Ocean Surface Area	Percent of Earth's Surface Area
Pacific	4028 m 13,216 ft	179.7×10^6 km^2 69.3×10^6 mi^2	723.8×10^6 km^3 173.3×10^6 mi^3	50.1	35.5
Atlantic	3332 m 10,932 ft	93.7×10^6 km^2 36.1×10^6 mi^2	312.2×10^6 km^3 74.8×10^6 mi^3	25.9	18.4
Indian	3897 m 12,786 ft	73.6×10^6 km^2 28.4×10^6 mi^2	286.8×10^6 km^3 68.7×10^6 mi^3	20.4	14.4
Arctic	1117 m 3665 ft	14.1×10^6 km^2 5.4×10^6 mi^2	15.7×10^6 km^3 3.8×10^6 mi^3	3.9	2.8
All Oceans	3795 m 12,451 ft	361.1×10^6 km^2 139.3×10^6 mi^2	1370.4×10^6 km^3 328.2×10^6 mi^3	100	70.8

The major flow of water into and out of the Arctic Ocean is through the North Atlantic Ocean. Use table 2.3 to help compare the four oceans.

> Name the oceans of the world.
>
> Which has the greatest surface area?
>
> Which has the least total volume of water?

Modern Navigation   2.11

Navigators may still use very accurate timepieces and wait for clear skies to "shoot" the Sun or stars with a sextant to determine positions at sea, but such measurements are used primarily to check their modern electronic navigational equipment. Chronometers are calibrated, or reset, by broadcast time signals. When vessels are near land, **radar** (radio detecting and ranging) can be used to bounce radio pulses off a target such as the shoreline or another vessel. The image on the radar screen is formed by energy that is sent out by a transmitter, reflected from an object, returned to the antenna, and then displayed on the screen. **Loran** (long-range navigation) can be used farther out at sea in areas that are within the receiving range of land-based transmitting stations. This electronic timing device measures the difference in arrival time of radio signals from pairs of land stations. A master station and up to 4 secondary stations transmit a synchronized pulse. The time differential between the master station pulse and the secondary pulses is measured.

From this data, a loran receiver can calculate the position of a vessel within 50 m or better within about 2000 km (1200 mi) of the coast of the United States. Modern loran receivers can be programmed with the latitude and longitude of a desired destination. The receiver can then read out the course to be sailed, the distance to the destination, the vessel's speed, and the estimated arrival time.

The **satellite navigation system** is an accurate and sophisticated navigational aid. Satellites orbiting the Earth transmit signals of a precise frequency that are picked up by a receiver on the ship. The ship's receiver monitors the frequency shift of the signal as the satellite passes overhead and determines the exact instant in time at which the frequency is correct. At this instant the ship's path and the satellite's orbit are at right angles. Given this information, a computer programmed with the satellite's orbital properties can determine the ship's position to within 30 m (100 ft) or less.

A more versatile and accurate method of finding one's position uses the U.S. Navstar **Global Positioning System (GPS).** The Global Positioning System is a worldwide radio-navigation system consisting of twenty-four navigational satellites and five ground-based monitoring stations (fig. 2.18). At any given time, from five to eight satellites are above the horizon from any point on the Earth. The system uses this constellation of satellites as reference points for calculating positions on the surface with an accuracy of about 10 m (33 ft) horizontally and 13 m (40 ft) vertically.

Each GPS satellite transmits a unique digital code. GPS receivers generate codes that are identical to those sent by the satellites at exactly the same time the satellites do. Because of the distance the satellite signals travel, there is a lag between the time when the receiver generates a specific signal and when it receives the same signal generated by the satellite. This time

(a)

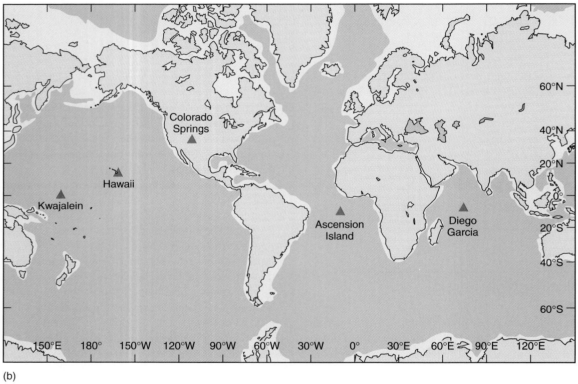

(b)

Figure 2.18

The Global Positioning System (GPS) consists of (a) twenty-four satellites and (b) five monitoring stations. The master control station is located at Falcon Air Force Base in Colorado Springs.

More Information: www.trimble.com/gps/index.htm

delay is a function of the distance between the receiver and the satellite. Precise measurements of distance between a receiver and four satellites will pinpoint the location of the receiver. In practice, there are a variety of sources of error in the measurements, so measurements are made between the receiver and as many satellites as it can detect to maximize location accuracy.

GPS can be used to measure the velocity of a moving receiver as well as its location.

GPS is used in a wide variety of applications including positioning and navigating marking the location of instruments, mapping land and sea-surface features, monitoring tides and currents, and measuring the motion of tectonic plates.

Oceanographers are currently developing new techniques for using GPS in underwater applications. A major problem in the use of GPS underwater is that GPS signals are radio waves, which are quickly absorbed by water. Underwater GPS systems that are being tested involve surface buoys containing GPS receivers that communicate with underwater targets using acoustic (sound) energy. Solutions to these problems will likely be developed rapidly because of the enormous benefit of being able to use GPS in underwater applications.

Shipboard computers can now store an electronic atlas that includes surface charts and seafloor bathymetry. Oceanographic vessels use devices that draw charts showing the vessel's changing position and, at the same time, conduct a seafloor bathymetric survey and keep track of water measurements (temperature and salt content) automatically as the ship moves along. Today's ocean scientists are able to return more and more accurately to the same places in the sea to repeat measurements and follow changes in ocean processes. Scientists have the ability to evaluate data as they are being taken, allowing changes to be made in a vessel's sampling pattern, thus insuring that the data satisfy a project's research requirements. Inexpensive, pocket-size GPS receivers are now available and are being used by recreational mariners, hikers, and back country adventurers to ensure that they know their position at all times, especially in the case of an emergency.

Why is electronic navigation more important to oceanographers than celestial navigation?

How do radar, loran, and satellite navigation systems differ from each other?

There is a fixed amount of water on Earth. Evaporation and precipitation move the water through the reservoirs of the hydrologic cycle. Seventy-one percent of the Earth's surface is covered by oceans. The Northern Hemisphere is the land hemisphere, and the Southern Hemisphere is the water hemisphere. The hypsographic curve is used to show land-water relationships of depth, elevation, area, and volume. The Earth has three large oceans extending north from the Southern Ocean and a fourth small ocean covering the North Pole region. Each has a characteristic surface area, volume, and mean depth.

Modern navigational techniques make use of radar, loran signals, computers, and satellite systems. The GPS allows more accurate position readings and maps land and ocean surface features.

Key Terms

Antarctic Circle, 34
Arctic Circle, 34
autumnal equinox, 34
bathymetric, 32
Big Bang, 25
contours, 32
Global Positioning System (GPS), 39
Greenwich Mean Time (GMT), 30
half-life, 27
hydrologic cycle, 34
hypsographic curve, 36
international date line, 30
isotope, 27
latitude, 30
longitude, 30
loran, 39

mean, 28
nebula, 25
physiographic, 32
prime meridian, 30
projection, 32
radar, 39
radiometric dating, 27
reservoir, 34
satellite navigation systems, 39
sublimation, 35
summer solstice, 34
temperate, 34
transpiration, 35
Tropic of Cancer, 34
Tropic of Capricorn, 34
Universal Time, 30
vernal equinox, 34
winter solstice, 34

Summary

The solar system began as a rotating cloud of gas. Over millions of years, the Earth heated, cooled, changed, and collected a gaseous atmosphere and an accumulation of liquid water.

Because it rotates, the Earth's shape is not perfectly symmetrical. Its surface is relatively smooth when compared to its size.

Reliable age dates for rocks are obtained by radiometric dating. The accepted age of the Earth is about 4.6 billion years. The geologic time scale is used to compile the Earth's history.

Latitude and longitude are used to locate positions on the Earth's surface. To determine longitudinal position requires accurate time measurement. Different types of map and chart projections have been developed. Bathymetric charts and physiographic maps use depth contours and elevation to depict the Earth's surface.

Natural cycles are based on the motions of the Earth and the Sun. Because of the tilt of the Earth's axis as it orbits the Sun, the Sun moves annually between $23\frac{1}{2}°$N and $23\frac{1}{2}°$S, producing the seasons.

Critical Thinking

1. Estimates of the age of the Earth have changed over the past centuries. Do you think the present estimate of the Earth's age will change in the future? Why or why not?
2. The route of a ship sailing a constant compass course on an equatorial cylindrical projection is indicated by a straight line that cuts all longitude lines at the same angle. This is a rhumb line. How does this line appear on a polar conic projection, a globe, and a tangent plane projection centered on the pole?
3. Use figure 2.13 to trace several routes for a water molecule moving between a mountain lake and an ocean. In which reservoirs would the molecule spend the greatest amounts of time and in which the least?
4. Use the values of area and volume given for all oceans in table 2.3 to find the ocean depth on a smooth Earth uniformly covered with water. Note: value will vary slightly from that cited in the text due to rounding off.
5. At what 1000 m depth interval in the world's oceans would the greatest change in ocean area occur? Use the hypsographic curve (fig. 2.16) to determine your answer.

Suggested Readings

Alper, J. 1994. Earth's Violent Birth. *Earth* 3(7):56–63.

Bowditch, N. 1984. *American Practical Navigator*, vol. 1, U.S. Defense Mapping Agency Hydrographic Center, Washington D.C., 1414 pp. History of navigation, chart projections, and navigation aids are covered in chapters 1–5 and 41–46.

Bjerklie, D., B. Hillenbrand, and J. O. Jackson. 1993. How Did Life Begin? *Time* 11 October:68–74.

Cone, J. 1994. Life's Undersea Beginnings. *Earth* 3(4):34–41.

Glick, D. 1994. Windows on the World. *National Wildlife* 32(2):6–13.

Gore, R. 1985. The Planets, between Fire and Ice. *National Geographic* 167(1):4–51.

Herring, T. 1996. The Global Positioning System. *Scientific American* 274(2):32–38.

Hoffmann, H. 2000. The Rise of Life on Earth. *National Geographic* 198(3):100–113. Discusses the Permian extinction.

Livermore, B. 1993. Bottoms Up. *Sea Frontiers* 39(3):40–45. Satellite images and seafloor mapping.

Macchetto, F. D., and M. Dickinson. 1997. Galaxies in the Young Universe. *Scientific American* 276(5):92–99.

Maranto, G. 1991. Way Above Sea Level. *Sea Frontiers* 37(4):16–23. A discussion of GPS.

Monastersky, R. 1998. The Rise of Life on Earth. *National Geographic* 193(3):54–81. Scientific theories about the first living organisms on earth.

Newcott, W. R. 1997. Hubble's Eye on the Universe. *National Geographic* 191(4):2–17.

Richelson, J. T. 1998. Scientists in Black. *Scientific American* 278(2):48–55.

Sawyer, K. 1999. Unveiling the Universe. *National Geographic* 196(4):8–41.

Sobel, D. 1995. *Longitude: The True Story of a Lone Genius Who Solved the Greatest Scientific Problem of His Time*. Walker Publishing Company, Inc., New York, 184 pp.

York, D. 1993. The Earliest History of the Earth. *Scientific American* 268(1):90–96.

internet references

worldwide websites

General Astronomy

AstroWeb—list of astronomy links
http://www.cv.nrao.edu/fits/www/astronomy.html

Cambridge Astronomy
http://www.ast.cam.ac.uk/

NASA High Energy Astrophysics Science Archive Research Center—general information and photographs
http://heasarc.gsfc.nasa.gov/

NASA's Origins Program
http://eis.jpl.nasa.gov/origins/index.html

National Space Science Data Center—photo gallery
http://nssdc.gsfc.nasa.gov/photo_gallery/photogallery.html

The Web Nebula—images and information
http://nineplanets.org/twn/

Solar System

Introduction to the Nine Planets—images and information
http://seds.lpl.arizona.edu/nineplanets/nineplanets/intro.html

The Solar System—images and information
http://emma.la.asu.edu/dsn_solarsyst.html

Planet Earth

Earth—images and information
http://nssdc.gsfc.nasa.gov/planetary/planets/earthpage.html

Earth and Moon Viewer—satellite images and information
http://www.fourmilab.ch/earthview/vplanet.html

National Imagery and Mapping Agency (NIMA)
http://164.214.2.59/nimahome.html

NIMA—Shape of the Earth
http://164.214.2.59/GandG/geolay/TR80003A.HTM#ZZ5

The Seasons
http://www.worldbook.com/fun/seasons/html/seasons.htm

The Hydrologic Cycle
http://www.epa.gov/seahome/groundwater/src/cycle.htm#cycle or
http://ga.water.usgs.gov/edu/

Geologic Time
United States Geological Survey—geologic time scale
http://vulcan.wr.usgs.gov/Glossary/geo_time_scale.html

Kentucky Geological Survey—important dates in Earth history
http://www.uky.edu/KGS/education/timelines.html

Milwaukee Public Museum—a walk through geologic time
http://www.mpm.edu/exhibit/third/tp4.html

Latitude and Longitude
Royal Observatory Greenwich
http://www.rog.nmm.ac.uk/

Latitude and Longitude—tutorials
http://www-spof.gsfc.nasa.gov/stargaze/Slatlong.htm or
http://www.hammondmap.com/latlong.html

Latitude and Longitude—look up the location of specific cities
http://www.bcca.org/misc/qiblih/latlong.html

Calculate the distance between two points
http://www.nau.edu/~cvm/latlongdist.html

Find Your Longitude exercise
http://www.pbs.org/wgbh/nova/shackleton/navigate/find.html

Time Zones
World Time zones
http://tycho.usno.navy.mil/tzones.html or
http:www.worldtimezone.com/

Time Around the World
http://swissinfo.net/cgi/worldtime/

Time Zone Converter
http://sandbox.xerox.com/stewart/tzconvert.cgi

Map Projections
Online Map Creation—construct your own maps with different projections
http://www.aquarius.geomar.de/omc/

Projection Animations—animated images of different projections
http://www.hammondmap.com/proani.html

Nautical Charts
National Ocean Service: National Oceanic and Atmospheric Administration Home Page
http://www.nos.noaa.gov

National Ocean Service: National Oceanic and Atmospheric Administration Charting and Navigation Page
http://www.noaa.gov/charts.html

Nautical Charts—historic and coastal charts, oceanographic maps
http://www.lib.berkeley.edu/EART/digital/nautic.html

Navigation
Global Positioning System—tutorials
http://www.colorado.Edu/geography/gcraft/notes/gps/gps_f.html or
http://www.trimble.com/gps/index.htm

Global Positioning System—location exercise
http://www.pbs.org/wgbh/nova/shackleton/navigate/gps.html

Analysis of GPS errors
http://www.cnde.iastate.edu/staff/swormley/gps/check_sa.html

National Imagery and Mapping Agency (NIMA) Marine Navigation Department
http://164.214.12.145/index/

National Ocean Service: National Oceanic and Atmospheric Administration Charting and Navigation Page
http://www.noaa.gov/charts.html

The Oceans
Smithsonian Ocean Planet—Fact sheets
http://seawifs.gsfc.nasa.gov:80/OCEAN_PLANET/HTML/oceanography_geography.html

Pacific Ocean—facts
http://oceanographer.navy.mil/pacific.html

Atlantic Ocean—facts
http://oceanographer.navy.mil/atlantic.html

Indian Ocean—facts
http://oceanographer.navy.mil/indian.html

Arctic Ocean—facts
http://oceanographer.navy.mil/arctic.html

chapter three

Plate Tectonics

The San Andreas Fault crossing the Carrizo Plain in California.

Learning Objectives

After reading this chapter, you should be able to

- Describe the interior of the Earth, and explain how this information has been obtained.
- Explain the relationship between the lithosphere and the asthenosphere.
- Trace the development of the theory of drifting continents.
- Relate mantle convection cells to seafloor spreading.
- List the major discoveries of the 1960s that provided the evidence for seafloor spreading.
- Explain the relationship between seafloor spreading and magnetic reversals.
- Define plate tectonics, and outline the main lithospheric plates.
- Describe the processes that occur at plate boundaries.
- Relate the direction and rate of motion of lithospheric plates.
- Describe a hot spot and its results on moving plates.
- Trace the breakup of Pangaea.
- Explain how deep-sea drilling and the use of submersibles have changed our ideas of the deep sea floor.

Recovery of Black Smokers

Listening to Seafloor Spreading

*m*any features of our planet have presented scientists with contradictions and puzzles. The remains of warm-water coral reefs are found off the coast of the British Isles; marine fossils occur high in the Alps and the Himalayas; and coal deposits that were formed in warm tropical climates are found in northern Europe, Siberia, and northeast North America. Great mountain ranges divide the oceans, volcanoes border the coasts of the Pacific Ocean, and deep-ocean trenches are found adjacent to long island arcs. No single coherent theory explained all these features until the new technology and new scientific discoveries of the 1950s and early 1960s combined to trigger a complete reexamination of the Earth's history.

The Interior of the Earth   3.1

Although we cannot directly observe the interior of the planet, scientists have been able to learn a great deal about the structure, composition, physical state, and behavior of the Earth's interior using indirect methods. The most detailed information we have about the interior has come from roughly a century of recording and studying the passage of seismic waves through the body of the Earth. Geologists and geophysicists monitor recording stations all over the surface of the Earth that measure the type, strength, and arrival time of seismic waves generated by earthquakes, volcanic eruptions, and deliberately caused detonations.

There are two basic kinds of **seismic waves:** surface waves travel along the surface of the Earth, and body waves travel through the Earth's interior. Most of the information we have about the Earth's interior comes from the study of body waves. There are two kinds of body waves. These are **P-waves,** or primary waves (so called because they travel faster than any other seismic waves and are the first to arrive at a recording station), and **S-waves,** or secondary waves (so called because they travel more slowly than P-waves and are the second waves to arrive at a station).

P-waves and S-waves produce different types of motion in the material they travel through. P-waves, also known as compressional waves, alternately compress and stretch the material they pass through, causing an oscillation in the same direction as they move. P-waves can travel through all three states of matter: solid, liquid, and gas (sound propagates through the air and the oceans as a compressional wave). S-waves, also known as shear waves, oscillate at right angles to their direction of motion (similar to a plucked string). S-waves propagate only through solids. The motion generated in materials by these two types of waves is shown in figure 3.1.

The velocity of seismic waves depends on the characteristics of the material they travel through including its chemistry, its density, and changes in the physical state of the material (solid, partially molten, or molten) caused by variations in pressure and temperature with depth. When seismic waves encounter boundaries separating layers that have significantly different characteristics, their speed and direction of travel will change. They may be reflected off a boundary, refracted along it, or transmitted through it. Detailed modeling of the paths taken by seismic waves and their expected travel times through the Earth, as a function of distance, has produced a simple Earth model consisting of four major layers as illustrated in figure 3.2. At the planet's center is the **inner core.** It is solid and nearly five times as dense as common surface rocks such as granite because of the tremendous pressure at that depth. The inner core is composed primarily of iron with lesser amounts of lighter elements that most likely include nickel, sulfur, and oxygen. Surrounding the inner core is a region of the same chemical composition that is at least partially if not totally molten called the **outer core.** The outer core would behave like a fluid even if as much as 30% of it were composed of suspended crystals of iron that had formed from the surrounding liquid. Material in the outer core flows rapidly compared to the layer above it, probably on the order of km/yr; this generates the Earth's magnetic field. S-waves, which only travel through solids, do not propagate through the outer core. The next layer, the **mantle,** comprises about 70% of the Earth's volume. It is less dense and cooler than the core and is composed of magnesium-iron silicates (rocky material rather than metallic like the core). Although the mantle is solid, some parts of

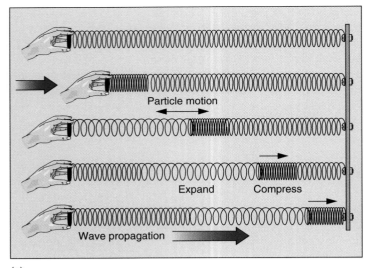

(a)

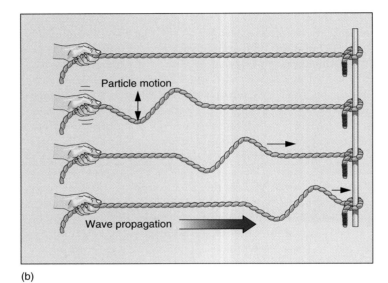

(b)

Figure 3.1

Particle motion in seismic waves. (a) P-wave motion can be illustrated with a sudden push on the end of a stretched spring. Vibration is parallel to the direction of propagation. (b) S-wave motion can be illustrated by shaking a rope to transmit a deflection along its length. Vibration is perpendicular to the direction of propagation.

More Information: www.geo.mtu.edu/UPSeis/waves.html

it are weaker than others. Material in the mantle flows very slowly in response to variations in temperature that create changes in density. Warmer, more buoyant material rises toward the surface while cooler, denser material sinks. The velocity of this motion is generally on the order of cm/yr, much slower than the flow in the liquid outer core. The Earth's outermost layer is the cold, rigid, thin surface layer called the **crust.** The boundary between the crust and the mantle is a

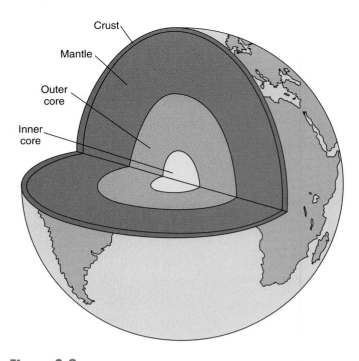

Figure 3.2

The four major layers of the Earth determined by studying the passage of seismic waves through the interior. The crust has not been drawn to scale. Its actual thickness at this scale would be roughly the thickness of the black lines used in the illustration.

chemical boundary. It is called the Mohorovičić discontinuity, known as the **Moho,** and is named for its discoverer, Andrija Mohorovičić, a Yugoslavian seismologist. There are two kinds of crust, continental and oceanic. Continental crust is relatively light and averages about 40 km (25 mi) in thickness. Its composition and structure are highly variable but it consists primarily of **granite**-type rock, which has a high content of sodium, potassium, aluminum, and silica. Oceanic crust is relatively dense, with an average thickness of about 7 km (4.3 mi). It is more homogeneous both chemically and structurally than continental crust and is composed primarily of **basalt**-type rock, which is low in silica and high in iron, magnesium, and calcium. See figure 3.2 and table 3.1 for a comparison of these layers and their properties.

> Describe the internal structure of the Earth.
>
> Describe the particle motion in P-waves and S-waves.
>
> What types of materials will P-waves and S-waves travel through?
>
> How does continental crust differ from oceanic crust?
>
> What and where is the Moho?

table 3.1 Layers of the Earth

Layer	Depth (km)	Thickness (km)	State	Composition	Density (g/cm^3)	Temperature (°C)
Crust						
Continental	0–65	40 (average)	Solid	Silicates rich in sodium, potassium, and aluminum	2.67	–89–1000
Oceanic	0–10	7 (average)	Solid	Silicates rich in calcium, magnesium, and iron	3.0	0–1100
Mantle	base of crust–2891	2866	Solid and mobile	Magnesium-iron silicates	3.4–5.6	1100–3200
Outer Core	2891–5149	2258	Liquid	Iron, nickel	9.9–12.2	3200
Inner Core	5149–6371	1222	Solid	Iron, nickel	12.8–13.1	4000–5500

The Lithosphere and Asthenosphere   3.2

More detailed study of the uppermost part of the Earth, the upper mantle and the crust, has shown that independent of the sharp chemical boundary marked by the Moho, it is possible to identify a different layered structure characterized by changes in the mechanical properties of the rock from rigid to ductile behavior. Rocks that behave rigidly do not deform or change shape when a force is applied. Rocks that have ductile behavior will deform, or flow, in response to an applied force. This has led to the identification of a strong, rigid surface shell called the **lithosphere,** which consists of crust and upper mantle material fused together. In oceanic regions the lithosphere thickens with increasing age of the sea floor. It reaches a maximum thickness of about 100 km (62 mi) at an age of 80 million years. In continental regions the thickness of the lithosphere is slightly greater than in the ocean basins, varying from about 100 km (62 mi) beneath the young marginal edges of continents to 200 km (124 mi) beneath ancient continental crust. The base of the lithosphere corresponds roughly to the region in the mantle where temperatures reach 650°C ± 100°C. At higher temperatures at these depths, mantle rock begins to lose its strength. The lithosphere is underlain by a weak, deformable region in the mantle called the **asthenosphere** where the temperature and pressure conditions lead to partial melting of the rock and loss of strength. The asthenosphere is often equated with a region of low seismic velocity called the Low Velocity Zone (LVZ). Seismic waves travel more slowly through the asthenosphere, indicating that it may be as much as 1% melt. The asthenosphere behaves in a ductile manner, deforming and flowing slowly when stressed. It behaves roughly the way hot asphalt does. With increasing depth, the increase in pressure results in greater strength once more in the lower mantle, sometimes called the **mesosphere,** where the material is once again solid but will convect slowly, moving upward in some regions and downward in others, because of temperature gradi-

ents and density differences. The depth to the base of the asthenosphere remains a matter of scientific debate. If it does correspond roughly to the LVZ, it would extend from the base of the lithosphere to about 350 km (217 mi). Some scientists believe it may extend to as much as 700 km (435 mi). The lithosphere and underlying asthenosphere are shown in figure 3.3 and are compared in table 3.2.

> Distinguish between the lithosphere and the crust; between the asthenosphere and the mantle.
>
> How and why does the speed of seismic waves in the lithosphere and the asthenosphere change?

History of a Theory   3.3

The shapes of the continents on either side of the Atlantic Ocean have intrigued observers for a long time. The possible "fit" of the bulge of South America into the bight of Africa was noted by the English scholar Francis Bacon (1561–1626), the French naturalist Georges de Buffon (1707–88), the German scientist and explorer Alexander von Humboldt (1769–1859), and others in later years. As scientists studied the Earth's crust, patterns of rock formation, fossil distribution, and mountain range placement, they recognized even greater similarities between continents. In a series of volumes published between 1885 and 1909, an Austrian geologist, Eduard Suess, proposed that the southern continents had once been joined into a single continent he called **Gondwanaland.** He assumed that portions of the continents sank and created the oceans between fragments of the original continents. At the beginning of this century, Alfred L. Wegener and Frank B. Taylor independently

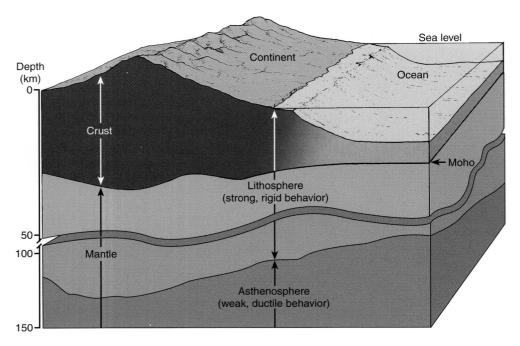

Figure 3.3

The lithosphere is formed from the fusion of crust and upper mantle. Notice that the Moho is relatively close to the Earth's surface under the ocean's basaltic crust but is depressed with the mantle under the granitic continents.

table 3.2 Upper Layers of the Earth, Based on Their Response to Applied Stress

Layer	Depth (km)	Thickness (km)	Characteristics
Lithosphere	0–100 (oceanic regions) 0 to 100–150 (continental regions)	0–150	Solid Rigid response
Asthenosphere	Base of lithosphere–350	200–350	Solid (~1% melt) Ductile response
Mesosphere	350–core-mantle boundary	4949	Solid Mobile

proposed that the continents were slowly moving about the Earth's surface. Taylor soon lost interest, but Wegener, a German meteorologist, astronomer, and Arctic explorer, continued to pursue this concept until his death in 1930.

Wegener's theory, often called **continental drift,** proposed the existence of a single supercontinent he called **Pangaea** (fig. 3.4). He thought that forces arising from the rotation of the Earth began Pangaea's breakup. First, the northern portion composed of North America and Eurasia, which he called **Laurasia,** separated from the southern portion formed from Africa, South America, India, Australia, and Antarctica, for which he retained the earlier name Gondwanaland. The continents as we know them today then gradually separated and moved to their present positions. Wegener based his ideas on the geographic fit of the continents, the scouring of landforms, and the way in which some of the older mountain ranges and

rock formations appeared to relate to each other when the landmasses were assembled as Pangaea. The presence of ancient coral reefs at high latitudes and traces of climate changes in the fossil record also supported his idea. He noted that fossils more than 150 million years old collected on different continents were remarkably similar, implying the ability of land organisms to move freely from one landmass to another. Fossils from different places dated after this period showed quite different forms, suggesting that the continents and their evolving populations had separated from one another.

Wegener's theory provoked considerable debate in the decade of the 1920s, but most geologists agreed that it was not possible to move the continental rock masses through the rigid basaltic crust of the ocean basins. The theory was not regarded very seriously in the scientific community and became a footnote in geology textbooks.

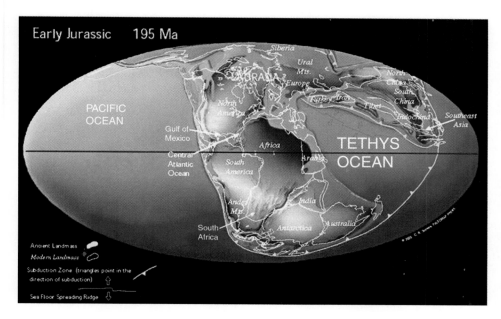

Early Jurassic 195 Ma

PACIFIC OCEAN

LAURASIA

Siberia

Ural Mts.

Europe

North China

South China

Turkey-Iran

Tibet

Indochina

Southeast Asia

North America

Gulf of Mexico

Central Atlantic Ocean

Africa

TETHYS OCEAN

Arabia

South America

India

Andes Mts.

South Africa

Antarctica

Australia

Ancient Landmass
Modern Landmass
Subduction Zone (triangles point in the direction of subduction)
Sea Floor Spreading Ridge

Figure 3.4

Pangaea in the early Jurassic, 195 Ma (million years before the present). Pangaea is composed of the two subcontinents, Laurasia and Gondwanaland, which will split apart with the opening of the central Atlantic Ocean. There is one great ocean, Panthalassa, that was the ancestral Pacific Ocean and a second small ocean, the Tethys, that would eventually close with the opening of the Indian Ocean. **More Information: www.scotese.com/jurassic.htm**

> Explain the evidence used by Wegener to propose the theory of drifting continents.
>
> What modern landmasses make up Pangaea, Laurasia, and Gondwanaland?

Evidence for a New Theory 3.4

Armed with new sophisticated instruments and technologies developed during World War II, Earth scientists returned to a study of the Earth's crust in the 1950s. Although the use of sound to examine the sea floor began in the 1920s, by the 1950s the equipment was greatly improved and much more readily available. (See chapter 4, Measuring the Depths, for a discussion of echo sounders and depth recorders.) In the 1950s a worldwide effort was made to survey the sea floor, and for the first time, scientists were able to examine, in detail, the deep-ocean floor. These surveys resulted in the construction of the first physiographic map of the world's oceans (fig. 3.5).

In the early 1960s, Harry H. Hess (1906–69) of Princeton University promoted the concept that there are convection cells within the Earth's mantle driven by heat from the Earth's core and natural radioactivity. When these upward-moving mantle currents reach the lithosphere, they move along under it, cooling as they do so until they become cool enough and dense enough to sink down toward the core again (fig. 3.6). There are currently two proposed

models of mantle convection. Some scientists believe that there are two sets of convection cells, one confined to the upper mantle above a depth of 700 km (435 mi) and the other in the lower mantle. Other scientists believe that convection occurs throughout the entire mantle from the base of the lithosphere to the core-mantle boundary; this model is called whole-mantle convection.

Upward-moving segments of mantle convection cells heat the overlying lithosphere, causing thermal expansion that produces great mountain ranges or **ridges** on the sea floor. These ridges are the sites of submarine volcanism as rapidly upwelling mantle material is partially melted to produce **magma.** New seafloor is created along the axis of the ridge as a result of this volcanism. Since there is no measurable change in the surface area of the Earth, there must be a mechanism to destroy old seafloor as it is being created at ridges. The deep, steep-sided **trenches** of the Pacific were proposed as areas where the older, colder, and denser oceanic lithosphere sinks back into the Earth's interior (fig. 3.6).

These mid-ocean ridges and deep trenches may be seen in figure 3.5. Follow the north-south ridge in the center of the Atlantic Ocean; it connects around Africa to the ridge in the central Indian Ocean. There is a ridge in the eastern South Pacific that can be followed south of Australia into the Indian Ocean and around South America's Cape Horn into the Atlantic. Conspicuous trenches may be seen along the west coast of South America, seaward of the Aleutian Islands and along the eastern coast of Asia into the western South Pacific.

Although the ascending magma breaks through the crust and solidifies, most of the rising material is turned aside under the rigid lithosphere. This allows the lithosphere to slide away from the spreading center toward the descending sides of the convection cells. The lateral movement of the crust produces **seafloor spreading** (fig. 3.6). Areas in which new crust is formed above rising magma are **spreading centers;** areas of descending older crustal material are **subduction zones.** In this model, the seafloor spreading mechanism provides the forces causing continental drift, with the continents carried as passengers on the lithosphere, like boxes on a conveyor belt.

Hess's model of the sea floor riding atop mantle convection cells has been modified and expanded in the decades since he first proposed it. Wegener's early ideas of continual drift and Hess's description of seafloor spreading have evolved into the theory of **plate tectonics,** which will be discussed in more detail in section 3.6.

> What causes convection cells in the Earth's mantle?
>
> How do convection cells cause seafloor spreading?
>
> Contrast the crust's movement at spreading centers with its movement at subduction zones.

Figure 3.5

A physiographic chart of the world's oceans.

World Ocean Floor *map from Bruce Heezen and Marie Tharp:* World Ocean Floor, *© Marie Tharp, 1977, South Nyack, NY 10960. Reproduced by permission.*

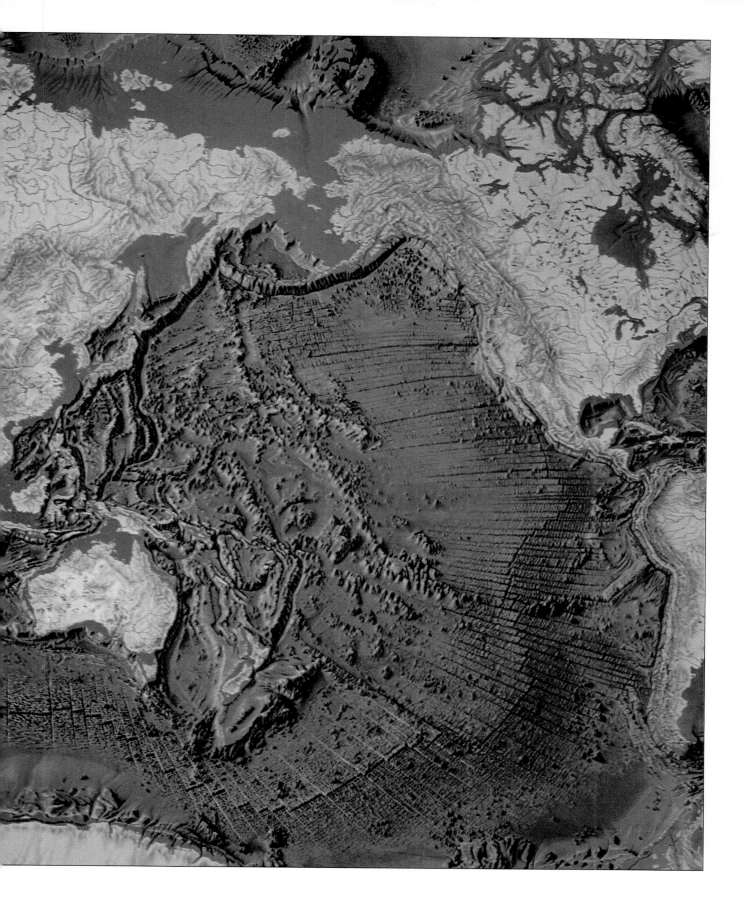

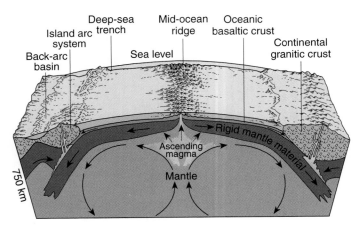

Figure 3.6

Seafloor spreading creates new crust at mid-ocean ridges and subducts old crust in deep-sea trenches. This process is shown being driven by convection cells in the mantle.

Evidence for Crustal Motion   3.5

As oceanographers, geologists, and geophysicists explored the Earth's crust on land and under the oceans, additional evidence to support the idea of seafloor spreading began to accumulate.

Epicenters of earthquakes were known to be distributed around the Earth in relatively narrow and distinct zones (fig. 3.7). An epicenter is the point on the Earth's surface directly above the earthquake source. These zones are found to correspond to the areas along the ridges, or spreading centers, and the trenches, or subduction zones. Relatively shallow earthquakes, above about 70 km (43 mi), occur in narrow bands along ridges and trenches. They also occur in more diffuse patterns throughout central and southern Eurasia and along the western coast of North America. Earthquakes that occur deeper than about 70 km (43 mi) can extend to depths as great as about 700 km (435 mi). These deep events are associated with subduction at ocean trenches. The deeper the earthquakes are, the more they are displaced away from the axis of the trench, beneath continents and island arcs.

Researchers sank probes into the sea floor to measure the heat from the interior of the Earth moving through the crust (fig. 3.8). The measured heat flow shows a regular pattern, highest in the vicinity of the mid-ocean ridges where the crust is thinner and younger, and decreasing as the distance from the ridge center increases (fig. 3.9). This heat-flow pattern is consistent with a model in which hot, buoyant mantle material rises beneath the ridge system and cools as it moves away from it.

Vertical cylindrical samples, or **cores,** were obtained by drilling through the **sediments** that cover the ocean bottom and into the basaltic rock of the oceanic crust. (A more detailed dis-

World Seismicity: 1977 - 1997

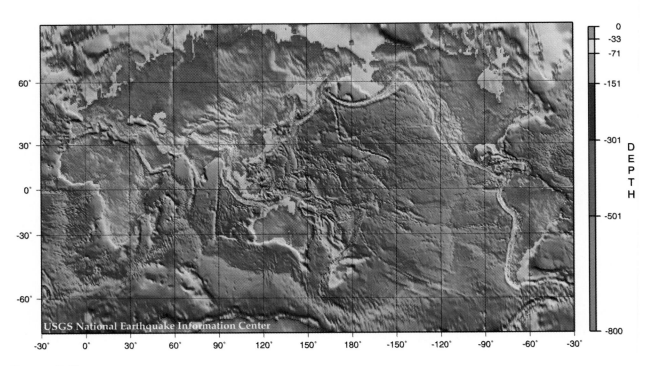

Figure 3.7

Earthquake epicenters, 1977–97. Epicenters are color-coded by the depth of the earthquake. Shallow earthquakes generally outline plate boundaries and regions of significant deformation in continents. Earthquakes deeper than about 70 km (green to red epicenters) are associated with ocean trenches and subduction.

More Information: www.neic.cr.usgs.gov/neis/general/seismicity/world.html and neic.usgs.gov/neis/bulletin/bulletin.html

Figure 3.8

Preparing to lower a heat flow probe. The probe consists of a cylinder holding the measuring and recording instruments above a thin probe that penetrates the sediment. Thermisters (devices that measure temperature electronically) at the top and bottom of the probe measure the temperature difference over a fixed distance equal to the length of the probe. This temperature change with depth is used to compute the heat flow through the sediment.

cussion of marine sediment is found in chapter 4.) This process required the development of a new technology and a new type of ship. In the late summer of 1968 the specially constructed drilling ship *Glomar Challenger* (see fig. 1.18a) went to sea on its first drilling expedition. One of the great challenges of deep-ocean drilling is keeping the drilling ship in a stable position at the surface. This is accomplished by using a specialized propulsion system including bow and stern thrusters, which are small propellers mounted perpendicular to the ship's keel that can be used to move the ship sideways or rotate it. The propulsion system responds automatically to computer-controlled navigation, using stationary acoustic beacons on the sea floor as stable reference positions. This allows the ship to remain for long periods of time in a nearly fixed position over a drill site in water too deep to anchor. An acoustic guidance system enabled it to replace drill bits and reenter the same bore holds in water about 6000 m (20,000 ft) deep. This method is illustrated in figure 3.10.

In 1983 the *Glomar Challenger* was retired, after logging 600,000 km (375,000 mi), drilling 1092 holes at 624 drill sites, and recovering a total of 96 km (60 mi) of deep-sea cores for study. A new deep-sea drilling program with an improved and more sophisticated ship, the *JOIDES Resolution* (see fig. 1.18b), began in 1985. The drill pipe comes in individual sections 9.5 m (31 ft) long that can be screwed together to make a single pipe up to 8200 m (27,000 ft) long.

The cores taken by the *Glomar Challenger* provided much of the data needed to establish the existence of seafloor spreading. No ocean crust older than 180 million years was found, and sediment age and thickness were shown to increase with distance from the ocean ridge system. Figure 3.11 shows that the sediments closest to the ridge system are thin over the new crust while older crust farther away from the ridge system is more deeply buried.

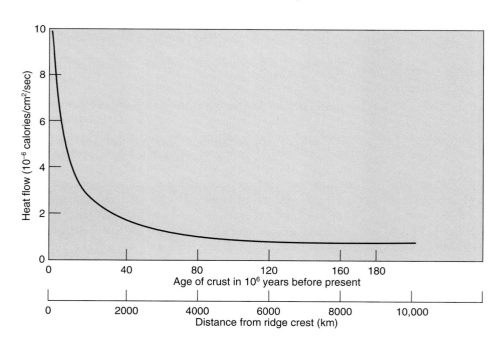

Figure 3.9

Heat flow through the Pacific Ocean floor. Values are shown against age of crust and distance from the ridge crest.

Source: Data from J. G. Sclater and J. Crowe, "On the Variability of Oceanic Heat Flow Average" in Journal of Geophysical Research *81 (17): 3004, June 10, 1970.*

Figure 3.10

Deep-ocean drilling technique. Acoustical guidance systems are used to maneuver the drilling ship over the bore hole and to guide the drill string back into the bore hole.

More Information: www-odp.tamu.edu/ or www.oceandrilling.org/

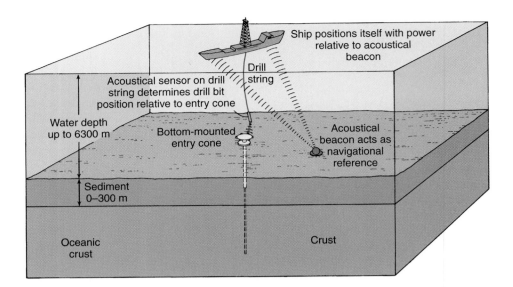

Figure 3.11

Age and thickness of seafloor sediments.

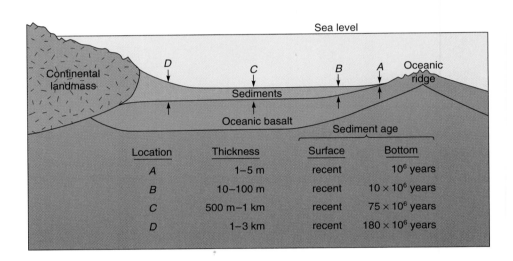

Location	Thickness	Sediment age	
		Surface	Bottom
A	1–5 m	recent	10^6 years
B	10–100 m	recent	10×10^6 years
C	500 m–1 km	recent	75×10^6 years
D	1–3 km	recent	180×10^6 years

Although each of these pieces of evidence fits the theory that the Earth's crust produced at the ridge system is new and young, the most elegant proof for seafloor spreading came from a study of the magnetic evidence locked into the oceans' floors.

The Earth's familiar north (90°N) and south (90°S) geographic poles mark the axis about which it rotates. The Earth also behaves as if it has a giant bar magnet embedded in its interior tilted roughly 11.5° away from its axis of rotation (fig. 3.12). The north magnetic pole is located near 79.3°N, 71.5°W in the Northwest Territories of Canada, and the south magnetic pole is almost directly opposite near 79.3°S, 108.5°E in the South Pacific Ocean. A magnetic field like the Earth's, with two opposite poles, is called a **dipole.** The Earth's magnetic field consists of invisible lines of magnetic force that are parallel to the Earth's surface at the magnetic equator and converge and dip toward the Earth's surface at the magnetic poles (fig. 3.12). A small, freely suspended magnet (such as a compass needle) will align itself with these lines

of magnetic force, with the north-seeking end pointing to the north magnetic pole and the south-seeking end pointing to the south magnetic pole. In addition, it will dip toward the Earth's surface by differing amounts depending on its distance from the magnetic equator. At the magnetic equator a small magnet would be horizontal, or parallel to the Earth's surface. In the magnetic northern hemisphere the north-seeking end of the magnet will point downward at an increasing angle until it pointed vertically downward at the north magnetic pole. In the magnetic southern hemisphere the north-seeking end of the magnet will point upward at an increasing angle until it pointed vertically upward at the south magnetic pole.

Most igneous rocks contain particles of a naturally magnetic iron mineral called magnetite. Particles of magnetite act like small magnets. These particles are particularly abundant in basalt, the type of rock that makes up the oceanic crust. When basaltic magma erupts on the sea floor along ocean ridges it cools and solidifies to form basaltic rock. During this time the

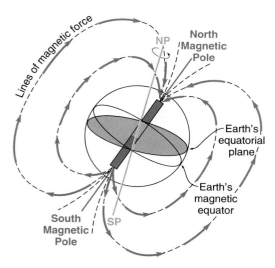

Figure 3.12

Orientation of the lines of magnetic force of the Earth's magnetic field.

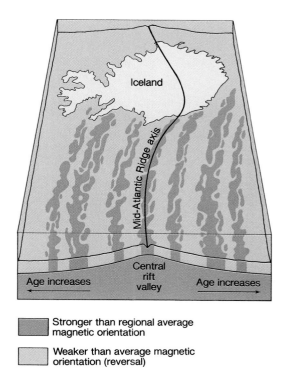

Stronger than regional average magnetic orientation

Weaker than average magnetic orientation (reversal)

Figure 3.13

Reversals in the Earth's magnetic polarity cause the symmetrically striped pattern centered on the Mid-Atlantic Ridge. The age of the sea floor increases with the distance from the ridge. The spreading rate along the Mid-Atlantic Ridge is about 2–3 centimeters per year.

magnetite grains in the rock become magnetized in a direction parallel to the existing magnetic field at that time and place. When the temperature of the rock drops below a critical level called the **Curie temperature,** or Curie point (named in honor of the twice Nobel prize-winning physicist Madame Marie Curie), roughly 580°C, the magnetic signature (magnetic strength and orientation) of these particles is "frozen" into them creating a "fossil" magnetism that will remain unchanged unless the rock is heated again to a temperature above the Curie temperature. In this manner, rocks can preserve a record of the strength and orientation of the Earth's magnetic field at the time of their formation. The investigation of fossil magnetism in rocks is the science of **paleomagnetism.**

Research done on age-dated layers of volcanic rock found on land shows that the polarity or north-south orientation of the Earth's magnetic field reverses for varying periods of geologic time. This means that at different times in the Earth's history the present north and south magnetic poles have changed places. Each time a layer of volcanic material solidified, it recorded the magnetic orientation and polarity of the time period in which it cooled. Nearly 170 magnetic reversals have been identified during the last 76 million years; our present magnetic orientation has existed for 710,000 years. The cause of reversals is not fully understood, but is thought to be associated with changes in the motion of the magnetic material of the Earth's liquid outer core. We do not know how long our present magnetic polarity will last.

When magnetometers were towed over the sea floor during the early 1960s, the resulting maps revealed a pattern of stripes that ran parallel with the mid-ocean ridge (fig. 3.13). Their significance was not understood until later in the decade, when Fred J. Vine (1940–) and Drummond H. Matthews (1931–) of Cambridge University proposed that these stripes represented a recording of the **polar reversals** of the Earth's

magnetic field, frozen into the sea floor. As the molten basalt rose along the crack of the ridge system and solidified, it locked in the direction of the existing magnetic field. Seafloor spreading moved this material off on either side of the ridge, to be replaced by more molten rock. Each time the Earth's magnetic field reversed, the direction of the magnetic field was recorded in the new crust. Vine and Matthews proposed that if such were the case, there should be a symmetrical pattern of magnetic stripes centered at the ridges and becoming older away from the ridges (fig. 3.13). The polarity and age of these stripes should correspond to the same magnetic field changes found in dated layers on land. Deep-sea core drilling confirmed their ideas and thus dramatically demonstrated seafloor spreading. The accumulation of magnetic stripe data across the world's oceans has been used to produce a map of the age of the sea floor. Study figure 3.14 and notice that seafloor age increases away from the oceanic spreading centers, providing verification of seafloor spreading. The present ocean basins are not old but new, created by seafloor spreading during the past 200 million years, or the last 5% of the Earth's history.

Paleomagnetic studies of fossil magnetism in continental rocks provide evidence of relative motion between tectonic plates and the Earth's magnetic poles through time. By carefully measuring the fossil magnetism in a rock, the location of

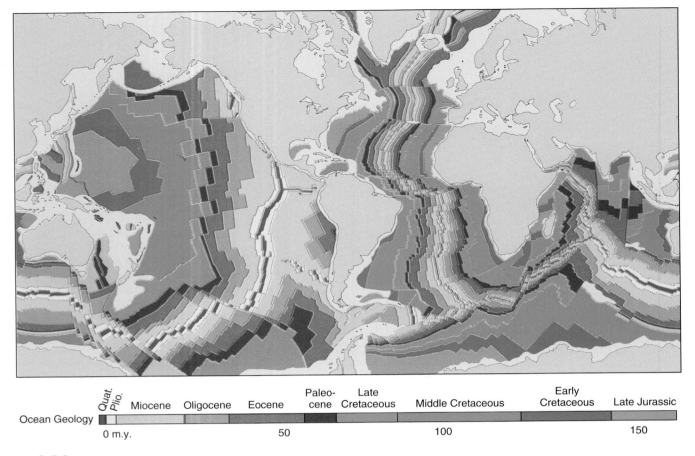

	Quat. Plio.	Miocene	Oligocene	Eocene	Paleo-cene	Late Cretaceous	Middle Cretaceous	Early Cretaceous	Late Jurassic

Ocean Geology

0 m.y.   50   100   150

Figure 3.14

The age of the ocean crust based on seafloor spreading magnetic patterns is shown in this color-shaded image. The age of the ocean floor increases with increasing distance from the mid-ocean ridge as a result of seafloor spreading. Oceanic fracture zones offset the ridge and the age patterns.

After *The Bedrock Geology of the World* by R. L. Larson, W. C. Pitman, III et al. W. H. Freeman.

More Information: www.usyd.edu.au/su/geosciences/geology/people/staff/dietmar/Agegrid/digit_isochrons.html

the north magnetic pole at the time of the rock's formation can be calculated. This is done by determining the direction and distance to the pole's position. The horizontal component of the rock's magnetism points in the direction of the pole. The angle of dip of the magnetic signature reveals the magnetic latitude at which the rock formed, which is directly related to the distance between the magnetic north pole and the rock's formation site.

By measuring the fossil magnetism in continental rocks of different ages that are all found on the same tectonic plate, it is possible to create a plot of the apparent location of the north magnetic pole through time. Such a plot is known as a **polar wandering curve** (fig. 3.15). A polar wandering curve is evidence that the relative positions of the plate and the north magnetic pole have changed with time. This may be either the result of the movement of the pole or the movement of the plate. While we know that the magnetic poles do move a little with time, their average location remains near the geographic poles. Consequently, a polar wandering curve records the motion of the plate with respect to a nearly stationary north magnetic pole. Because all plates do not move along the same path,

each plate produces its own polar wandering curve, which converges with all other polar wandering curves at the current location of the north magnetic pole. If we look at the polar wandering curves constructed for the North American and Eurasian plates we see that they have the same general shape but diverge with increasing age of the rocks (fig. 3.15). The divergence of these two polar wandering paths is a record of the opening of the North Atlantic Ocean and the drifting apart of the North American and European plates. Think of the polar wandering curves as the open blades of a pair of scissors with one handle attached to North America and the other to Europe. If the scissor blades are closed to make the polar wandering curves coincide, the handles move Europe and North America toward each other, joining the continents, as the polar wandering curves approach each other. This demonstrates the case for the time between 500 and 200 million years before the present, when North America and Europe moved as one landmass across the Earth's surface. Two hundred to one hundred million years ago the polar wandering curves diverged, indicating that the continents began to move independently, drifting apart and forming the North Atlantic Ocean. This plan is supported

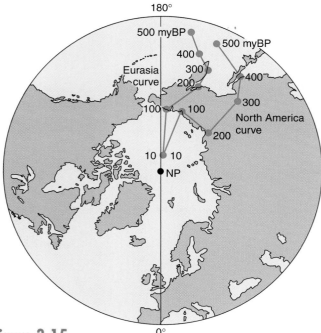

Figure 3.15

Lines of magnetic force surround the Earth and converge at the magnetic poles. A freely suspended magnet will align itself with these lines of magnetic force, the north-seeking end of the magnet pointing to the north magnetic pole and the south-seeking end pointing to the south magnetic pole. The magnet would hang parallel to the Earth's surface at the magnetic equator and would dip at an increasing angle as it approached the magnetic poles. NP and SP indicate the north and south geographic poles respectively.

More Information: antwrp.gsfc.nasa.gov/apod/ap991019.html and www.agso.gov.au/geophysics/geomag/information/faq.html

by geological evidence including the joining of ancient mountain ridges and fault systems.

> How do earthquake patterns, seafloor heat flow, age of the Earth's crust, and thickness of seafloor sediments help to explain seafloor spreading?
>
> What is a magnetic reversal and how is it recorded in the rock of the sea floor?
>
> How do the seafloor patterns of magnetic reversals verify seafloor spreading?
>
> How and why is a polar wandering curve constructed?

Plate Tectonics   3.6

When the ideas of continental drift and seafloor spreading were joined, it led to the formation of a single unified concept of the fragmentation and movement of the outer rigid shell of the Earth; this concept is known as plate tectonics. This rigid shell is the layer we call the lithosphere (you may want to review section 3.2 and figure 3.3). The lithosphere is fragmented into seven major plates along with a number of smaller ones (fig. 3.16 and table 3.3). Each lithospheric plate consists of the upper roughly 80 to 100 km (50 to 62 mi) of rigid mantle rock capped by either oceanic or continental crust. Lithosphere capped by oceanic crust is often simply called oceanic lithosphere and, in a similar fashion, lithosphere capped by continental crust is often referred to as continental lithosphere. Some plates, such as the Pacific Plate, consist entirely of oceanic lithosphere but most plates, like the South American Plate, consist of variable amounts of both oceanic and continental lithosphere with a gradual transition from one to the other along the margins of continents. The plates move with respect to one another, sliding on top of the ductile asthenosphere below. As the plates move, they interact with one another along their boundaries, producing the majority of the earthquake and volcanic activity that occurs on Earth.

There are three basic kinds of plate boundaries. These are defined by the type of relative motion between the plates they separate. Each of these boundaries is associated with specific kinds of geological features and processes (table 3.4 and fig. 3.17). Plates move away from each other along **divergent plate boundaries.** Ocean basins and the sea floor are created along divergent boundaries. Plates move toward one another and collide along **convergent plate boundaries.** Convergent boundaries are associated with the destruction of sea floor, the closing of ocean basins, and the creation of massive mountain ranges. The third type of plate boundary occurs where two plates are neither converging nor diverging, but are simply sliding past one another. These are called **transform boundaries** and are marked by large faults called **transform faults.**

table 3.3   Approximate Plate Area (10^6 km^2)

Major Plates	
Pacific Plate	105
African Plate	80
Eurasian Plate	70
North American Plate	60
Antarctic Plate	60
South American Plate	45
Australian Plate	45
Smaller Plates	
Nazca Plate	15
Indian Plate	10
Arabian Plate	8
Philippine Plate	6
Caribbean Plate	5
Cocos Plate	5
Scotia Plate	5
Juan de Fuca Plate	2

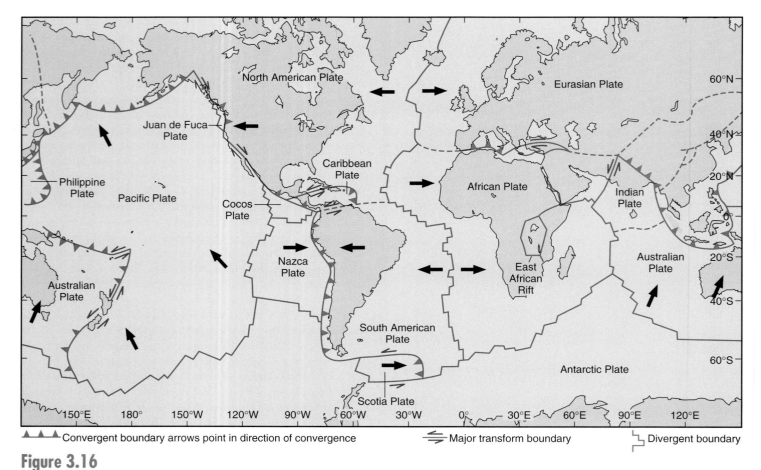

▲▲▲ Convergent boundary arrows point in direction of convergence   ⇋ Major transform boundary   ⌐⌐ Divergent boundary

Figure 3.16

The Earth's lithosphere is broken into a number of individual plates. The edges of the plates are defined by three different types of plate boundaries. Arrows indicate direction of plate motion. **More Information: geology.er.usgs.gov/eastern/plates.html**

table 3.4   Types and Characteristics of Plate Boundaries

Plate Boundary	Types of Lithosphere	Geologic Process	Geologic Feature	Earthquakes	Volcanism	Examples
Divergent (Move apart)	Ocean-Ocean	New seafloor created, Ocean basin opens	Mid-ocean ridge	Yes, shallow	Yes	Mid-Atlantic Ridge, East Pacific Rise
	Continent-Continent	Continent breaks apart, New ocean basin forms	Continental rift, Shallow sea	Yes, shallow	Yes	East African Rift, Red Sea, Gulf of Aden, Gulf of California
Convergent (Move together)	Ocean-Ocean	Old seafloor destroyed by subduction	Ocean trench	Yes, shallow to deep	Yes	Aleutian, Mariana, and Tonga Trenches (Pacific Ocean)
	Ocean-Continent	Old seafloor destroyed	Ocean trench	Yes, shallow to deep	Yes	Peru-Chile and Middle-America Trenches (Eastern Pacific Ocean)
	Continent-Continent	Mountain building	Mountain range	Yes, shallow to intermediate	No	Himalaya Mountains, Alps
Transform (Slide past each other)	Ocean	Seafloor conserved (neither created nor destroyed)	Transform fault (offsets segments of ridge crest)	Yes, shallow	No	Mendocino and Clipperton, (Eastern Pacific Ocean)
	Continent	Seafloor conserved (neither created nor destroyed)	Transform fault (offsets segments of ridge crest)	Yes, shallow	No	San Andreas Fault, Alpine Fault (New Zealand), North and East Anatolian Fault (Turkey)

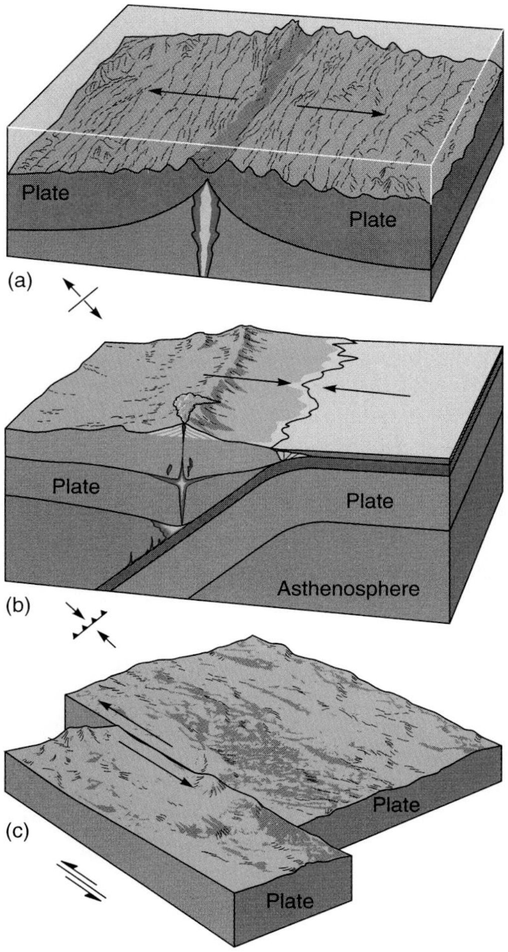

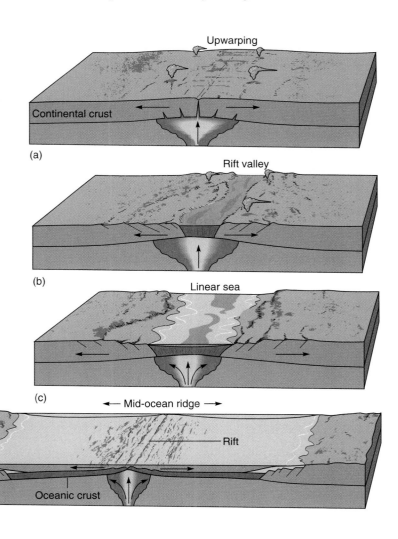

Define the term "plate tectonics."

Locate the boundaries of the seven major plates (listed in table 3.2) in figure 3.5.

What are the three types of plate boundaries and what is the relative motion of plates along each of them?

What geologic features would we expect to find along each type of plate boundary?

Figure 3.17

The three basic types of plate boundaries include: (a) divergent boundaries where plates move apart, (b) convergent boundaries where plates collide, and (c) transform boundaries where plates slide past one another.

More Information: scign.jpl.nasa.gov/learn/plate4.htm

Figure 3.18

(a) Continental rifting begins as rising magma heats the overlying continental crust, making it dome upward and thin as tensional forces cause it to stretch. (b) Stretching and pulling apart of the crust produces a rift valley with active volcanism. (c) Continued spreading results in the formation of new sea floor along the boundary and the creation of a young, shallow sea. (d) Eventually, a mature ocean basin and spreading ridge system are created.

Divergent Boundaries   3.7

Ocean basins are created along divergent boundaries that break apart continental lithosphere (fig. 3.18). A series of successive divergent boundaries were responsible for the breakup of Pangaea, producing the individual continents and ocean basins we see today. Upwelling of hot mantle rock is thought to be responsible for the creation of divergent boundaries in continental lithosphere. Mantle upwelling heats the base of the lithosphere, thinning it and causing it to dome upward (fig. 3.18a). This weakens the lithosphere and produces extensional forces and stretching that lead to faulting and volcanic activity. Faulting along the boundary will create rifting in the continental crust, forming a **rift valley** along the length of the boundary (fig. 3.18b). A present-day example of this early rifting is the continuing development of the East African Rift Valley, stretching from Mozambique to Ethiopia (fig. 3.19).

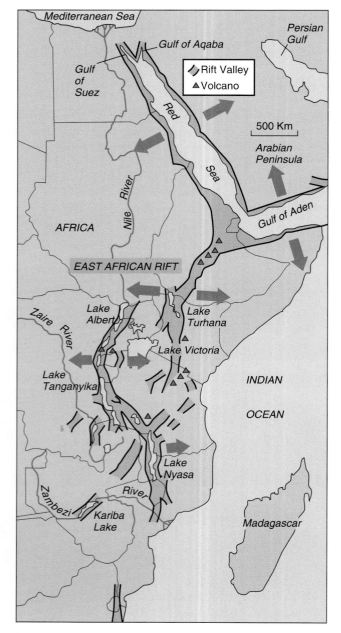

Figure 3.19

An active continent-continent divergent boundary, still in its early stage of development, has created the East African Rift system. This same boundary is more fully developed to the north where it has created two shallow seas, the Red Sea and the Gulf of Aden.

More Information: bilbo.clpccd.cc.ca.us/lpc/science/rhanna/puzzle/rift/rift.htm

Volcanoes in the rift include Kilimanjaro and Mount Kenya. The next "stage" in continental rifting can be seen further to the north where two shallow seas, the Red Sea and the Gulf of Aden, have formed (fig. 3.18c). If this rift system remains active in the future, East Africa will separate from the rest of the continent and a new ocean basin will form with seafloor spreading and a central spreading ridge (fig. 3.18d).

Most divergent boundaries are located along the axis of the mid-ocean ridge system (fig. 3.20). Mantle upwelling beneath the ridge heats the overlaying oceanic lithosphere, causing it to expand, creating a submarine mountain range. As the plates on either side of the boundary move apart, cracks form along the crest of the ridge, allowing molten rock to seep up from the mantle and flow onto the ocean floor. In the early stages of a volcanic eruption, basaltic magma can flow rapidly onto the sea floor in relatively flat flows called sheet flows that may extend several kilometers away from their source. As the eruption rate decreases, the magma is often extruded more slowly onto the sea floor creating rounded flows called pillow lavas or **pillow basalts** (fig. 3.21). The magma solidifies to form new ocean crust along the edges of each diverging plate. The thickness of the oceanic crust remains relatively constant from the time of its formation at the ridge. As the two plates diverge, however, the mantle rock immediately beneath the crust cools, fuses to the base of the crust, and begins to behave rigidly, causing the thickness of the oceanic lithosphere to increase with age and distance from the ridge. Plates move apart at different rates along the length of the ridge system.

> What causes continental rifting?
>
> How are ocean basins formed?
>
> Where are most divergent plate boundaries located and what major geological feature is associated with them?

Transform Boundaries 3.8

The ocean ridge system is divided into segments that are offset by transform faults (fig. 3.22). As two plates move away from the ridge crest they will simply slide past one another along the transform fault. In this manner the ridge crest segments and transform faults form a continuous plate boundary that alternates between being a divergent boundary and a transform boundary. Differences in the age and temperature of the plates across a transform boundary can create significant and rapid changes in the elevation of the sea floor (look at fig. 3.22 and imagine walking across the transform fault along a line that begins very near the ridge crest in Plate A and ends far from the ridge crest in plate B). These changes in elevation are propagated into each plate by seafloor spreading where they are preserved as "fossil" transform faults. The two symmetrical fossil transform faults in each plate together with the active transform

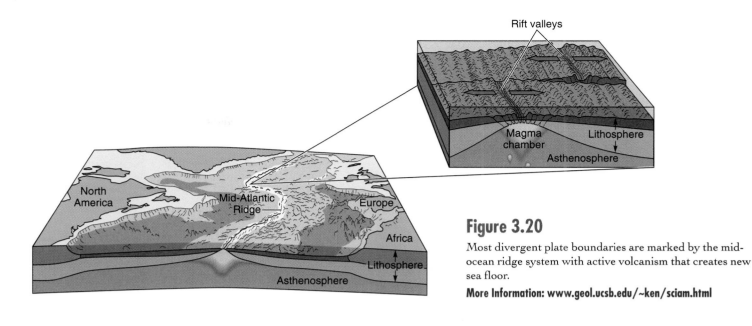

Figure 3.20

Most divergent plate boundaries are marked by the mid-ocean ridge system with active volcanism that creates new sea floor.

More Information: www.geol.ucsb.edu/~ken/sciam.html

Figure 3.21

Pillow lavas are mounds of elongate lava "pillows" formed by repeated oozing and quenching of extruding basalt magma on the sea floor. First, a flexible glassy crust forms around the newly extruded lava, forming an expanded pillow. Next, pressure builds within the pillow until the crust breaks and new basalt extrudes like toothpaste, forming another pillow.

More Information: volcanoes.usgs.gov/Products/Pglossary/PillowLava.html

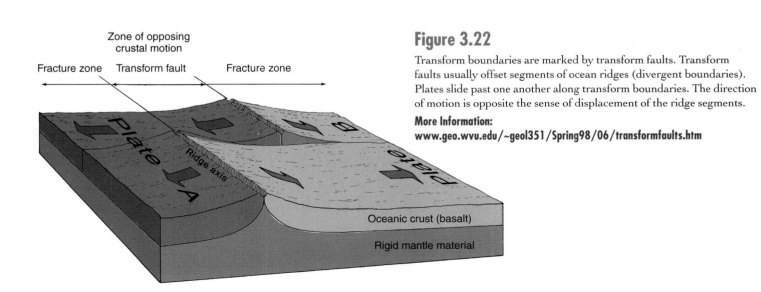

Figure 3.22

Transform boundaries are marked by transform faults. Transform faults usually offset segments of ocean ridges (divergent boundaries). Plates slide past one another along transform boundaries. The direction of motion is opposite the sense of displacement of the ridge segments.

More Information:
www.geo.wvu.edu/~geol351/Spring98/06/transformfaults.htm

fault between the adjacent segments of ridge crest form a single linear feature called a **fracture zone** that is roughly perpendicular to the ridge crest. The longest fracture zones can be up to 10,000 km (6200 mi) long, extending deep into each plate on either side of the ridge. It is important to remember that there is no movement or earthquake activity along the fracture zone where it extends into either plate. However, there is relative motion and seismicity along the transform fault between adjacent segments of ridge crest. The direction of motion of the two plates on either side of a transform fault is determined by the direction of seafloor spreading in each plate and is opposite the sense of displacement of the ridge segments. While most transform faults join two divergent boundaries (segments of ridge crest), they may also join different combinations of other types of plate boundaries. Transform faults can join two convergent boundaries (ocean trenches) as they do between the Caribbean and North American Plates, or a convergent boundary and a divergent boundary as they do on either side of the Scotia Plate (see fig. 3.16). The diversity of settings in which transform faults play a role in marking the boundaries of plates can be seen along the west coast of the United States (fig. 3.23). Much of the boundary between the North American and Pacific Plates is marked by a long transform fault; its southern portion is known as the San Andreas fault (see this chapter's opening photograph), cutting through continental crust from the Gulf of California to Cape Mendocino, California. Offshore of Cape Mendocino a transform fault known as the Mendocino fault

links a spreading center and a trench, forming a boundary between the Pacific and Juan de Fuca Plates. Further to the north, additional transform faults offset segments of the Juan de Fuca Ridge.

> What is the difference between a fracture zone and a transform fault?
>
> Most transform faults join what type of plate boundaries together?

Convergent Boundaries 3.9

Plates collide along convergent plate boundaries. The result of plate collision depends largely on whether the colliding edges of the plates are oceanic or continental lithosphere. There are three possible combinations, and hence three different outcomes to plate convergence; ocean-continent, ocean-ocean, and continent-continent convergence (fig. 3.24). The geologic features and processes that are characteristic of these different types of convergent boundaries are determined by the relative density of continental and oceanic lithosphere. Roughly 20% to

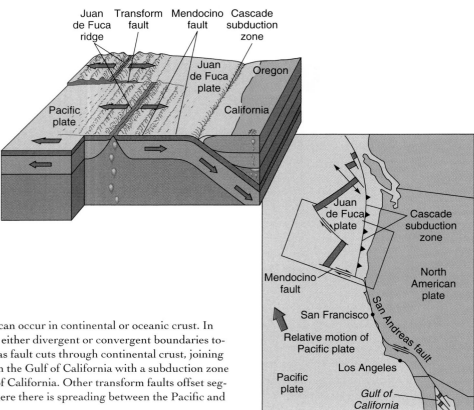

Figure 3.23

Transform boundaries can occur in continental or oceanic crust. In addition, they may join either divergent or convergent boundaries together. The San Andreas fault cuts through continental crust, joining a divergent boundary in the Gulf of California with a subduction zone off the northern coast of California. Other transform faults offset segments of ridge crest where there is spreading between the Pacific and Juan de Fuca Plates.

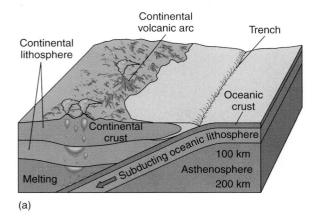

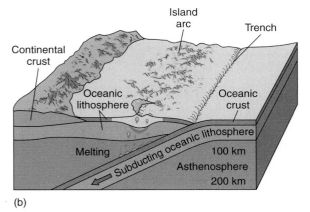

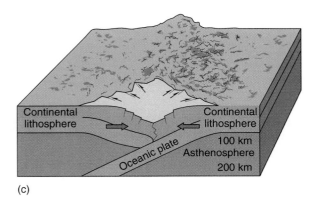

Figure 3.24

The three different types of convergent plate boundaries are
(a) ocean-continent convergence, (b) ocean-ocean convergence, and
(c) continent-continent convergence.

40% of the thickness of continental lithosphere consists of continental crust. Because continental crust has a relatively low density, continental lithosphere is very buoyant and will not sink into the underlying mantle. The density of oceanic lithosphere is greater than continental lithosphere and changes with age. The density of oceanic lithosphere increases with increasing age as seafloor spreading carries it away from the ridge

crest and it cools and thickens. Roughly 90% of old oceanic lithosphere consists of cold, dense mantle rock. Old oceanic lithosphere can actually be more dense than the hot mantle material below it, making it relatively easy for it to be subducted.

Along ocean-continent convergent boundaries the dense oceanic lithosphere will sink into the mantle, creating an ocean trench on the sea floor marking the boundary between the two plates (fig. 3.24a). Some of the marine sediments will be scraped off the descending plate by the edge of the continental plate. The remainder of the sediments will be carried into the mantle with the descending plate. As the oceanic plate descends it will gradually heat up. When it reaches a depth of about 100 to 150 km (roughly 60 to 100 mi) the temperature will be high enough to drive water and other volatiles out of the subducted sediments and into the surrounding mantle rock that the plate is penetrating. The addition of water to the mantle will reduce its melting temperature and it will begin to partially melt, producing a basaltic magma that will rise beneath the edge of the continent. This magma may partially melt the overlying continental crust as well, resulting in a magma with a mixed composition between basalt and granite called andesite. Eventually, this rising magma will produce active volcanoes along the edge of the continental plate. These volcanoes often erupt explosively. Examples of active continental volcanic chains include the Andes along the west coast of South America and the Cascade Range of the Pacific Northwest, including Mount St. Helens (fig. 3.25). Subducting oceanic plates are also associated with intense earthquake activity as they bend and descend into the mantle. Dipping zones of earthquakes, called **Benioff zones** in honor of seismologist Hugo Benioff, mark the location of the subducted plate. Some of the deepest earthquakes in the world occur in Benioff zones extending to depths of as much as 650 km (400 mi) in the western Pacific Ocean.

Ocean-ocean convergent boundaries are also marked by ocean trenches where one of the plates is subducted beneath the other (fig. 3.24b). The plate that is subducted will generally be the one whose convergent edge is the oldest, and hence the furthest from the ridge where the sea floor was first created. Just as in the case of ocean-continent convergence, the subducted plate will partially melt producing a basaltic magma. This magma will rise beneath the overriding oceanic plate and produce a line of active volcanoes on the ocean floor that may eventually break the surface to form a vocanic **island arc.** Island arcs are most commonly found in the Pacific Ocean; examples include the Aleutian, Mariana, and Tonga Islands. There are only two island arcs in the Atlantic Ocean, the Sandwich Islands in the South Atlantic and the Lesser Antilles on the eastern side of the Caribbean Sea. In some cases island arcs can form on pieces of continental crust that have broken away from the mainland. Examples of this include Japan and the Philippines.

Continent-continent collision is the end result of the closing of an ocean basin by ocean-continent convergence (fig. 3.24c). As the ocean basin closes, the continental lithosphere of the subducting plate will move progressively closer to the trench

Figure 3.25

Mount St. Helens erupted violently on May 18, 1980. The mountain lost nearly 4.1 cubic kilometers (1 cu mi) from its once symmetrical summit, reducing its elevation from 2950 meters to 2550 meters (a 1312-ft change). The force of the lateral blast blew down forests over a 594 square kilometer (229 sq mi) area. Huge mud flows of glacial meltwater and ash moved down the mountain.

More Information: vulcan.wr.usgs.gov/Volcanoes/MSH/framework.html

and the continental edge of the overriding plate. When the oceanic lithosphere is completely subducted and the ocean basin has closed, the continental lithosphere of the subducting plate will collide with the continental edge of the overriding plate. The subducted oceanic plate will tear away and continue to descend. The subduction process will cease however, along with the associated volcanic activity, since continental lithosphere is too buoyant to descend into the mantle. Along the edges of the two colliding continents, the continental crust will buckle, fracture, and thicken. What was once a well-defined convergent boundary marked by a narrow ocean trench between the two plates will become a broad zone of intense deformation with no single discrete boundary. This will produce relatively shallow earthquakes over a large area.

Continent-continent convergence created the Himalayas as India collided with Asia. India continues to collide with Asia at the rate of about 5 cm (2 in) per year and the Himalayas are still rising in elevation. Other examples of mountain ranges created in this manner are the Appalachians, Alps, and Urals. Continent-continent collision can trap some of the marine sediment of the old ocean basin and incorporate it into the mountains that are built. Old marine fossils and sediments, including limestone remains of coral reefs, are found in the summits of peaks in the Himalayas and the Alps.

What is a Benioff zone and what does it mark?

What is the source of the magma associates with ocean-ocean and ocean-continent convergent boundaries?

Why can oceanic lithosphere be subducted but continental lithosphere cannot?

How does plate tectonics explain fossils of marine organisms found high in the Alps and Himalayas?

Continental Margins   3.10

When a continent rifts and moves away from a spreading center, the resultant continental margin is known as a **trailing,** or **passive margin.** These are also frequently referred to as Atlantic-style margins since they are found on both sides of the Atlantic Ocean as well as around Antarctica, the Arctic Ocean, and in the Indian Ocean. Continental and oceanic lithosphere are joined along passive margins so there is no plate boundary at the margin. As passive margins move away from the ridge the oceanic lithosphere cools, increases its density, thickens, and subsides. This causes the edge of the continent to slowly subside as well. While passive margins begin at a divergent plate boundary, they end up in a midplate position as a result of seafloor spreading and the opening of the ocean basin. Old passive margins are not greatly modified by tectonic processes because of their distance from the ridge. These margins are often broad and shallow and have thick sedimentary deposits, as along the eastern coast of the United States.

When a plate boundary is located along a continental margin, the margin is called a **leading,** or **active margin.** Active continental margins are often marked by ocean trenches where oceanic lithosphere is subducted beneath the edge of the continent. These margins are typically narrow and steep with volcanic mountain ranges, as along the west coast of South America as well as Oregon and Washington. Active margins are found primarily in the Pacific Ocean.

Studies of the North American crust indicate that the core of the continent was assembled about 1.8 billion years ago by collisions with four or five large pieces of even older continental land. Crustal fragments with properties and histories distinct from adjoining crust and added by collisions are known as **terranes.** Terranes may be pieces of island arc systems, seamount volcanoes, seafloor plateaus, or parts of other continental landmasses. The terranes of Alaska and the West Coast of the United States and Canada appear to have arrived from the south about 70 million years ago. Fossils collected between Virginia and Georgia indicate that a long section up and down the East Coast was formed somewhere adjacent to an island arc system; the fossils point to a past European connection. It is estimated that 25% of North America was formed from terranes; see figure 3.26 for examples. The subcontinent of India is considered by some to be a single, giant terrane; it arrived from Antarctic latitudes and is now firmly attached to the continent of Asia. North of the Himalayas there appear to be terranes that became a part of the continent before the arrival of India.

> What is the difference between "passive" and "active" continental margins?
>
> Is the east coast of South America a passive or an active continental margin? What about the west coast?
>
> What does the presence of terranes tell us about the past geological history of an area?

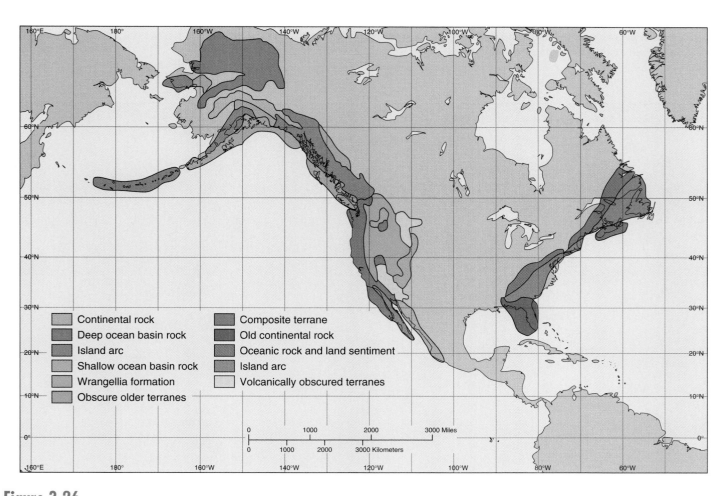

Figure 3.26

Terranes of the west and east coasts of North America.

Rates of Motion 3.11

The sea floor, acting like a conveyor belt, moves away from the ridge system where it is created by volcanic activity. The rate at which each plate moves away from the axis of the ridge is commonly known as the half-spreading rate while the rate at which the two plates move away from each other is called the full spreading rate, or simply the **spreading rate.** Spreading rates vary between about 1 and 20 cm (0.4–8 in) per year but are generally between about 2 and 10 cm (0.8 to 4 in) per year. Although these rates are slow by everyday standards, about as fast as fingernails grow, they produce large changes over geologic time. For example, if a plate moved at the rate of just 1.6 cm (0.6 in) per year it would take 100,000 years for it to travel 1.6 km (1 mi); therefore, in the 200 million years since the breakup of Pangaea it could move more than 3200 km (2000 mi), which is more than half the distance between Africa and South America. The spreading rate affects the physical structure of divergent plate boundaries. Slow spreading rates are found along ridges that have steep profiles and deep central valleys, like the Mid-Atlantic Ridge. Fast spreading rates produce ridges with gentler slopes and shallower, or non-existent, central valleys, like the East Pacific Rise. Spreading rates are estimated at about 2.5 to 3 cm (1–1.2 in) per year for the Mid-Atlantic Ridge and about 8 to 13 cm (3–5 in) per year for the East Pacific Rise. The structure of ridges will be discussed in more detail in chapter 4. Keep in mind that the process of spreading does not occur smoothly and continuously; it goes on in "fits and starts," with varying rates and time periods between occurrences.

Iceland is the only large island lying across a mid-ocean ridge and rift zone; many of the processes of seafloor spreading can here be seen on land. Spreading in Iceland occurs at rates similar to those found at the crest of the mid-ocean ridge. Northeastern Iceland had been quiet for one hundred years, until volcanic activity began in 1975; in six years this activity widened by 5 m (17 ft) along an 80 km (50 mi) stretch of the ridge.

> What is the average rate of seafloor spreading?
>
> Compare spreading rates across ridges in the Atlantic and Pacific.
>
> Relate the spreading rate of Iceland since 1975 to the average spreading rate of the Mid-Atlantic Ridge.

Hot Spots 3.11

Scattered around the Earth are approximately forty specific fixed areas of isolated volcanic activity known as **hot spots** (fig. 3.27). They are found under continents and oceans, in the

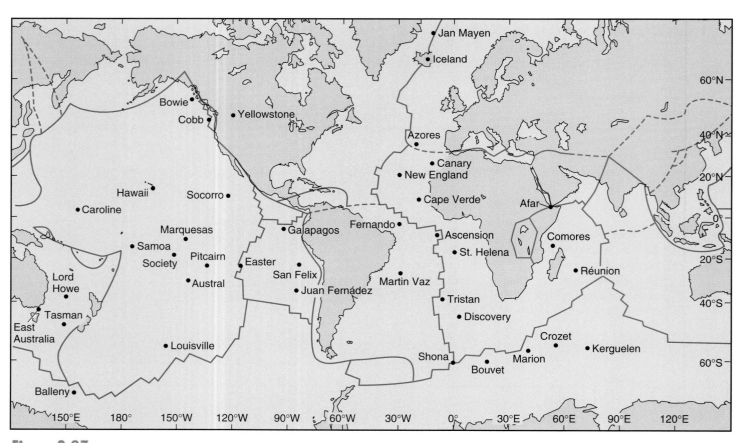

Figure 3.27
Location of major hotspots.

center of plates, and at the mid-ocean ridges. These hot spots periodically channel hot material to the surface from deep within the mantle, possibly from the core-mantle boundary. The life span of a typical hot spot appears to be about 200 million years. At these sites, mantle material may force its way through the lithosphere and form a seamount directly above. The hot spot may also raise a broad swelling of the ocean floor that subsides as the crust moves away from the magma source and cools.

As an oceanic plate moves over a hot spot, successive, and usually nonexplosive, eruptions can produce a linear series of peaks or **seamounts** with the youngest peak above the hot spot and the seamounts increasing in age as the distance from the hot spot increases (fig. 3.28). For example, in the islands and seamounts of the Hawaiian Islands system, the island of Hawaii with its active volcanoes is presently over the hot spot. The newest volcanic seamount in this series is Loihi, found erupting in this area in 1981. Loihi lies 45 km (28 mi) east of Hawaii's southernmost tip, rising more than 2450 m (8000 ft) above the sea floor. At its current rate of growth it should become an island sometime between 50,000 and 100,000 years from now.

The Hawaiian seamount system continues to the west, until just west of Midway Island, the chain changes direction and stretches to the north, indicating that the crust over the hot spot moved in a different direction some 40 million years ago. This is the Emperor Seamount Chain, volcanic peaks that once were above the sea surface as islands but have eroded and subsided over time. **Subsidence** occurs when the plate supporting the seamounts slides down and away from the bulge of the hot spot. This motion, in addition to the contraction and increasing density of the cooling plate and the weight of the seamount, depresses the mantle and carries the seamounts below the surface (fig. 3.28). Refer to figure 3.5 to follow this seamount chain. Recently, the ages of sequentially related seamounts from hot spots have been used to reconstruct the opening of the Atlantic and Indian Oceans.

Define "hot spot."

Discuss the formation of the Hawaiian Islands.

What do the Hawaiian Islands and the Emperor Seamount Chain tell us about the motion of the Pacific plate?

The Breakup of Pangaea   3.13

Figure 3.29 traces the recent plate movements that led to the configuration of the continents and oceans as we know them today. In the early Triassic when the first mammals and dinosaurs appeared, the continents were all joined in a single landmass we call Pangaea (fig. 3.29a). Most of the rest of the globe at this time was covered by the massive Panthalassic Ocean, also known as Panthalassa. A second, much smaller ocean called the Tethys occupied an indentation in Pangaea between what would eventually become present-day Australia and Asia. About 200 million years ago, Pangaea began to break apart into Laurasia and Gondwana. By about 150 million years ago, when the dinosaurs were flourishing, Laurasia and Gondwana were separated by a narrow sea that would grow to be the central Atlantic Ocean (fig. 3.29b). At the same time, India and Antarctica were beginning to move away from South America and Africa. Water flooded the spreading rift between South America and Africa about 135 million years ago and the

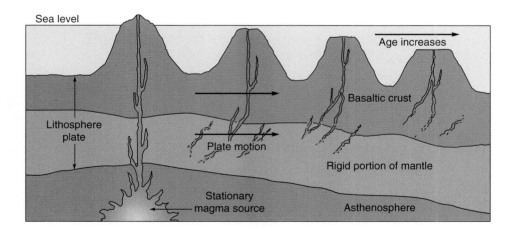

Figure 3.28

Chains of islands and seamounts are produced on the sea floor when an oceanic plate moves over a stationary hot spot. The islands and seamounts subside as they are carried away from the heat of the hot spot by seafloor spreading.

More Information: pubs.usgs.gov/publications/text/hotspts.html

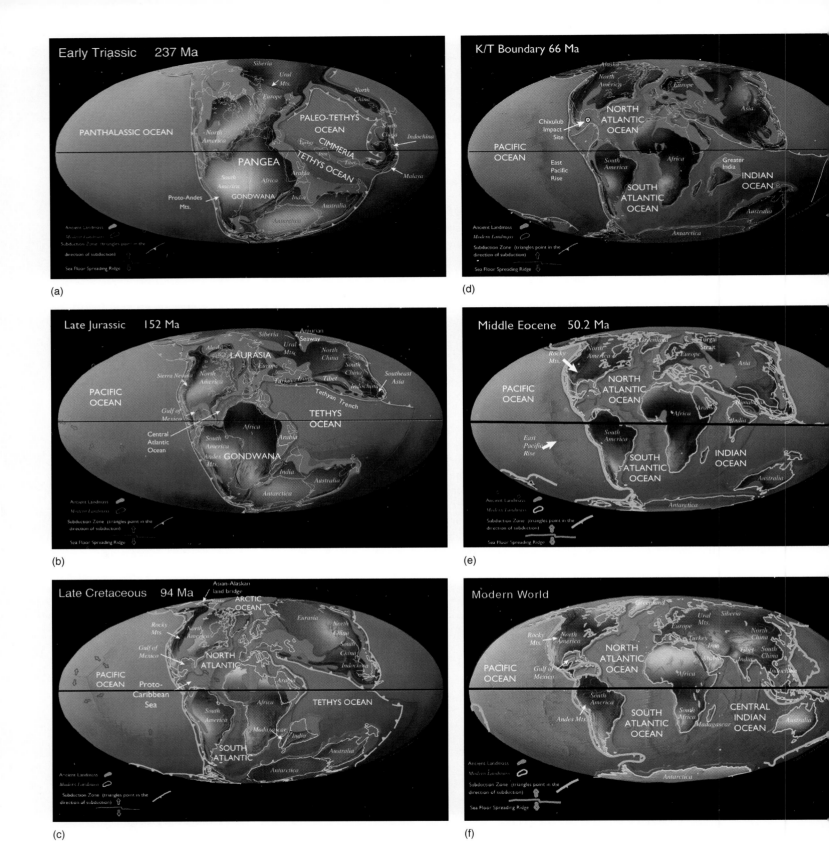

Figure 3.29

The configuration of landmass and oceans from Pangaea to the present. Geologic Period (or Epoch) and time before the present in millions of years (Ma) are given in the upper left-hand corner of each illustration.

More Information: www.scotese.com/

South Atlantic Ocean began to form. As seafloor spreading opened the Atlantic Ocean, Panthalassa (what we now call the Pacific Ocean) was growing progressively smaller. At the same time, India was moving north and the southern Indian Ocean was forming as the Tethys Ocean was closing. By the late Cretaceous, 94 million years ago, the North Atlantic Ocean was opening and the Caribbean Sea was beginning to form. India was continuing to move toward Asia and the Tethys Ocean was being consumed from the north as the Indian Ocean was expanding from the south (fig. 3.29c). By the end of the Cretaceous, shortly before the meteor impact that led to the extinction of the dinosaurs and many other species, the Atlantic Ocean was well-developed and Madagascar had separated from India (fig. 3.29d). During the Middle Eocene, 50 million years ago, Australia had separated from Antarctica, India was on the verge of colliding with Asia, and the Mediterranean Sea had formed (fig. 3.29e). About 40 million years ago India collided with Asia initiating a continent-continent convergent boundary, destroying the last of the Tethys Ocean and beginning the formation of the Himalayas. Twenty million years ago, Arabia moved away from Africa to form the Gulf of Aden and the new and still opening Red Sea.

Since the Earth is about 4.6 billion years old, there is no reason to believe that Pangaea's breakup was the first or only time during which the sea floor spread or the continents changed position and recombined. Refer to the discussion of terranes in section 3.10. It has been suggested that there is an orderly, cyclic, 500-million-year pattern of assembling and disassembling the landmasses powered by the heat from radioactive decay in the Earth's interior. Although production of heat is continuous, this model proposes that the heat is lost at varying rates that are related to the movement of the continents. As the continents shift, coalesce, and compress, the sea level drops. When sufficient heat accumulates under a continental mass, the rifting process begins unstitching the continents, and as the continents move apart, cool, and subside, the sea level rises and covers their borderlands.

> Trace the breakup of Pangaea over the past 200 million years.
>
> Compare the histories of the Red Sea and the Mediterranean Sea.
>
> Compare possible fates of the North Atlantic and the North Pacific after the next 500 million years.

Research Projects 3.14

The Ocean Drilling Program (ODP) is an international effort that began in 1983. It is presently guided by the Joint Oceanographic Institutions for Deep Earth Sampling (JOIDES) lo-
cated at Texas A&M University and uses the specialized drilling ship *JOIDES Resolution* (refer back to fig. 1.18b). Since 1985 the *JOIDES Resolution* has conducted drilling in all of the world's ocean basins seeking to obtain answers to questions related to global climate change, the creation and destruction of ocean basins, the occurrence of great submarine earthquakes, and more.

The *JOIDES Resolution* makes six scientific expeditions, or legs, each year. Each cruise is approximately two months in length and has specific scientific goals chosen through a careful review process. Each leg is staffed with thirty shipboard scientists (including graduate students) from the ODP member institutions. Upon completion of a cruise, the recovered cores are transported to one of four repositories for curation, storage, and future research. ODP scientists are able to use these repositories much as the general public uses a library. Scientists can also access the ODP databases containing vast amounts of data gathered during each cruise.

In 1997 ODP legs were carried out in the Southern Ocean off Antarctica to study global climate change, off the tip of South Africa to study the arid southern African climate, in the North Atlantic to study sea water circulation in the ocean crust, and off the coast of New Jersey to study the effect of the ice ages on sea level. In 1998 cruises were conducted in the southwest Pacific and Indian Oceans to study unusual cool-water coral reefs and climate change. In addition, cores were obtained along large faults in a young spreading basin near Papua New Guinea to learn more about submarine earthquakes. To improve the detection of earthquakes in remote areas an ocean bottom seismometer package, for the detection and recording of earthquakes, was placed in a drill hole in the central Indian Ocean. The 1999 legs included studies of earthquake and subduction processes in the Japan and Mariana trenches and an investigation of the properties of a large igneous plateau roughly one-third the size of the United States in the southern Indian Ocean called the Kerguelen Plateau. The ODP is scheduled to end in 2003 and plans are now underway for a follow-up drilling program and the construction of two new drilling vessels.

Geological oceanographers and geophysicists have descended into the mid-ocean ridges' rift valleys to study the production of new oceanic crust. In the summer of 1973, French oceanographers aboard the submersible *Archimède* made the first visual study of the rift valley of the Mid-Atlantic Ridge. The following summer, the U.S. submersible *Alvin* (fig. 3.30) joined in the dive program.

In 1977 and 1979, expeditions from the Woods Hole Oceanographic Institution explored the Galápagos Rift between the East Pacific Rise and the South American mainland, west of Ecuador. The most surprising discovery of these expeditions was the unexpected and perplexing presence of large communities of animals concentrated in areas of hot water rising from **hydrothermal vents** along the rift. In 1979, *Alvin* made twenty-four dives to the 2500-meter-deep (8000-ft) rift to study these animal populations, which are discussed in chapter 12.

Figure 3.30

The submersible *Alvin* being lowered to the sea surface for a dive along the Juan de Fuca Ridge in the northeastern Pacific. The *Alvin* carries two scientists and a pilot. It is equipped with cameras inside and out, as well as baskets for samples taken with its mechanical claws. It dives at a speed of 1.3 meters per second to a depth of 3000 m (10,000 ft). It requires 3 hours for the round trip and spends 4 to 5 hours on the bottom.

Also in 1979, the Mexican-French-American research program RISE (Riviera Submersible Experiment) investigated seafloor hot springs near the tip of Baja California. Two thousand meters (6500 ft) down, *Alvin* found mounds and chimney-shaped vents 20 m (65 ft) high ejecting hot (350°C) streams of black mud containing particles of lead, cobalt, zinc, silver, and other minerals. Here, too, the vents are surrounded with rich animal life.

Hydrothermal vent activity is now known to occur at seafloor spreading centers worldwide. Cold, dense seawater circulating through magma chambers close to the seafloor is heated and then released through cracks in the newly created upper lithosphere. Some of these vents discharge low-temperature (up to 30°C [80°F]), clear waters; others, known as white smokers, give rise to milky discharges (200 to 330°C [392 to 626°F]); and still others, black smokers, release jets of sulfur-blackened water (300 to 400°C [571 to 752°F]). This hydrothermal vent activity is thought to go on for years or decades. Rapid, unpredictable temperature fluctuations have been recorded on a time scale of days to seconds, indicating a very unstable environment surrounding the vents. During dives on the Juan de Fuca Ridge, submersibles collected samples (fig. 3.31) and an electronic still camera detected thermal radiation coming from seafloor vents (fig. 3.32). This long-wave-length light is undetectable by the human eye and is known as "vent glow."

As researchers' abilities to monitor the activity of the seafloor increase, additional and unexpected discoveries continue to be made. Oceanographers studying the Juan de Fuca Ridge in 1986 discovered a huge mass of warm water about 1000 m (3300 ft) above the ridge. Analysis of the water mass

Figure 3.31

A submersible uses its mechanical arm to collect a sample.
Courtesy of Dr. Verena Tunnicliffe, Univ. of Victoria, B.C. Canada.

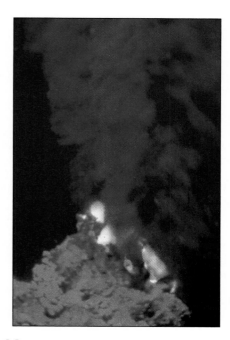

Figure 3.32

A color enhanced photo from the Juan de Fuca Ridge showing the glowing vent light produced by a hot vent.

What kind of information do the *JOIDES Resolution* drilling projects produce?

What kinds of discoveries have been made by deep-sea submersibles, and how do their observations help us to better understand the processes of plate tectonics?

Describe an area of hydrothermal vent activity.

What is a megaplume?

(estimated at 100 million cubic meters) showed that it had recently been formed by a large hydrothermal discharge. It was given the name "megaplume."

The Vents and RIDGE (Ridge Inter-Disciplinary Global Experiments) programs are two major U.S. projects associated with identifying and understanding processes along the mid-ocean ridges. The Vents program is exploring the source and strength of hydrothermal discharges as well as the time intervals between discharges. The program investigates the pathways followed by the materials issuing from vents, and its researchers have set up monitors to discover the impact of vent discharges on the chemistry of the water.

The RIDGE program includes a wide spectrum of science, from the physics and chemistry of the deep mantle; the formation of new ocean crust; the chemistry and physics of hydrothermal vents and their associated plumes; to the unique, complex ecosystems sustained by hydrothermal activity and, perhaps, the origin of life itself. One of the major parts of the RIDGE program is the Mantle ELectromagnetic and Tomography (MELT) Experiment, which is designed to provide observational constraints on the nature of the melting region in the mantle and the pattern of upwelling beneath a mid-ocean ridge that produces the magma that ultimately forms the oceanic crust.

RIDGE has installed sampling devices, recording cameras, sensors to monitor temperature and magnetic properties, and seismic event detectors in the Juan de Fuca Ridge Rift area and is making use of the underwater sound surveillance system installed by the U.S. Navy in the early 1950s; see "Listening to Seafloor Spreading" at the end of this chapter.

Summary

The Earth is made up of a series of layers: the crust, the mantle, the liquid outer core, and the solid inner core. Evidence for this structure comes from the ways in which seismic waves change speed and direction as they move through the Earth.

Continental crust is dominantly granite, which is less dense than oceanic crust formed of basalt. The boundary that separates the crust and the mantle is the Moho. The top of the mantle is fused to the crust to form the rigid lithosphere. The lithosphere floats on the deformable upper mantle or asthenosphere.

Alfred Wegener's theory of drifting continents was based on the geographic fit of the continents and the similarity of fossils collected on different continents. The discovery of the mid-ocean ridge system and the idea of convection cells in the asthenosphere led to seafloor spreading as the mechanism for continental drift. New oceanic lithosphere is formed at the ridges, or spreading centers; old oceanic lithosphere descends into trenches at subduction zones. Evidence for plate motion includes the match of the earthquake zones to spreading centers and subduction zones, greater crustal heat flow near ridges, age measurement of seafloor rocks, age and thickness measurements of sediment from deep-sea cores, and the magnetic stripes in the seafloor basalt on either side of the ridge system. Polar wandering curves are also used to trace plate movement. Rates of seafloor spreading range from 1 to 20 centimeters per year. The concept of continental drift and the theory of seafloor spreading combined to produce plate tectonics, in which the plates are made up of continental and oceanic lithosphere bounded by ridges, trenches, and faults.

There are three basic types of plate boundaries. Plates move apart along divergent plate boundaries. Divergent boundaries in oceanic lithosphere are marked by the mid-ocean ridge system and the creation of new sea floor. Divergent boundaries in continental lithosphere create rift zones that may expand to create new ocean basins. Plates collide along convergent boundaries. Convergence of oceanic lithosphere with either continental lithosphere or other oceanic lithosphere will produce an ocean trench and subduction of the oceanic plate into the mantle with associated volcanism. The convergence of two continental plates will produce buckling and thickening of the continental crust and the formation of a large mountain range. Plates simply slide past one another along transform boundaries, marked by transform faults.

Rifting has separated landmasses and produced new ocean basins. Passive or trailing continental margins move away from spreading centers to become midplate margins; they are not greatly modified by tectonic processes. Active or leading continental margins move toward subduction zones to be modified by the subduction process. Plate collisions produce earthquakes, island arcs, volcanic activity, and mountain building. Plate movements traced over the last 225 million years show the breakup of Pangaea to form the present continents.

Deep-sea drilling programs investigate tectonic processes, the evolution of oceanic crust, and the sediment records of past ages. Submersibles have been used to explore the rift valley of the Mid-Atlantic Ridge and the hydrothermal vents and animal communities found along the Galápagos Rift, Baja California, and off the Oregon coast. New programs use undersea robots, seafloor samplers, and sound systems to detect and monitor mid-ocean ridge dynamics.

Key Terms

active margin, 65	plate tectonics, 49
asthenosphere, 47	polar reversal, 55
basalt, 46	polar wandering curve, 56
continental drift, 48	P-waves, 45
Gondwanaland, 47	ridge, 49
granite, 46	rift valley, 60
hot spot, 66	seafloor spreading, 49
hydrothermal vent, 69	seamount, 67
inner core, 45	sediment, 52
Laurasia, 48	seismic wave, 45
lithosphere, 47	spreading center, 49
magma, 49	subduction zone, 49
mantle, 45	subsidence, 67
Moho (Mohorovičić	S-waves, 45
discontinuity), 46	terrane, 65
outer core, 45	transform fault, 57
Pangaea, 48	trench, 49
passive margin, 64	

Critical Thinking

1. Compare Wegener's theory of continental drift with today's concept of seafloor spreading. Are there differences? What are their similarities?
2. What had to be learned about the Earth to allow the acceptance of plate tectonics?
3. Should India be considered a terrane? Explain.
4. If the plates continue to move in the directions they move today for another 500 million years, propose a configuration of oceans and continents at the end of that time.
5. What technologies of the recent past have contributed to our increased understanding of seafloor processes?

Suggested Readings

Alper, J. 1994. Earth's Violent Birth. *Earth* 3(7):56–63.

Appenzeller, T. 1996. Travels of America. *Discover* 17(9):890–87. (Terranes.)

Baker, E. T. 1991/92. Megaplumes. *Oceanus* 34(4):84–91.

Ballard, R. 1975. Dive into the Great Rift. *National Geographic* 147(5):604–15. Project FAMOUS and photographs from *Alvin*.

Ballard, R. D., and J. F. Grassle. 1979. Return to Oases of the Deep. *National Geographic* 156(5):689–705.

Bambach, R. K., C. R. Scotese, and A. M. Ziegler. 1980. Before Pangaea: The Geographies of the Paleozoic World. *American Scientist* 68(1):26–38.

Bloxham, J., and D. Gubbins. 1989. The Evolution of the Earth's Magnetic Field. *Scientific American* 261(6):68–75.

Bonatti, E. 1994. The Earth's Mantle Below the Oceans. *Scientific American* 270(3):44–51.

Bonatti, E. and K. Crane. 1984. Oceanic Fracture Zones. *Scientific American* 250(5):40–51.

Broad, W. J. 1997. The Hot Dive. *Earth* 6(4):26–33.

Dalziel, I. 1995. Earth Before Pangaea. *Scientific American* 272(1):58–63.

Davidson, K., and A. Williams. 1996. Under Our Skin — Hot Theories on the Center of the Earth. *National Geographic* 189(1):100–11.

Dietz, R. S. 1977. San Andreas: An Oceanic Fault that came Ashore. *Sea Frontiers* 23(5):258–266.

Fuller, M., C. Laj, and E. Herrero-Bervera. 1996. The Reversal of the Earth's Magnetic Field. *American Scientist* 84(6):552–61.

Hoffman, K. A. 1988. Ancient Magnetic Reversals: Clues to the Geodynamo. *Scientific American* 258(5):76–83.

Hopkins, R. L. 1997. Land Torn Apart. *Earth* 6(1):36–41.

Jeanloz, R., and T. Lay 1993. The Core-Mantle Boundary. *Scientific American* 268(5):48–55.

Lonsdale, P., and C. Small. 1991/92. Ridges and Rises: A Global View. *Oceanus* 34(4):26–35.

Macdonald, K. C., and P. J. Fox. 1990. The Mid-Ocean Ridge. *Scientific American* 262(6):72–95.

Maxwell, A. E. 1993/94. An Abridged History of Deep Ocean Drilling. *Oceanus* 36(4):9–27.

Murphy, J. B., and R. D. Nance. 1992. Mountain Belts and the Supercontinent Cycle. *Scientific American* 266(4):84–91.

Nance, R. D., T. R. Worsley, and J. B. Moody. 1988. The Supercontinent Cycle. *Scientific American* 259(1):72–79.

Oceanus. 1992. 34(4). Special issue devoted to mid-ocean ridges.

Parks, N. 1994. Exploring Loihi: The next Hawaiian Island. *Earth* 3(5):56–63.

———.1997. Loihi Rumbles to Life. *Earth* 6(2):42–49.

Schneider, D. 1997. Hot-spotting. *Scientific American* 276(4):22–24.

Shaping the Earth — Tectonics of Continents and Oceans, Readings from Scientific American. 1990. W. H. Freeman, New York. 206 pp.

Taira, A. 1993/94. When Plates Collide. *Oceanus* 36(4):95–98.

Taylor, S. R., and S. M. McLennan. 1996. The Evolution of Continental Crust. *Scientific American* 274(1): 76–81.

The MELT Experiment. 1998. Several articles under the section 'Reports,' *Science* 280:1215–38.

Tivey, M. K. 1991/92. Hydrothermal Vent Systems. *Oceanus* 34(4):68–74.

Toomey, D. R. 1991/92. Tomographic Imaging of Spreading Centers. *Oceanus* 34(4):92–99.

Vink, G. E., W. J. Morgan, and P. R. Vogt. 1985. The Earth's Hot Spots. *Scientific American* 252(4):50–57.

Vogel, S. 1995. Inner Earth Revealed. *Earth* 4(4):42–49.

White, R. S., and D. McKenzie. 1989. Volcanism at Rifts. *Scientific American* 261(1):62–71.

Winter, S. (photog.). 1997. Iceland's Trial by Fire. *National Geographic* 191(5):58–71.

Wyession, M. 1995. The Inner Workings of the Earth. *American Scientist* 83(2):134–147.

Wyession, M. E. 1996. Journey to the Center of the Earth. *Earth* 5(6):46–49.

York, D. 1994. The Earliest History of the Earth. *Scientific American* 268(1):90–96.

Zimmer, C. 1996. The Light at the Bottom of the Sea. *Discover* 17(11):62–73.

internet references

 ## worldwide websites

Earth's Interior
Explanation of seismic waves
http://www.geo.mtu.edu/UPSeis/waves.html

Earthquakes and Volcanoes
National Earthquake Information Center—Home Page
http://neic.usgs.gov/

Earthquakes, Volcanoes, and Plate Tectonics
http://vulcan.wr.usgs.gov/Glossary/PlateTectonics/
Maps/map_quakes_volcanoes_plates.html

Submarine Volcanoes—Descriptions
http://vulcan.wr.usgs.gov/Glossary/Submarine
Volcano/description_submarine_volcano.html

Plate Tectonics
Major Tectonic Plates of the World
http://geology.er.usgs.gov/eastern/plates.html

Illustrations of landmass and ocean configurations with time
http://www.scotese.com

Animations of Plate Boundaries
http://scign.jpl.nasa.gov/learn/plate4.htm

Understanding plate motions
http://pubs.usgs.gov/publications/text/
understanding.html

Spreading Rate Calculator
http://triton.ori.u-tokyo.ac.jp/~intridge/calc.html

Relative Plate Motion Calculator
http://manbow.ori.u-tokyo.ac.jp/tamaki-html/
nuvel1a.html

Sea floor magnetic anomaly patterns
http://www.usyd.edu.au/su/geosciences/geology/
people/staff/dietmar/Agegrid/digit_isochrons.html

Ocean Ridge Structure
http://www.geol.ucsb.edu/~ken/sciam.html

VENTS program Home Page
http://www.pmel.noaa.gov/vents/index.html

RIDGE program Home Page
http://ridge.oce.orst.edu/

Pillow lavas and pillow basalts
http://volcanoes.usgs.gov/Products/Pglossary/
PillowLava.html

East African Rift system
http://bilbo.clpccd.cc.ca.us/lpc/science/rhanna/
puzzle/rift/rift.htm

NASA images of East Africa Rift
http://daac.gsfc.nasa.gov/DAAC_DOCS/
geomorphology/GEO_2/GEO_PLATE_T-35.HTML

Fracture Zones and Transform Faults
http://www.geo.wvu.edu/~geol1351/Spring98/
06/transformfaults.htm

Mount St. Helens
http://vulcan.wr.usgs.gov/Volcanoes/MSH/
framework.html

Hotspots
http://pubs.usgs.gov/publications/text/hotspots.html

Deep Sea Hydrothermal vents—journey to the seafloor and hot vents
http://www.ocean.washington.edu/people/grads/
scottv/explo raquarium/vent/intro.htm

Seafloor Topography from Satellite Altimetry
http://www.ngdc.noaa.gov/mgg/image/seafloor.
html

Ocean Drilling
Ocean Drilling Program
http://www-odp.tamu.edu/ and

http://www.oceandrilling.org/

Ocean Drilling Links and Data
http://www.ngdc.noaa.gov/mgg/geology/drill.
html

Paleomagnetism
Earth's magnetic poles
http://antwrp.gsfc.nasa.gov/apod/ap991019.html

http://www.agso.gov.au/geophysics/geomag/
information/faq.html

http://www.agu.org/sci_soc/campbell.html

In 1998 an ambitious expedition was undertaken by scientists at the University of Washington and the American Museum of Natural History in cooperation with the Canadian Coast Guard to recover black smokers from an active hydrothermal vent system on the Endeavour Segment of the Juan de Fuca Ridge. The Endeavour, located about 300 km (180 mi) off Washington State and Vancouver Island, is one

fluid flowing out. Neither chimney had many large organisms attached to it. The remaining two chimneys recovered, Roane and Gwenen, were venting diffuse fluids at temperatures of 20°–200°C (68°–392°F) and had diverse communities of organisms living on them, including tube worms, limpets, and snails (fig. 5). These two structures remained hot when brought up from a water depth of over

Recovery of Black Smokers

of the most active hydrothermal vent regions ever discovered. There are at least four major hydrothermal vent fields along the Endeavour Segment with hundreds of black smokers, some of which emit fluids at temperatures of 400°C (750°F). One giant smoker, Godzilla, stood roughly 45 m (150 ft) tall before it collapsed in 1996. The southernmost field in this area, the Mothra Hydrothermal Field, was chosen as the recovery site because it contains abundant steep-sided smokers that host diverse macrofaunal communities. Diffusely venting sulfide structures in this area are up to 24 m (79 ft) high and extend linearly for over 400 m (1300 ft). The tops of black smokers targeted for recovery were initially identified during a preliminary survey conducted in 1997 using the remotely operated vehicle Jason and the manned submersible Alvin. The smokers identified as possible targets for retrieval were relatively small; the largest was about 3 m (10 ft) tall and weighed roughly 6800 kg (15,000 lb).

The expedition used the Canadian remotely operated vehicle ROPOS to recover four sulfide chimneys from a depth of about 2250 m (7400 ft). ROPOS works out of a cage that acts as a garage for the vehicle while it is being lowered to the bottom or brought back to the surface (fig. 1). The fiber-optic tether that connects the cage to the ship is also used to transmit images from cameras on ROPOS and other data, including piloting commands to the vehicle. After the cage has been lowered to the target area, ROPOS "swims" out on a separate tether connected to the cage. Because ROPOS is on its own tether, it is unaffected by any motion of the cage caused by ship motion at the surface.

Once the expedition had arrived at the site, ROPOS was lowered from the University of Washington's research vessel Thomas G. Thompson to the bottom, where it moved to the first target, an inactive black smoker called Phang. Its initial task was to take pictures of this structure to document its characteristics for before-and-after studies. It then cinched cables around the smoker using a recovery cage that had been specially designed to fit over it. The next step was for ROPOS to approach the smoker and cut into its base using a chain saw with carbide and diamond-embedded blocks in the chain (fig. 2). After a series of cuts had been made, ROPOS was backed away from the chimney. The recovery cage and cables were attached to a line brought over by ROPOS from a previously deployed recovery basket on the bottom holding 2400 m (8000 ft) of line. ROPOS was then brought on deck, and the recovery line was floated to the surface, where it was brought on board the Canadian Coast Guard's vessel John P. Tully and attached to a winch. The Tully then took up slack on the line, broke the smoker off its base, and brought it to the surface (fig. 3), where the structure was cut in half in preparation for geological and microbiological studies.

This process was repeated three times during the course of the expedition to recover a total of four sulfide chimneys (fig. 4). The sulfide edifices were chosen in part for their diversity. Phang, the dead chimney, did not have any water flowing through it, but another chimney, Finn, was very active, with 304°C (579°F) hydrothermal

Figure 1

The Canadian remotely operated platform ROPOS (Remotely Operated Platform for Ocean Science) in its deployment cage.

2000 m (nearly 1.5 mi); Roane had temperatures of 90°C (194°F) and Gwenen had temperatures of 60°C (140°F). Their temperatures were measured from within the structures when they were on deck. The largest chimney recovered is 1.5 m (5 ft) tall and weighs 1800 kg (4000 lbs); it is on display at the American Museum of Natural History in New York City.

The smokers will be studied in detail by geologists, biologists, and chemists to learn more about the extreme chemical and thermal gradients that characterize these environments, the conditions under which sulfide structures grow and evolve, and how nutrients may be delivered to the organisms that live on and within the structures. Such studies are likely to find new species of microorganisms that thrive in these high-temperature, sunlight-free environments. Preliminary studies on these samples indicate that microorganisms living within the structures at temperatures of 90°C (194°F) derive their energy from carbon-bearing species in the hydrothermal fluids and from mineral-fluid reactions within the rocks. These and similar findings are exciting in that they show that microorganisms are capable of living in the absence of sunlight in volcanically active water-saturated regions of our planet. Other hydrothermally active planets may harbor similar life forms.

To Learn More about Recovery of Black Smokers

Holden, C. 1998. Deep-Sea Curios. *Science* 281 (5377): 639.

NOVA Online: Into the Abyss—retrieval of black smokers from the sea floor
http://www.pbs.org/wgbh/nova/abyss/

REVEL Project
http://www.ocean.washington.edu/outreach/revel/

(a)

Figure 2

With the recovery cage in place, ROPOS approaches the base of the chimney in preparation for cutting with a carbide and diamond-embedded chain saw.

(b)

Figure 4

(a) Several segments of recovered chimneys secured on deck.
(b) Section of a chimney that was cut in half for study of its interior.

Figure 3

A sulfide chimney breaking the surface during recovery, with the recovery cage surrounding it.

Figure 5

Tube worm colony attached to one of the recovered chimneys.

Volcanic events on land are often predicted in advance using seismometers to record the weak, low-frequency Earth tremors associated with moving magma and tilt meters to detect the gradual swellings of the Earth's crust prior to eruptions. Volcanism on land is easily recognized and documented as it occurs, but seafloor volcanism is rarely visible, and monitoring equipment has been installed in

Internet Resources

Acoustic Detection of a Seafloor Spreading Episode
http://www.pmel.noaa.gov/pubs/outstand/fox1526/abstract.shtml

Listening to Seafloor Spreading

very few undersea areas. The discovery of an undersea volcanic event in progress has depended on chance, for observers must be in the right place at the right time.

SOund SUrveillance Systems (SOSUS) are networks of underwater hydrophones (listening devices) installed by the U.S. Navy in the early 1950s; they were set up primarily on continental slopes and seamounts to detect sound signals from submarines and monitor coastal vessel traffic. However, the system also detects nearby, weak, low-frequency Earth tremors (Richter magnitude 2 or less) and can locate the area where the tremors are formed. On June 22, 1993 NOAA began acoustic monitoring of the SOSUS network near the Juan de Fuca Ridge, off the Washington and Oregon coasts. On June 26th scientists detected a burst of low-level seismic activity that migrated northward for the next 40 hours to a site along the Juan de Fuca Ridge where more than 600 seismic events were measured during the next three weeks. The intensity, frequency, and duration of the signals indicated that a seafloor spreading episode had occurred forming a magma dike about 60 km (36 mi) long as well as a possible seafloor extrusion of lava (fig. 6).

Good fortune made available two oceanographic research vessels, the NOAA Ship *Discoverer* and the Canadian Research Vessel *Tully*, to respond to these events. Both proceeded to the area of disturbance and began looking for changes in the sea floor and the seawater over the area. The *Discoverer* detected a recent lava flow covering 3.4×10^5 m^2 (0.13 mi^2) and decreasing water depth by about 29 m (95 ft). The *Tully* deployed a robotic device that observed an area of fresh glassy lava 2.5 km (1.5 mi) long and 300 m (0.2 mi) wide. It also observed nearly continuous diffuse venting of warm water (<30°C). Much of the new lava flow was covered with orange flocculent material that was shown to contain disorganized aggregations of bacteria and debris. Water samples south of the new lava flow showed elevated levels of manganese, iron, calcium, silica, and lithium but depleted levels of magnesium. In October the area was revisited by the *Alvin* and additional samples were taken. Between 1 July and 2 August 1993 towed sensors detected three huge (1.3 – 4 $\times 10^{10}$ m^3 [0.3 – 1.0 mi^3]) megaplumes or masses of warm water.

The origin of the burst of seismic activity at the north end of the rift zone is still unclear. The presence of megaplumes appears to point to a preexisting hydrothermal system for which there is no crustal evidence at this time. Analysis of the 1993 data and monitoring of the SOSUS net continue as oceanographers listen for more signals of volcanic events. Detecting mid-ocean ridge events as they occur allows scientists to be in the right place at the right time to observe and investigate the dynamics of seafloor spreading.

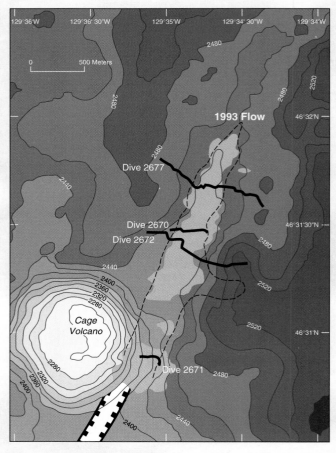

Figure 6

A bathymetric chart of the 1993 Juan de Fuca volcanic event. Fresh lava is shown in pink. The edge of the flow is shown by dashes. Dive tracks of the *Alvin* are shown in black and identified by dive numbers.

From H. Paul Johnson, "Alvin Magnetic Survey of Zero-Age crust: CoAxial Segment Eruption, Juan de Fuca Ridge, 1993" in Geophysical Research Letters, *Vol. 22 (2) 171–174, January 15, 1995.*

chapter four

The Sea Floor

The fog continued ... very light breeze, before which we ran to the eastward, literally feeling our way along. The lead was ... every two hours and the gradual change from black mud to sand, showed that we were approaching Nantucket south shoals. On Monday morning, the increased depth and deep blue color of the water and the mixture of shells and white sand which we brought up, upon ... in the channel, and nearing ... head was put directly to the ... with perfect confidence in the ... taken an observation for two ... difference of an eighth of a ... the ... prevailed ... at eight ... which we passed, told us we ... before ... carried us well along; and at four o'clock, ... ourselves to the northward of Race Point, we hauled upon the wind and stood into the bay, north-north-west, for Boston light, and commenced firing guns for a pilot.

Richard Henry Dana, Jr.,
from Two Years Before the Mast

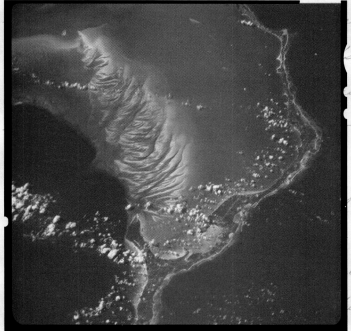

Eleuthera Island in the Bahamas from the Space Shuttle Columbia. The light blue of the shallow water Bahama Bank contrasts with the dark blue of the deep ocean.

Learning Objectives

After reading this chapter, you should be able to

- Describe the divisions and features of the continental margin.
- Describe the formation of a submarine canyon.
- Describe the features of the ocean basin floor.
- Understand the use of echo sounding to map the sea floor.
- Compare ocean sediments by source, properties, and distribution.
- Relate sediment particle size and sinking rate.
- Describe the instruments used to sample seafloor sediments.
- Give examples of seafloor mineral resources.
- Understand the political, legal, and economic factors related to harvesting seafloor mineral resources.

*t*he sea floor was an unknown environment to early mariners and the first curious scientists. They believed that the oceans were large basins or depressions in the Earth's crust, but they did not conceive that these basins held features that were as magnificent as the mountain chains, deep valleys, and great canyons of the land. As maps were created in greater detail and as ocean travel and commerce increased, it became essential to map seafloor features in the shallower regions, but it was not until the 1950s that improvements in technology made it possible to sample the deep sea floor routinely and in detail.

The geography and the geology of the sea floor are the products of processes that occur on both human and geological time scales. The main features of the sea floor are compelling evidence for plate tectonics as discussed in chapter 3.

The Sea Floor   4.1

The mountain ranges under the sea are longer, the valley floors are wider and flatter, and the canyons are often deeper than those found on land. Land features are continually eroded by wind, rain, and ice, and are affected by changes in temperature and rock chemistry. The erosion of seafloor features is slow, and occurs only by way of waves and currents along the shore and on steep underwater slopes. Beneath the surface the water dissolves away certain materials, changing the rock chemistry, but the features retain their shape and are only slowly modified as they are covered by a constant rain of particles or **sediment** falling from above. Computer-drawn profiles of elevations across the United States and the Atlantic Ocean are compared in figure 4.1. Notice that the profiles of the mountain ranges on land and ridge systems of the sea floor are similar.

The Continental Margin   4.2

The region of the sea floor that is closest to land is called the **continental margin.** Continental margins are the edges of the landmasses at present below the ocean surface and their steep slopes that descend to the deep sea floor. The two different types of continental margins, passive and active, are discussed in section 3.10. Passive continental margins can be subdivided into four distinct regions: the continental shelf, shelf break, slope, and rise (fig. 4.2). Along active continental margins, the continental slope typically continues into a deep ocean trench except where large amounts of sediment carried off the continent have filled the trench.

The **continental shelf** is a nearly flat land-border of varying width that slopes very gently toward the ocean basins. Shelf widths average about 65 km (40 mi) and vary from only a few tens of meters to more than 100 km (60 mi).

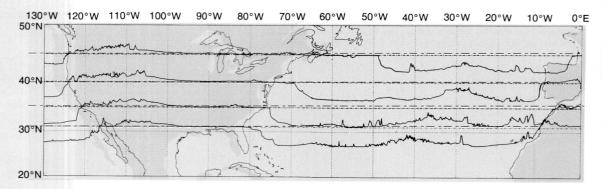

Figure 4.1

Computer-drawn topographic profiles from the west coast of Europe and Africa to the Pacific Ocean. The elevations and depths above and below 0 meters are shown along a line of latitude by using the latitude line as zero elevation. For example, the ocean depth at 40°N and 60°W is 5040 m (16,500 ft). The vertical scale has been exaggerated about one hundred times the horizontal scale. If both the horizontal and vertical scales were kept the same, a vertical elevation change of 5000 m would measure only 0.05 mm (0.002 in).

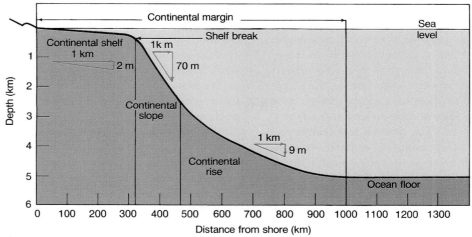

Figure 4.2

A typical profile of a passive continental margin. Notice both the vertical and horizontal extent of each subdivision. The average slope is indicated for the continental shelf, slope, and rise. The vertical exaggeration is one hundred times greater than the horizontal scale.

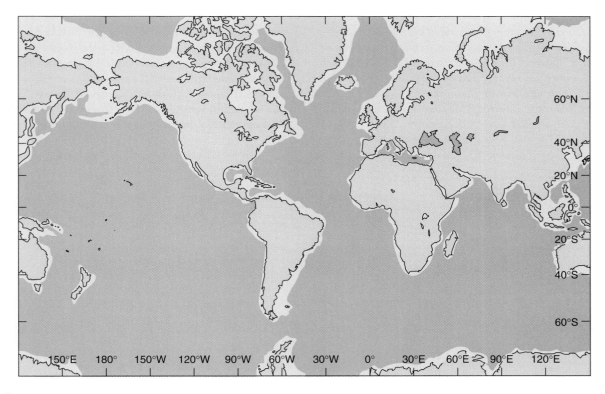

Figure 4.3

Distribution of the world's continental shelves. The seaward edges of these shelves are at an average depth of approximately 130 m (430 ft).

The distribution of the world's continental shelves is shown in figure 4.3. The shelf is narrow along the mountainous coasts of active continental margins and wide along low-lying land at passive continental margins. Passive margin continental shelves are often further modified by sea-level changes and storm waves that erode the edges of the continents and, in some cases, natural dams on the shelves trap sediments between the offshore dam and the coast. Such dams include reefs, volcanic barriers, upthrust rock, and salt domes. Notice the narrow shelf along the west coasts of North and South America and the wide shelves off the eastern and northern coasts of North America, Siberia, and Scandinavia.

The continental shelves are geologically part of the continents, and during past ages, they have been covered and uncovered by fluctuations in sea level. When the sea level was low during the ice ages (periods of increased ice on land), erosion deepened valleys, waves eroded previously submerged land, and rivers left their sediments far out on the shelf. When the ice melted and sea level rose, these areas were flooded and sediments built up in areas close to the new shore. Although presently submerged, these shelf areas still show the scars of old riverbeds and glaciers, features they acquired when exposed as part of the continent. Some continental shelves are covered with thick deposits of silt, sand, and mud derived from the land; for example, the Mississippi and Amazon Rivers deposit large amounts of sediments at their mouths. Other shelves are bare of sediments, such as where the Florida Current sweeps the tip of Florida, carrying the shelf sediments northward to the deeper water of the Atlantic Ocean.

The boundary of the continental shelf on the ocean side is determined by an abrupt change in slope, leading to a more rapid increase in depth. This change in slope is referred to as the **continental shelf break,** while the steeper slope extending to the ocean basin floor is known as the **continental slope.**

The angle and extent of the slope vary from place to place. The slope may be short and steep—for example, the depth may increase rapidly from 200 m (650 ft) to 3000 m (10,000 ft), as in figure 4.2 or it may drop as far as 8000 m (26,000 ft) into a deep sea trench, as it does off the west coast of South America. The continental slope may show rocky outcroppings, and it is often relatively bare of sediments because of its steepness.

The most outstanding features found on the continental slopes are **submarine canyons.** These canyons sometimes extend up, into, and across the continental shelf. A submarine canyon is steep-sided and has a V-shaped cross section, with tributaries similar to those of river-cut canyons on land. Figure 4.4a shows the Monterey and Carmel canyons off the coast of California. Figure 4.4b compares a profile of the Monterey Canyon with a profile of the Grand Canyon of the Colorado River.

Many of these submarine canyons are associated with existing river systems on land and were apparently cut into the shelf during periods of low sea level, when the rivers flowed across the continental shelves. Ripple marks on the floors of canyons and sediments fanning out at the ends of these canyons suggest that they have been formed by moving flows of sediment and water called **turbidity currents.** Turbidity currents are fast-moving avalanches of mud, sand, and water caused by earthquakes or the overloading of sediments on steep slopes. They flow down the slope, eroding and picking up sediment as they gain speed, and over time they erode the slope and excavate the submarine canyon. The abundance of submarine canyons can be seen along the continental slopes in the chart of the sea floor shown earlier in figure 3.5. As the turbidity current flow reaches the sea floor, it slows and spreads, and the sediments settle. During this settling process, coarse materials drop out and are overlaid by successive layers of decreasing particle size, and **turbidites** are formed; figure 4.5 shows a size-graded deposit preserved in a shore cliff. Turbidity currents have never been directly observed, although similar but smaller

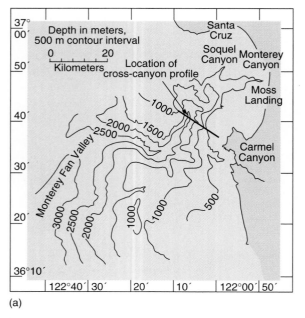

(a)

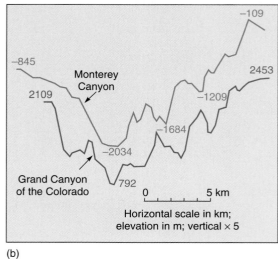

(b)

Figure 4.4

(a) Depth contours depict the Monterey and Carmel canyons off the California coast as they cut across the continental slope and up into the continental shelf. (b) Cross-canyon profile of the Monterey Canyon shown in (a). Compare this profile with that of the Grand Canyon drawn to the same scale.

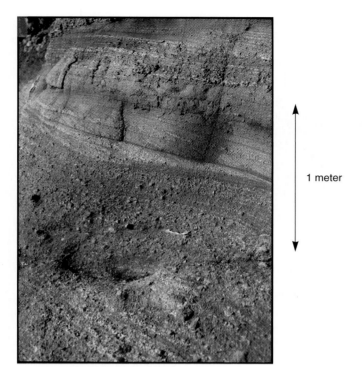

1 meter

Figure 4.5

This beach cliff shows a series of ancient turbidite deposits that have been uplifted and then exposed by wave erosion. Each individual turbidite is a graded deposit, with the largest particles in the deposit at the bottom of the turbidite and the smallest particles at the top.

and more continuous flows, such as sand falls, have been observed and photographed. Compelling evidence of a large turbidity current in the ocean was observed in 1929 when a series of trans-Atlantic telephone cables south of Newfoundland along the Grand Banks were broken after an earthquake occurred beneath the continental slope. The cables closest to the earthquake broke when the earthquake occurred. The other cables broke in sequence with increasing distance from the earthquake. Scientists concluded that the earthquake triggered a turbidity current that moved down the continental slope at speeds of as much as 80 km (50 mi) per hour and across the less steep continental rise at about 24 km (15 mi) per hour, breaking cables as it reached them. This turbidity current traveled a total distance of about 650 km (400 mi).

At the base of the steep continental slope, there may be a gentle slope formed by the accumulation of sediment. This portion of the sea floor is the **continental rise** (fig. 4.2); it is a region of sediment deposition by turbidity currents, underwater landslides, and any other processes that carry sand, mud, and silt down the continental slope. The continental rise is a conspicuous feature in the Atlantic and Indian Oceans and around the Antarctic continent. Few continental rises occur in the Pacific Ocean; here great seafloor trenches are often at the base of the continental slope, where material from the active margin is being subducted.

The Ocean Basin Floor   4.3

The deep sea floor, between 4000 and 6000 m (13,000–20,000 ft), covers more of the Earth's surface (30%) than do the continents (29%). In many places, the ocean floor is a flat plain, known as an **abyssal plain.** Abyssal plains form as sediment deposits (from turbidity currents and the waters above) cover the irregular topography of the sea floor. Large wavelike undulations have been discovered in these sediment layers; these "mud waves" are formed by deep-ocean currents flowing across the abyssal plain. **Abyssal hills** and seamounts are scattered across the sea floor in all the oceans. Abyssal hills are volcanic features less than 1000 m (3300 ft) high, and seamounts are steep-sided volcanoes that are formed by local vulcanism and over hot spots; they rise abruptly toward the sea surface. Sometimes seamounts pierce the sea surface to become islands. These features are shown in figure 4.6. Abyssal hills are probably the Earth's most common topographic feature. They are found over 50% of the Atlantic sea floor and about 80% of the Pacific floor; they are also abundant in the Indian Ocean.

In the warm waters of the Atlantic, Pacific, and Indian Oceans, coral reefs and coral islands are formed in association with seamounts. Reef-building coral is a warm-water animal that requires a place of attachment and grows in intimate association with a single-celled, plant-type organism. It is therefore confined to sunlit, shallow, warm waters. When a seamount pierces the sea surface to form an island, it provides a base on which the coral can grow. The coral grows to form a **fringing reef** around the island. If the seamount sinks or subsides slowly enough, the coral continues to grow upward at the same rate as the rising water, and a **barrier reef** with a lagoon between the reef and the island is formed. If the process continues, eventually the seamount disappears below the surface and the coral reef is left as a ring, or **atoll.** This process is illustrated in figure 4.7.

These steps required to form an atoll were suggested by Charles Darwin, based on his observations during the voyage of the *Beagle* in 1831–36. Darwin's ideas have been proved substantially correct by more recent expeditions that drilled through the coral debris of lagoon floors and found the basalt peak of the seamount that once protruded above the sea surface.

Submerged flat-topped seamounts, known as **guyots,** or **tablemounts** (fig. 4.6), are found most often in the Pacific Ocean. These guyots are 1000 to 1700 m (3300 to 5600 ft) below the surface, with many at the 1300 m (4300 ft) depth.

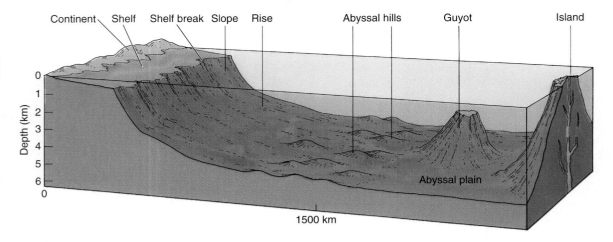

Figure 4.6

An idealized portion of ocean basin floor with abyssal hills (less than 1000 m of elevation), a guyot (a flat-topped seamount), and an island on the abyssal plain. The island was previously a seamount before it reached the surface. Seamounts and guyots are known to be volcanic in origin (vertical × 100).

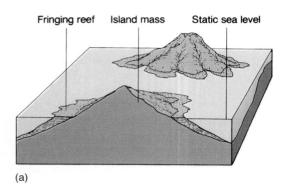

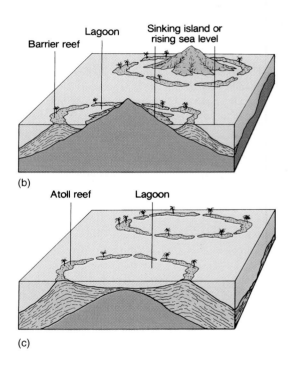

Figure 4.7

Types of coral reefs and the steps in the formation of a coral atoll are shown in profile: (a) fringing reef; (b) barrier reef; and (c) atoll reef.

Many guyots show the remains of shallow marine coral reefs and the evidence of wave erosion at their summits. This indicates that at one time they were surface features and that their flat tops are the result of rain and wave erosion, past coral reef growth, or both. Guyots have dropped far below sea level because their weight has helped depress the oceanic crust and because the plate supporting them has moved away from the hot spot or ridge where they were created; refer to section 3.11. They have also been submerged by rising sea level during periods in which the land ice has melted.

List the features of the ocean basin floor.

Explain the series of events that produce an atoll.

How are seamounts and guyots the same, and how are they different?

The Ridges and Rises   4.4

The most remarkable features of the ocean floor are the mid-ocean **ridge and rise systems** that form at the spreading boundaries of the lithospheric plates. This series of great continuous underwater volcanic mountain ranges stretches for 65,000 km (40,000 mi) around the world and runs through every ocean; see figure 4.8 and refer to figure 3.5. These ridge systems are about 1000 km (600 mi) wide and 1000 to 2000 m (3500 to 7000ft) high. If the slopes of these mountain ranges are steep, they are referred to as ridges (such as the Mid-Atlantic Ridge and the Mid-Indian Ridge); if the slopes are more gentle, they are called rises (such as the East Pacific Rise). Along portions of the system's crest is a central **rift valley,** 15 to 50 km (9 to 30 mi) wide and 500 to 1500 m (1500 to 5000 ft) deep. The rift valley is volcanically very active and bordered by rugged rift mountains. Fracture zones with steep sides run perpendicular to the ridges and rises, connecting offset sections of the mid-ocean ridges to make a stair-step pattern. The faults between the offset sections of rift valley are the transform faults; refer to section 3.8. The mid-ocean ridge and rise systems and their smaller lateral extensions separate the ocean basins into sub-basins, in which bodies of deep water are isolated from each other. Refer back to figure 3.5.

Distinguish between a ridge and a rise.

Where would you find examples of each? Use figure 3.5.

Use figures 3.5 and 4.8 to identify South Atlantic ocean basins formed by the Mid-Atlantic Ridge and its lateral extensions.

The Trenches   4.5

The Mid-Atlantic Ridge is the dominant feature of the floor of the Atlantic Ocean, but the narrow, steep-sided, deep-ocean trenches characterize the Pacific. Trenches occur at converging plate boundaries; refer to section 3.9 to review their formation, and see figures 4.9 and 3.5 for their locations. Some of these trenches are located on the seaward side of chains of volcanic islands known as **island arcs.** For example, the Japan-Kuril Trench, the Aleutian Trench, the Philippine Trench, and the deepest of all ocean trenches, the Mariana Trench, are all associated with island arc systems. The Challenger Deep, a portion

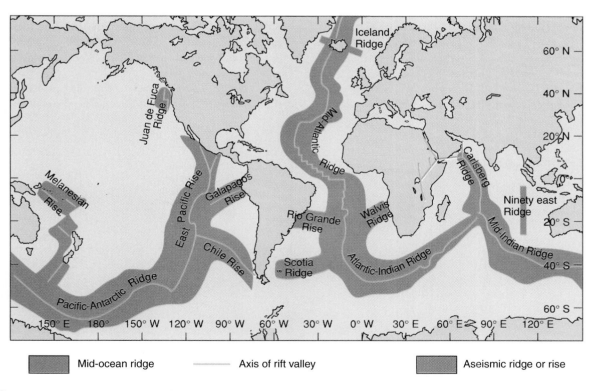

| Mid-ocean ridge | —— Axis of rift valley | Aseismic ridge or rise |

Figure 4.8

The mid-ocean ridge and rise system of divergent plate boundaries. Locations of major aseismic (no earthquakes) ridges and rises have been added. Aseismic ridges and rises are elevated linear features thought to be created by hot spot activity.

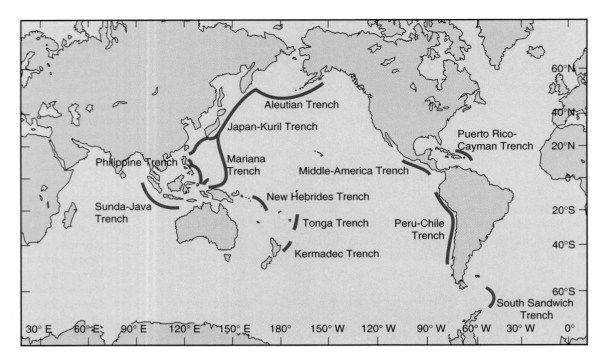

Figure 4.9

Major ocean trenches of the world. The deepest ocean depth is 11,020 m, east of the Philippines in the Mariana Trench. It is known as the Challenger Deep.

of the Mariana Trench, has a depth of 11,020 m (36,150 ft), making it the deepest known spot in all the oceans. The longest of the trenches is the Peru-Chile Trench, stretching 5900 km (3700 mi) along the west side of South America. To the north, the Middle-America trench borders Central America. The Peru-Chile and Middle-America Trenches are not associated with volcanic island arcs, but are bordered by land volcanoes.

In the Indian Ocean the great Sunda-Java Trench runs for 4500 km (2800 mi) along Indonesia, while in the Atlantic there are two comparatively short trenches, the Puerto Rico-Cayman Trench and the South Sandwich Trench, both associated with chains of volcanic islands.

Figure 4.10 summarizes the topography of the land and the bathymetry of the sea floor as percents of the Earth's sur-

Figure 4.10

The Earth's main topographic features shown as percents of the total Earth's surface and as percents of the land and of the oceans.

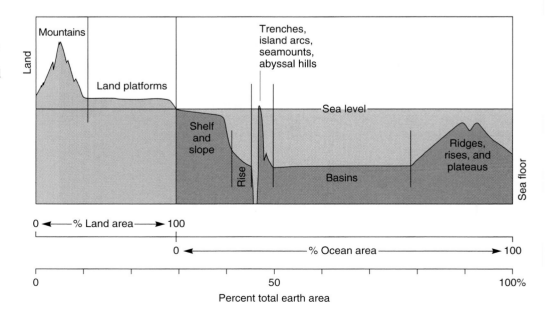

face area. Compare the tectonically active areas of ridges and trenches. Compare the area of land platforms to the area of ocean basins.

What land features are associated with deep-sea trenches?

Where are the major deep-sea trenches of the world's oceans? Use figures 3.5 and 4.9.

Measuring the Depths 4.6

Early sailors made their soundings, or depth measurements, using a hemp line or rope marked in equal distances (usually in **fathoms,** a 6 ft unit) with a greased lead weight at the end. The change in line tension when the weight touched bottom indicated depth, and the particles from the bottom adhering to the grease confirmed the contact and brought a bottom sample to the surface. This method was quite satisfactory in shallow water, and the experienced captain used bottom samples with depth measurements to aid in navigation, particularly at night or in heavy fog.

Later, piano wire with a cannonball attached was used for deep-water soundings. The heavy weight of the ball made it easier to sense the bottom, but the time (eight to ten hours) and effort to winch the wire out and in for each measurement were so great that by 1895 over all the world's oceans only about 7000 depth measurements had been made in water deeper than 2000 meters and only 550 measurements had been made of depths greater than 9000 meters. It was not until the 1920s, when the **echo sounder** or **depth recorder** was invented, that deep-sea depth measurements became routine. The depth recorder measures the time required for a sound pulse to leave the surface vessel, reflect off the bottom, and return. It allows continuous measurements to be made easily and quickly while the ship is under way. The behavior of sound in seawater and its use as an oceanographic tool are discussed in chapter 5. Today data from depth recorders (fig. 4.11) and underwater photography and television, as well as direct observation from submersibles, are used to produce even more detailed knowledge of the sea floor. See "Undersea Robots" at the end of this chapter to learn more about their role in exploring the sea floor.

Very large scale sea floor surveys use satellite measurements of changes in sea surface elevation caused by changes in the Earth's gravity field due to seafloor bathymetry. These changes in sea surface elevation can be detected by radar altimeters that measure the distance between the satellite and the sea surface. The sea surface is not flat even when it is perfectly calm. Changes in gravity caused by seafloor topography create gently sloping hills and valleys on the sea surface. The excess mass of features such as seamounts and ridges creates a gravitational attraction that draws water toward them, resulting in

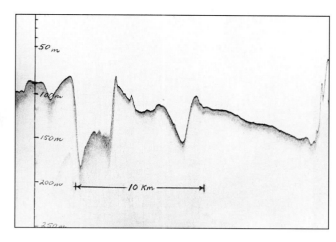

Figure 4.11

Depth recorder trace. Sound pulses reflected from the ocean floor trace a depth profile as the ship sails a steady course. The horizontal scale depends on ship speed.

an elevation of the sea surface. Conversely, the deficit of mass along deep ocean trenches, and subsequent weaker gravitational attraction, results in a depression of the sea surface as water is drawn away toward surrounding areas with greater gravitational attraction. Sea level is elevated over large seamounts by as much as 5 m (16 ft), over ocean ridges by about 10 m (33 ft), and is depressed over trenches by about 25–30 m (80–100 ft). These changes in elevation occur over tens to hundreds of kilometers so the slopes are very gentle. Because these changes in sea surface elevation are related to sea floor topography it is possible to use them to reconstruct what the bathymetry must look like to produce the observed variations in sea surface topography (fig. 4.12).

What kinds of equipment have oceanographers used to develop the detailed seafloor chart shown in figure 3.5?

How was figure 4.12 obtained?

Sediments 4.7

The margins of the continents and the ocean basin floors receive a continuous supply of particles from many sources. Regardless of where these particles are found, or what their origin is, they are called sediment when they accumulate on the sea floor. Oceanographers study the rate at which sediments accumulate, the distribution of type and abundance of sediments on the sea floor, their chemistry, and the history they record. In order to describe and catalog the sediments found on the ocean floor, geological oceanographers classify sediments by particle size, location, origin, and chemistry.

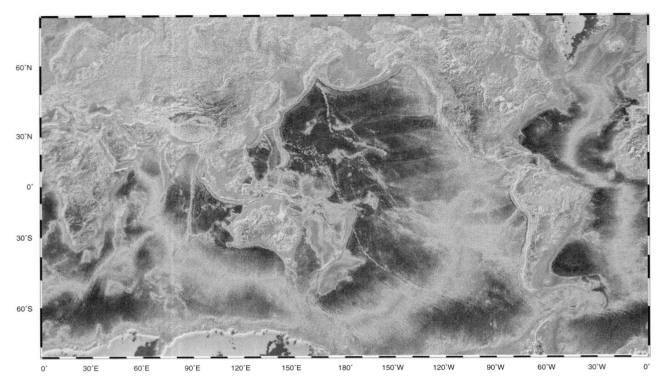

Figure 4.12

Color-shaded relief image of the bathymetry of the world's ocean basins modeled from satellite altimetry data and checked against ship depth soundings.
More Information: topex.ucsd.edu/marine_topo/text/topo.html

Size

Sediment particles are classified by size as indicated in table 4.1. Familiar terms such as gravel, sand, and mud are used to identify broad size ranges of large, intermediate, and small particles respectively. Within each of these ranges particles are further ranked to produce a more detailed scale from boulders to the very smallest clay-sized particles that can only be seen with a microscope. Note that a boulder is a particle whose di-

ameter is greater than 256 mm or 10 in; much smaller than what we commonly envision when we hear the word "boulder."

The individual particles in a sediment sample can be sorted using a series of sieves that have woven mesh of decreasing size. A sample is said to be well sorted if it is nearly uniform in particle size and poorly sorted if it is made up of many different particle sizes. Particle size influences the deposition of sediment by the horizontal distance it is transported before settling out of the water and the rate at which it sinks. In general, it takes more energy to transport large particles than it does small particles. In the coastal environment, when poorly sorted sediment is transported by wave or current action, the larger particles will settle out and be deposited first while the smaller particles may be carried farther away from the coast and deposited elsewhere. In the open ocean, the variation in sinking rate between large and small particles has a tremendous influence on how long it takes for a particle to sink to the deep sea floor, and hence, how far the particle may be transported by deep horizontal currents while it is settling (table 4.2). Sand-sized particles may settle to the deep sea floor in a matter of days while it may take clay-sized particles over 100 years to make the same journey. The speed of deep horizontal currents in the oceans is generally quite slow but even at a speed of 5 cm (2 in)/s a clay-sized particle could theoretically be transported around the world five times before it reached the deep sea floor. Smaller soluble particles also have time to dissolve as they slowly sink in the deep ocean.

table 4.1	Sediment Size Classifications	
Descriptive Name		**Diameter (mm)**
Gravel	Boulder	>256
	Cobble	64–256
	Pebble	4–64
	Granule	2–4
Sand	Very Coarse	1–2
	Coarse	0.5–1
	Medium	0.25–0.5
	Fine	0.125–0.25
	Very fine	0.0625–0.125
Mud	Silt	0.0039–0.0625
	Clay	<0.0039

table 4.2 Sediment Sinking Rate and Distance Travelled

Sediment Size	Approximate Sinking Rate (m/s)	Time for a Vertical Fall of 4 km (days)	Horizontal Distance Travelled in a 5 cm/s Current (km)
Very fine sand	9.8×10^{-3}	4.7	20.4
Silt	9.8×10^{-5}	470	2040
Clay	9.8×10^{-7}	47,000	204,000

The sinking rate of a particle depends on its density, shape, and diameter. These rates are based on the assumption that the particles are spherical and have a density similar to that of quartz. Estimates of the speed of deep currents vary. A conservative estimate of 5 cm/s is chosen for purposes of illustration.

More Information: www.filtration-and-separation.com/settling/settling.htm

If small particles are differently charged, they will attract each other, forming larger particles that sink more rapidly. Also, when predators eat tiny plants and animals, they process the organic material to produce energy for life and package the inorganic remains, often microscopic shells called **tests,** and expel them as larger fecal pellets, which sink more rapidly. It is estimated that as many as 100,000 tests of small organisms can be packaged in a single fecal pellet. These processes help to decrease the time it takes for small particles to sink to the sea floor from years to just a few days, minimizing their horizontal displacement by water movements.

Location

Marine sediments are classified as either **neritic** (*neritos* = of the coast) or **pelagic** (*pelagios* = of the sea) based on where they are found (fig. 4.13). Neritic sediments are found near conti-nental margins and islands and have a wide range of particle sizes. Most neritic sediments are eroded from rocks on land and transported to the coast by rivers. It is estimated that as much as 15 billion metric tons of sediments are carried to the oceans each year by rivers. Once they enter the ocean they are spread across the continental shelf and down the slope by waves, currents, and turbidity currents. The largest particles are left near coastal beaches while smaller particles are transported farther from shore. Accumulation rates of neritic sediments are quite variable. In river estuaries, sedimentation rates can reach several meters per year; the rivers of Asia contribute more than one-quarter of the world's marine sediments eroded from the continents each year. In quiet bays, the deposit rate may be 500 cm (200 in)/1000 years, while on the continental shelves and slopes, values of 10 to 40 cm (4 to 16 in)/1000 years are typical, with the flat continental shelves receiving the larger amounts.

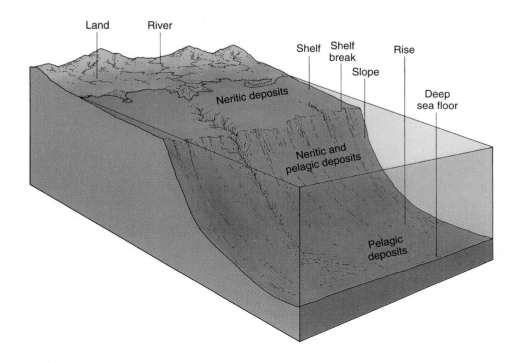

Figure 4.13

Classification of sediments by location of deposit. The distribution pattern is partially controlled by the sediments' proximity to source and rate of supply.

Pelagic sediment deposits are fine-grained sediments, which collect slowly on the deep sea floor. The thickness of pelagic sediments is related to the length of time they have been accumulating, or the age of the sea floor they cover. Consequently, their thickness tends to increase with increasing distance from mid-ocean ridges (review fig. 3.11). An average accumulation value for sediments in the deep oceans is 0.5 to 1.0 cm (0.2 to 0.4 in)/1000 years. Although deep-sea sedimentation rates are extremely slow, there has been plenty of time during geologic history to accumulate the average deep-sea sediment thickness of approximately 500 to 600 m (1600 to 2000 ft). At a rate of 0.5 cm (0.2 in)/1000 years, it takes only 100 million years to accumulate 500 m of sediment.

Source and Chemistry

Marine sediments are also classified by the source of the particles that comprise the sediment. Sedimentary particles may come from one of four different sources; pre-existing rocks, marine organisms, the seawater itself, or from space.

Sediments derived from pre-existing rocks are classified as **lithogenous** (*lithos* = stone, *generare* = to produce) **sediments.** These are also commonly called **terrigenous** (*terri* = land, *generare* = to produce) **sediments** since the majority of them come from landmasses. Active volcanic islands in the ocean basins are also an important source of lithogenous sediment. Rocks on land are weathered and broken down into smaller particles by wind, water, and seasonal changes in temperature that result in freezing and thawing. The resulting particles are transported to the oceans by water, wind, ice, and gravity. Windblown dust from arid areas of the continents, ash from active volcanoes, and rock picked up by glaciers and embedded in icebergs are additional sources of lithogenous materials. The chemical composition of lithogenous sediments is controlled by the chemistry of the rocks they came from and their resistance to chemical and mechanical weathering. Most lithogenous sediments have high concentrations of quartz because it is one of the most abundant and stable minerals in continental rocks. Quartz is very resistant to both chemical and mechanical weathering so it can easily be transported long distances from its source. Lithogenous material can be found everywhere in the oceans. It is the dominant type of neritic sediment because the supply of lithogenous particles from land simply overwhelms all other types of material. Lithogenous deposits in pelagic sediments on the deep sea floor are called **abyssal clay,** if they are composed of at least 70% by weight clay-sized particles. Abyssal clay accumulates very slowly. It is important to understand that where it is the dominant pelagic sediment, it is only because of the lack of other types of material that would otherwise dilute it, not because of an increase in the supply of clay particles. This is generally the case in regions where there is little marine life in the surface waters above. This fine rock powder, blown out to sea by wind and swept out of the atmosphere by rain, may remain suspended in the water for many years. These clays are often rich in iron that oxidizes in the water and turns a reddish brown color, hence they are frequently

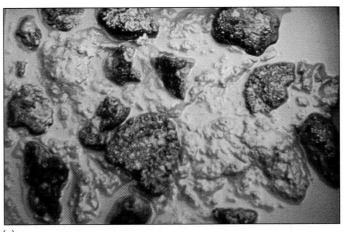

(a)

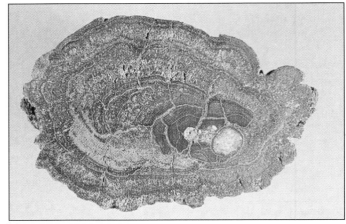

(b)

Figure 4.14

(a) Manganese nodules resting on red clay photographed on deck in natural light. Nodules are 2–8 cm in diameter. (b) A cross section of a manganese nodule showing concentric layers of formation.

called **red clay** (fig. 4.14a). The distribution of red clay is illustrated in figure 4.15.

Sediments derived from organisms are classified as **biogenous** (*bio* = life, *generare* = to produce) **sediments.** These may include shell and coral fragments as well as the hard skeletal parts of single-celled plants and animals that live in the surface waters. Pelagic biogenous sediments are composed almost entirely of the shells, or tests, of single-celled organisms (fig. 4.16). The chemical composition of these tests is either calcereous (calcium carbonate: $CaCO_3$, or seashell material) or siliceous (silicon dioxide: SiO_2, clear and hard). If pelagic sediments are more than 30% biogenous material by weight, the sediment is called an **ooze;** specifically either a **calcareous ooze** or **siliceous ooze** depending on the chemical composition of the majority of the tests found in the sediment. The distribution of calcereous and siliceous oozes on the sea floor is related to the supply of organisms in the overlying water, the rate at

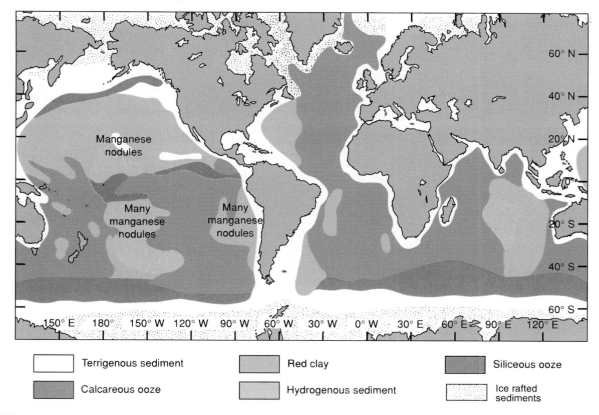

Figure 4.15

Distribution of the principal sediment types on the deep sea floor. Sediments are usually a mixture but are named for their major component.

Map legend:
- Terrigenous sediment
- Calcareous ooze
- Red clay
- Hydrogenous sediment
- Siliceous ooze
- Ice rafted sediments

Map labels: Manganese nodules, Many manganese nodules, Many manganese nodules

which the tests dissolve as they descend through the water, the depth at which they are deposited, and dilution with other sediment types.

Calcareous tests are created by small, single-celled plants called coccolithophores, small snails called pteropods, and minute amoeba-like animals called foraminifera (fig. 4.16a and c). Calcareous oozes are the dominant pelagic sediments (fig. 4.15). The dissolution, or destruction, rate of calcium carbonate varies with depth and is different in different ocean basins. Calcium carbonate generally dissolves more rapidly in cold, deep water that characteristically has a higher concentration of CO_2 and is slightly more acidic. The depth at which calcareous skeletal material first begins to dissolve is called the **lysocline.** Below the lysocline, there is a progressive decrease in the amount of calcareous material preserved in the sediment. The depth at which the amount of calcareous material preserved falls below 20% of the total sediment is called the **carbonate compensation depth (CCD).** Calcareous ooze tends to accumulate on the sea floor at depths above the CCD and it is generally absent at depths below the CCD. The CCD has an average depth of about 4500 m (14,800 ft), or roughly mid-way between the depth of the crests of ocean ridges and the deepest regions of the abyssal plains. In the Pacific, the CCD is generally at depths of about 4200 to 4500 m (13,800 to 14,800 ft). An exception to this is the deepening of the CCD to

about 5000 m (16,400 ft) in the equatorial Pacific where high rates of biological productivity result in a large supply of calcareous material. In the North Atlantic and parts of the South Atlantic, it is at or below depths of 5000 m (16,400 ft).

Siliceous tests are created by small, single-celled photosynthetic organisms called diatoms, and animals called radiolaria (fig. 4.16b). The pattern of dissolution of siliceous tests is opposite to that of calcareous tests. The oceans are undersaturated in silica everywhere so siliceous material will dissolve at all depths, but it dissolves most rapidly in shallow, warm water. Siliceous oozes are only preserved below areas of very high biological productivity in the surface waters (fig. 4.15). Even in these areas, an estimated 90% or more of the siliceous tests produced are dissolved, either in the water or on the sea floor, before they can be preserved.

Sediments derived from the water are classified as **hydrogenous** (*hydro* = water, *generare* = to produce) **sediments.** Hydrogenous sediments are produced in the water by chemical reactions. Most of them are formed by the slow precipitation of minerals onto the sea floor, but some are created by the precipitation of minerals in the water column as is the case in the plumes of hot water at hydrothermal vents along the ocean ridge system that create sediments rich in metal sulfides as discussed in chapter 3. Hydrogenous sediments include some **carbonates** (limestone-type deposits), **phosphorites** (phosphorous

(a)

(b)

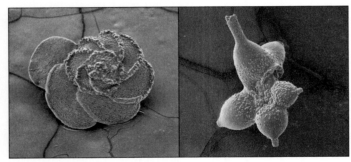

(c)

Figure 4.16

Scanning electron micrographs of biogenous sediments: (a) Diatoms and coccolithospheres. The small disks are detached coccoliths. (b) Radiolarians. (c) Foraminifera.

Fig. 4.16c images courtesy of Steve Nathan and R. Mark Leckie.

in phosphate form in crusts and nodules), and **manganese nodules.** Manganese nodules are deposits of manganese, iron, copper, cobalt, and nickel in the form of black or brown rounded masses scattered across the ocean floor and lying on top of other sediments (fig. 4.14). They were first recovered from the ocean floor in 1873 during the *Challenger* expedition. Nodules generally range from pea-sized to potato-sized. They develop very slowly (1–200 mm per million years, or 1000 times slower than other pelagic sediments), accumulating layer upon layer, often around a hard skeletal piece such as a shark's tooth or a fish bone. The nodules form in areas of very little sediment supply from other sources or where rapid bottom currents prevent them from being buried. If a nodule is buried by some other type of sediment it will cease to grow because it will no longer react chemically with the seawater. Manganese nodules have been mapped in all oceans except the Arctic. They are most abundant in the central Pacific north and south of the biogenous oozes along the equator (fig. 4.15). In the Atlantic and Indian oceans there are higher rates of lithogenous and biogenous sedimentation and consequently fewer deposits of manganese nodules.

Sediments derived from space are classified as **cosmogenous** (*cosmos* = universe, *generare* = to produce) **sediments.** Particles from space constantly bombarded the Earth. Most of these particles burn up as they pass through the atmosphere, but roughly 10% of the material reaches the surface of the Earth. Cosmogenous particles are very small and those that survive the passage through the atmosphere and fall in the ocean stay in suspension in the water long enough that they dissolve before they reach the sea floor. The particles typically have characteristic rounded or teardrop shapes due to partial melting as they pass through the atmosphere.

Oceanic sediments are laid down in layers, characterized by changes in color, particle size, and kinds of particles (fig. 4.17). Layering occurs if the sediment type varies over time or if the rate at which a sediment type is supplied varies. Over long periods of geologic time, climatic changes such as the ice ages have altered the populations of organisms that produce sediments and so have left their record in the sediment layers.

A summary of the major sediment types is given in table 4.3.

What are the principal sources of ocean sediments?

What can you learn about the formation of manganese nodules from figure 4.14?

Relate the distribution of calcareous oozes to seafloor features.

Explain the factors responsible for the seafloor location of siliceous oozes and red clays.

Give examples of sediments classified by their place of deposit.

Explain the layering of the core shown in figure 4.17.

Relate sediment size and sinking rate to the accumulation patterns found on the sea floor.

Sampling the Sediments 4.8

In order to analyze sediments, the geological oceanographer must have an actual bottom sample to examine. A variety of devices have been developed to take a sample from the sea floor for laboratory analysis. **Dredges** are net or wire baskets that are dragged across the bottom to collect loose bulk material, surface rocks, and shells; see figure 4.18. **Grab samplers** are hinged devices that are spring- or weight-loaded to snap shut when the sampler strikes the bottom. Figure 4.19 shows examples of this kind of device.

A **corer** is a hollow pipe with a sharp cutting end (fig. 4.20). The free-falling pipe is forced down into the sea floor, by its weight (fig. 4.20a and c) or, for longer cores, by a piston device within the core barrel that uses water pressure to help drive the corer into the sediments (fig. 4.20b). The use of a piston corer results in a cylinder of sediment, up to 20 m (60 ft) long containing undisturbed sediment layers (like those in fig. 4.17). Box corers are used when a large and nearly undisturbed sample of surface sediment is needed. These corers drive a rectangular metal box into the sediment, with doors that close over the bottom of the box before the sample is retrieved (fig. 4.20d). Longer cores may be obtained by drilling, as discussed in chapter 3.

Discuss the advantages and disadvantages of sediment sampling devices.

Figure 4.17

Layered sediments are seen in this deep-sea core obtained by the drilling ship *Glomar Challenger*.

table 4.3 **Sediment Summary**

Type	Source	Areas of Significant Deposit	Examples
Lithogenous (terrigenous)	Eroded rock, volcanoes, airborne dust	Dominantly neritic, pelagic in areas of low productivity	Coarse beach and shelf deposits, turbidites, red clay
Biogenous	Living organisms	Regions of high surface productivity, areas of upwelling, dominantly pelagic, some beaches, shallow warm water	Calcareous ooze (above the CCD), siliceous ooze (below the CCD), coral
Hydrogenous	Chemical precipitation from seawater	Mid-ocean ridges, areas starved of other sediment types, neritic and pelagic	Metal sulfides, manganese nodules, phosphates, some carbonates
Cosmogenous	Space	Everywhere but in very low concentration	Meteorites, space dust

Figure 4.18

A geological dredge with heavy-duty chain bag. The dredge picks up loose rock and breaks off fragments from rock outcroppings on the sea floor.

Seabed Resources 4.9

Mineral Resources

People began to exploit the materials of the seabed long ago. The ancient Greeks extended their lead and zinc mines under the sea, and sixteenth-century Scottish miners followed veins of coal under the Firth of Forth at Culross. As technology has developed, and as people have become concerned about the depletion of onshore mineral reserves, interest in seabed minerals and mining has grown. The development of a seabed source depends largely on international markets, needs for strategic materials, and whether offshore production costs can compete with onshore costs.

The largest superficial seafloor mining operations are for sand and gravel, used in cement and concrete for buildings, for landfills, and to construct roads and artificial beaches. This is a high-bulk, low-cost material tied to the economics of transport and the distance to market. Annual world production is approximately 1.2 billion metric tons; the reported potential reserve is more than 800 billion tons. Sand and gravel mining is

Figure 4.19

Vanveen (left) and orange peel (right) grab samplers are shown in open positions. Grabs take surface sediment samples.

the only significant seabed mining done by the United States at this time. It is estimated that the United States has a reserve of 450 billion tons of sand off its northeast coast. Along the Gulf coast, shell deposits are mined for use in the lime and cement industries and as a gravel substitute. Iron-rich sediments are dredged in Japan with an estimated reserve of 36 million tons, and tin is recovered from sands found from northern Thailand to Indonesia.

Phosphorite, which can be mined to produce the phosphates needed for fertilizers, is found in shallow waters as phosphorite muds and sands and as nodules on the continental shelf and slope. Large deposits are known to exist off Florida, California, North Carolina, Mexico, Peru, Australia, Japan, and northwestern and southern Africa.

In the Gulf of Mexico, sulfur is present in mineable quantities. Today, the cheaper and easier recovery of sulfur waste from pollution-control equipment has replaced all but one of the area's mining operations. Millions of tons of sulfur reserves exist in the Gulf of Mexico and the Mediterranean Sea.

Coal deposits under the sea floor are mined when the coal is present in sufficient quantity and quality to make the operation worthwhile. In Japan, undersea coal deposits are reached by shafts that stretch under the sea from the land or descend from artificial islands.

Oil and gas represent over 95% of the mineral value presently taken from the sea floor. The proportion of oil and gas production from offshore areas has increased from about 10 to 20% in the 1980s to over 30% in the 1990s, in large part because of advances in technology. In 1965 there were no offshore wells in water deeper than 100 m (300 ft), but today wells are drilled in as much as 1000 m (3000 ft) of water. Major offshore oil fields are found in the Gulf of Mexico, the Persian Gulf, the North Sea, and off the northern coast of Australia, the southern coast of California, and the coasts of the Arctic Ocean. During the 1990s new deep-water fields have been opened off Louisiana, west of the Shetland Islands in the North

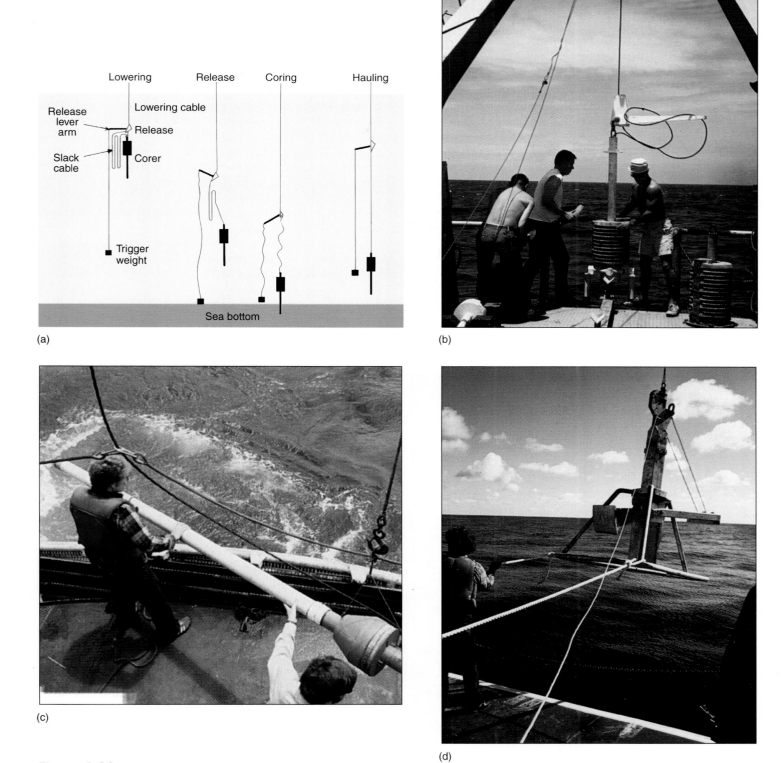

Figure 4.20

(a) The Phleger corer is a free-fall gravity corer. The weights help to drive the core barrel into the soft sediments. Inside the core is a plastic liner. The sediment core is removed from the corer by removing the plastic tube, which is capped to form a storage container for the core. (b) Loading a piston corer with weights to prepare it for use. (c) A gravity corer ready to be lowered. (d) A box corer is used to obtain large, undisturbed seafloor surface samples.

Diagram labels in (a): Lowering, Release, Coring, Hauling; Release lever arm, Lowering cable, Release, Slack cable, Corer, Trigger weight, Sea bottom.

Atlantic, and off the western and southern African coasts. Canada's $4.2 billion Hibernia field off Nova Scotia completed its final construction in the summer of 1997 (fig. 4.21). International companies are showing new interest in the Caspian Sea and the Russian coast, as well as recently licensed tracts around the Falkland Islands in the southern Atlantic Ocean. The giant Natuna gas field off Indonesia is expected to open the first of 216 wells by 2001.

Continental margin deposits account for nearly one-third of the world's estimated gas and oil resources. Although the cost of drilling and equipping an offshore well is three to four times greater than a similar venture on land, the large size of the deposits allows offshore ventures to compete successfully. The gas and oil potential of the deeper areas of the sea floor is still relatively unknown, but the deeper the water in which the drilling must be done, the higher the cost. Though legal restraints, environmental concerns, and political uncertainties

continue to slow the development of offshore deposits, petroleum exploration and development is expected to continue as the main focus of ocean mining in the near future.

Manganese nodules, the hydrogenous pelagic deposits found scattered across the world's deep ocean floors, have been the focus of intense research and development of mining and extraction techniques over the last twenty-five years. The mineral content varies from place to place, but the nodules in some areas contain 30% manganese, 1% copper, 1.25% nickel, and 0.25% cobalt, much higher concentrations than are usually found in land ores. Cobalt is of particular interest since it is classified as being of "strategic" importance to the United States, and hence essential to the national security. This is because it is an important component in the manufacture of strong alloys used in tools and aircraft engines. The nodules grow very slowly but they are present in huge quantities. An estimated 16 million additional tons of nodules accumulate each year.

Since the 1960s, large multinational consortia and mining corporations have spent over $600 million to locate the highest nodule concentrations and develop technologies for their collection. However, expectations of rapid development have been disappointing; some of these consortia have withdrawn completely, while others have become dormant. In the early 1990s, Australia and New Zealand with twelve small South Pacific island nations continued to support survey cruises. A 1996 feasibility study on the recovery of manganese nodules in the waters surrounding the Cook Islands was inconclusive and the financial difficulties of the Cook Islands' government have made any further efforts uncertain. However, China, Korea, and Japan have continued research and exploration for deep seabed mineral deposits in the Okinawa region, and India is exploring the deep waters of the Indian Ocean. See the following section on mining laws and treaties for a discussion of the difficulties surrounding the ownership and harvesting of pelagic nodules. Cobalt enriched manganese crusts, or hard coatings on other rocks, were discovered in relatively shallow water on the slopes of seamounts and islands within United States territorial waters in the 1980s. The concentration of cobalt in these deposits is roughly twice that found in typical pelagic manganese nodules and about one and one-half times that found in known continental deposits. These crusts are not being actively mined because of the relatively low cost and continued availability of continental sources.

Expeditions to the rift valleys of the East Pacific Rise near the Gulf of California, the Galápagos Ridge off Ecuador, and the Juan de Fuca and Gorda Ridges off the northwestern United States have found deposits of minerals combined with sulfur to form sulfides of zinc, iron, copper, and possibly silver, lead, chromium, gold, and platinum. Molten material from beneath the Earth's crust rises along the rift valleys, fracturing and heating the rock. Seawater percolates into and through the fractured rock forming these mineral-rich deposits. Deposits may be tens of meters thick and hundreds of meters long. Too little is presently known about these deposits to know whether they might be of economic importance at some future time. There is no technology to retrieve them at this time, and like the manganese nodules these deposits are found outside na-

Figure 4.21

The oil-drilling platform, Hibernia, located 200 miles offshore Newfoundland, produced its first oil in November 1997.

tional economic zones presenting ownership problems; see the next section.

In recent years there has been increasing interest in gas hydrates trapped in deep marine sediments. Gas hydrates are a combination of natural gas, primarily methane (CH_4), and water that form a solid, ice-like structure under pressure at low temperatures. Drill cores of marine sediment have recovered samples of gas hydrates that melt and bubble as the natural gas escapes. These melting samples will burn if lit. Gas hydrates are a subject of intense interest for three reasons; they are a potential source of energy, they may contribute to slumping along continental margins, and they may play a role in climate change.

Scientists have mapped two relatively small areas, each the size of the State of Rhode Island, off the coasts of North and South Carolina that are thought to contain more than 1300 trillion cubic feet of methane gas, an amount that is more than 70 times the 1989 gas consumption of the United States.

A second reason gas hydrates are significant is their effect on sea floor stability. Along the southeastern coast of the United States a number of submarine landslides, or slumps, have been identified that may be related to the presence of gas hydrates. The hydrates may inhibit normal sediment consolidation and cementation processes, creating a weak zone in the sediments. Alternately, the lowering of sea level during the last glacial period may have reduced the pressure on the sea floor enough to allow some of the gas to escape from the hydrates and accumulate in the sediment, decreasing its strength.

A final reason for studying gas hydrates is their potential link to climate changes. The amount of methane stored in hydrates is believed to be about 3000 times the amount currently present in the atmosphere. Since methane is a greenhouse gas, its release from hydrates could effect global climate.

> What seabed resources have been identified with commercial value, and which of these are being exploited at the present time?

Mining Laws and Treaties

Because of the potential value of deep-sea minerals, specifically manganese nodules, and because the nodules are found in international waters, the developing nations of the world think they have as much claim to this wealth as those countries that are presently technically able to retrieve the nodules. The developing nations want access to the mining technology and a share in the profits. For nearly ten years, the United Nations worked to produce the United Nations Convention on the Law of the Sea (UNCLOS). One section of this treaty regulated deep-ocean exploitation, including mining. The treaty recognized the manganese nodules as the heritage of all humankind, to be regulated by a UN seabed authority, which would license private companies to mine in tandem with a UN company. The quantities removed would be limited and the profits would be shared. The Law of the Sea Treaty was completed in April 1982. The United States and some other industrialized countries chose not to sign the treaty at that time.

During the past several years, negotiations have been underway to resolve issues related to the part of the convention that deals with seabed mining. The United States announced that it intends to sign a new agreement that changes the provisions for managing the mining of deep-sea resources, when UNCLOS came into force in 1994. The new agreement and convention was forwarded to the U.S. Senate for consideration and ratification in the fall of 1994. In 1998, as part of the International Year of the Ocean, the Department of Commerce and the Department of the Navy brought leaders from government, academia, industry, environmental groups, and other concerned organizations together for a National Ocean Conference. The purpose of the conference was to stimulate a focused discussion of national ocean problems and solutions. One of the clear recommendations that emerged from this conference was that the United States move swiftly to join the 1982 U.N. Convention on the Law of the Sea. It may take several years of deliberation before the United States ratifies the agreement and convention.

In any case, it is unlikely that any significant deep-sea mining will occur until well into the twenty-first century. The high costs of sea mining, low international metal market prices, and still undeveloped land sources combine to make rapid commercialization unlikely.

> Explain the present situation concerning the commercial exploitation of deep-sea resources.

Summary

The features of the ocean floor are as rugged as the features of the land but erode more slowly. The continental margin includes the continental shelf, shelf break, slope, and rise. Submarine canyons are major features of the continental slope and, in some cases, the continental shelf. Some canyons are associated with rivers; others are believed to have been cut by turbidity currents. Turbidity currents deposit sediments known as turbidites.

The flat abyssal plains are interrupted by scattered abyssal hills, volcanic seamounts, and flat-topped guyots. In warm shallow water, corals have grown up around the seamounts to form fringing reefs. A barrier reef is formed when a seamount subsides while the coral grows. An atoll results when the seamount's peak is fully submerged. Great, continuous volcanic mountain ranges, the mid-ocean ridges and rises, extend through all the oceans. Rift valleys run along the ridge and rise crests; fracture zones form perpendicular to the ridge system. Trenches are long, deep depressions in the ocean floor that are associated with volcanic island arcs and are found mainly in the Pacific Ocean.

Depth soundings were made first with hand lines, then with wire lines, and are now made with echo sounders. Bathymetric charts can now be produced from satellite data.

Sediment classifications are based on their size, location, origin, and chemistry. Sediment particles are broadly categorized in order of decreasing size as gravel, sand, and mud. Within each of these categories particles can be further subdivided by size. The sinking rate and distance traveled in the water column are related to sediment size. Small particles sink more slowly than large particles. The very smallest particle sizes, silts and clays, sink so slowly that they may be transported large distances while falling to the sea floor. The sinking rate of particles is increased by clumping and incorporation into fecal pellets of small organisms. Coarse sediments are concentrated close to shore; finer sediments are found in quiet offshore or nearshore environments.

Sediments that accumulate on continental margins and the slopes of islands are called neritic sediments. Sediments of the deep sea floor are pelagic sediments. In general, pelagic sediments accumulate very slowly and neritic sediments accumulate more rapidly.

Sediments formed from pre-existing rocks are called lithogenous sediments. Since lithogenous sediments are typically derived from the land, they are also known as terrigenous sediments. Pelagic lithogenous sediment is dominated by red clay. Red clay dominates marine sediments only in regions that are starved of other sources of sediment. Biogenous sediments come from living organisms. Sediments composed of at least 30% biogenous material are called oozes; this material accumulates in regions of high biological productivity. Siliceous sediments are subjected to dissolution everywhere in the oceans while calcareous sediments dissolve rapidly in deep, cold water below the CCD. Sediments that precipitate directly from the water are called hydrogenous sediments. These include manganese nodules on the deep sea floor and metal sulfides along mid-ocean ridges. Sediments that originate in space are called cosmogenous sediments.

Sediments are sampled with dredges, grab samplers, corers, and drilling ships.

Seabed resources include sand and gravel, used in construction and landfills. Sediments that are rich in mineral ores are mined. Phosphorite nodules have potential as fertilizer. Oil and gas are the most valuable of all seabed resources. Manganese nodules, rich in several metals, are present on the ocean floor in huge numbers. The status of manganese-nodule mining is clouded by disputes over international law with reference to mining claims and shared technology. Sulfide mineral deposits have been discovered along rift valleys; their economic importance is unknown.

Large deposits of gas hydrates are now being studied to determine their potential as economically important sources of methane gas. These deposits are ice-like accumulations of natural gas and water that form at low temperature and high pressure on the sea floor. Scientists are also studying their possible role in the occurrence of submarine landslides and what effect they may have on global climate.

Key Terms

Critical Thinking

1. How are continental shelf types related to active and passive plate margins and on which plate margins, and why, would you expect to find submarine canyons?

2. Imagine that you are in a submersible on the ocean bottom. You leave New York and travel across the North Atlantic to Spain. Draw a simple ocean-bottom profile showing each major bathymetric feature you see as you move across the ocean. Name each feature. Do the same for the South Pacific between the coast of Chile and the west coast of Australia. Compare the two profiles. Did your depth scale differ from your horizontal scale? How much?

3. What can geologists determine from seafloor cores? Consider information in both chapters 3 and 4.

4. Why are calcareous oozes common in the southern Pacific Ocean but uncommon in the northern Pacific Ocean?

5. Consider the future commercial development and exploitation of deep-sea mineral resources. What factors need to be discussed. How soon do you think such exploitation might occur? Which mineral resources are most likely to be developed?

Suggested Readings

Seafloor Geology

Bambach, R. K., C. R. Scotese, and A. M. Ziegler. 1980. Before Pangaea: The Geographies of the Paleozoic World. *American Scientist* 68(1):26–38.

Emery, K. O. 1969. The Continental Shelves. In *Ocean Science, Readings from Scientific American* (1977), 221(3):32–44.

Kunzig, R. 1996. The Sea Floor from Space. *Discover* 17(3):58–64.

Mapping the Margins, 1997. *Smithsonian* 28(1):68–71. (New images of continental shelves.)

Menard, H. W. 1969. The Deep-Sea Floor. In *Ocean Science, Readings from Scientific American* (1977) 221(3):55–64.

Normark, W. R., and D. Piper. 1993/4. Turbidite Sedimentation. *Oceanus* (36)4:107–110.

Pratson, L., and W. Haxby. 1997. Panoramas of the Seafloor. *Scientific American* 276(6):82–87. (Sonar mapping of continental margins.)

Pratson, L. F., and W. F. Haxby. 1996. What is the Slope of the U.S. Continental Slope? *Geology* 24(1):3–6.

Rea, D. K. 1993/94. Terrigenous Sediments in the Pelagic Realm. *Oceanus* (36)4:103–106.

Rice, A. L. 1991. Finding Bottom. *Sea Frontiers* 37(2):28–33.

Smith, W. H. F., and D. T. Sandwell. 1997. Global Sea Floor Topography from Satellite Altimetry and Ship Depth Soundings. *Science* 277:1956–62.

Seabed Resources

Broadus, J. M. 1987. Seabed Materials. *Science* 235(4791):853–60. (Discussion of seabed resources.)

Cruickshank, M. J. 1999. Mining: A Significant Non-Action for 1998. *Sea Technology* 40(1):48–50.

Cruickshank, M. J. 1991. Ocean Mining: For the Future a Good Omen. *Sea Technology* 32(1):38–39.

Pagano, S. S. 1999. Offshore Oil & Gas: Amid Uncertainty, Survival Strategies, and Continued Activity. *Sea Technology* 40(1):13–16.

Rona, P. 1986. Mineral Deposits from Seafloor Hot Springs. *Scientific American* 254(1):84–92.

Weldon, C. 2000. Ocean Action 1999: Some — But Still a Long Way to Go. *Sea Technology* 41(1):47–49.

internet references
worldwide websites

Sea Floor and Sediments

World Data Center for Marine Geology and Geophysics
http://www.ngdc.noaa.gov/mgg/aboutmgg/

Sea Floor Bathymetry from Satellite Altimetry
http://topex.ucsd.edu/marine_topo/text/topo.html

Predicted and Measured Sea Floor Topography
http://www.ngdc.noaa.gov/mgg/announcements/images_predict.HTML

Calculate Sediment Particle Settling Rates
http://www.filtration-and-separation.com/settling/settling.htm

Coastal Ocean Sediment Transport
http://penguin.whoi.edu/leo15.html

HUGO-Hawaii Undersea Ge-Observatory
http://www.soest.hawaii.edu/HUGO/hugo.html

Gas Hydrates

Blake Ridge Gas Hydrates
http://obs.er.usgs.gov/BlakeRidge95.html

USGS Fact Sheet on Gas Hydrates
http://marine.er.usgs.gov/fact-sheets/gas-hydrates/title.html

Gas Hydrates—A Possible Future Energy Source
http://walrus.wr.usgs.gov/resources/hydrate.html

Gas Hydrate Studies at Woods Hole
http://atlantic.er.usgs.gov/hydrates/hydrate.htm

Law of the Sea

Oceans and Law of the Sea—United Nations
http://www.un.org/Depts/los/index.htm

Law of the Sea Convention—Table of Contents
http://www.greenpeace.org/~intlaw/lscouts.html

Law of the Sea Convention and U.S. Policy
http://www.cnie.org/nle/mar-16.html

Historically, deep-sea oceanographic data has been gathered by surface ships that towed instruments and samplers or sent equipment down thousands of meters to collect data from scattered areas. Since the 1970s manned submersibles, such as the *Alvin* (refer to chapter 3), have allowed researchers to actually experience the deep sea. A single dive in the *Alvin* takes more than eight hours from launch to retrieval, but only four hours are spent at the target depth. The cost of the dive and support costs for the vessel and crew at the surface are about $25,000 per day. In order to explore larger areas faster, decrease the costs, and decrease the risks, dozens of deep-diving robotic devices are presently in use or are being built by the world's centers for oceanographic research. These devices act as eyes, samplers, and manipulators for oceanographers confined to submersibles or surface vessels.

Remotely Operated Vehicles

Remotely Operated Vehicles or ROVs are tethered to a surface vessel, a manned submersible, or a below-the-surface system that isolates the ROV from ship motion at the sea surface. The tether is the umbilical cord between ROV and operator; it supplies power and transmits information over an electrical or fiber-optic cable. ROVs use video, electronic-digital and still cameras accompanied by lights to explore their environment. They are equipped with mechanical hands to manipulate objects and depending on their mission, other sensors such as sonar and temperature and salinity monitors. The operator "flies" the ROV over the sea floor, and the ROV sends data to the operator and receives back directions to change position, to manipulate or retrieve objects, or to use cameras and other sensors. Full-ocean-depth ROVs, required to perform a variety of tasks over long periods, weigh several tons and must be towed slowly because of the stresses on their long cables. Small, limited-use, shallow-water vehicles that may weigh less than 45 kg (100 lbs) can be towed rapidly by small surface vessels. These smaller robotic devices are used to service, inspect, and aid in construction of offshore oil and gas facilities, bridge piers, and dam footings. Mkl ROV is shown in figure 1.

ROV *Jason* is a tethered vehicle operated by the Woods Hole Oceanographic Institution. *Jason* weighs about 1350 kg (3000 lbs), operates to a depth of 6000 m (20,000 ft), and provides simultaneous use of up to four color video channels, sonar, electronic cameras, and manipulators. A smaller ROV, *Jason Jr.* or *JJ*, operating from the submersible *Alvin*, was used to inspect the wreck of the *Titanic* and photograph its interior spaces (see fig. 2). In 1992 ROPOS or Remotely Operated Platform for Ocean Science was used by a team of Canadian and U.S. scientists to map and sample a hydrothermal vent field on the Juan de Fuca Ridge off the Oregon coast. It also successfully retrieved data packages from previously installed samplers. Unfortunately during a severe storm in October 1996 ROPOS broke loose from its tether and was lost in the North Pacific.

In March 1995 the Japanese ROV, *Kaiko*, descended 10,920 m (36,000 ft) into the Mariana Trench. *Kaiko* is the world's deepest diving ROV and the only ROV capable of descending below 6500 m (20,000 ft). This ROV is equipped with five television cameras, still cameras, sonar, manipulators, and other sensors. Plans for *Kaiko* include detail mapping of active faults, examining the sediment on top of subducting plates, and observing the ocean's deepest life-forms.

The U.S. Navy has been working with an Underwater Security Vehicle or USV, another ROV, that can be operated from small support craft or land stations. Its mission is to search for, track, and intercept underwater targets that may represent a threat to ships, submarines, and naval shore stations. These ROVs use special acoustical sensors to home in on their targets and then use visual sensors to evaluate and identify them. This ROV may be equipped with response capability to be used if the target is considered a security threat.

ROVs are much less costly than manned submersibles, for there is no need to provide safety for humans. Figure 3 shows the *Ventana* ROV. Supplied with continuous power through their tethers they can operate to deeper depths, work under water for long periods of time, and explore areas of high risk such as vent and volcanic areas. The ROV's two-dimensional video images are a drawback, for they do not measure up to the three-dimensional image seen by a human in a submersible. Therefore, ROVs cannot yet react in the same way as a manned submersible. Also they do require the expensive support of surface vessels. Efforts to further reduce costs and extend deep-sea exploration time has led to the development of Autonomous Underwater Vehicles.

Undersea Robots

Figure 1

Mkl ROV, a MiniRover, is an example of a small "low-cost" system.

Figure 2

The camera-equipped ROV, *Jason Jr.*, peers into a cabin of the luxury liner *Titanic*, 3800 m (12,500 ft) below the surface. A tether connecting *Jason Jr.* to the submersible *Alvin* is seen at left.

Autonomous Underwater Vehicles

AUVs or Autonomous Underwater Vehicles are independent, mobile, instrument platforms with sensors. They are not tethered to another vessel, and therefore, their range of movement is greater than that of ROVs. AUVs may be linked acoustically to an operator in a surface vessel, or they may be programmed to complete surveys and take samples with little or no human supervision. Many of these vehicles are small, less than 25 kg (50 lbs) and have specialized capabilities and restricted depth ranges. Others work effectively in deep water and in inaccessible regions such as under sea ice. A typical AUV is diagrammed in figure 4.

Figure 3

The *Ventana* ROV is owned and operated by Monterey Bay Aquarium Research Institute. This ROV is used to study marine life in Monterey Canyon.

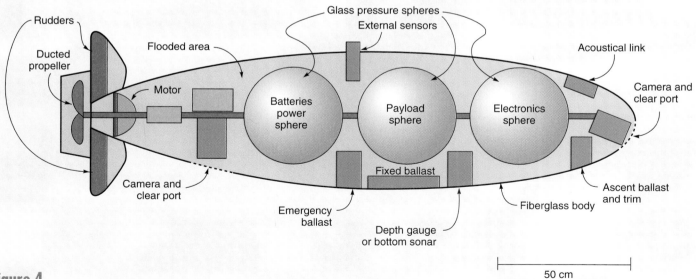

Figure 4

A typical AUV design has a streamlined, external, fiberglass shell surrounding pressurized internal functional units. Electronics, payload (sensor controls, data recording, etc.), and batteries are housed in glass spheres; each camera and sensor has its own pressure housing. The electrical junctions and drive motor are oil-filled; internal oil pressure changes with depth to equal external pressure. The space between the external shell and the pressure-protected internal units is flooded when the AUV submerges.

The Navy's Advanced Unmanned Search System or AUSS locates unidentified objects on the sea floor at depths from 5 m (17 ft) to 6 km (20,000 ft); it is a small vehicle, 23 m (77 ft) long, 78 cm (31 in) in diameter, and weighs 1270 kg (2800 lbs). It follows a predefined search pattern while transmitting sonar images to the surface. Its acoustic link may order suspension of the search pattern, require the AUSS to take a closer look at an object, and then resume its search pattern. The Autonomous Benthic Explorer or ABE is designed to survey deep-sea hydrothermal vent areas for up to one year, making a daily survey of its area. It monitors currents, takes pictures, and makes temperature and other measurements; then it returns to its mooring platform on the sea floor and remains quiet to conserve energy.

An AUV can communicate its data acoustically to a surface vessel, or it can dock with a buoy or beacon, transfer its data to the buoy system, receive new instructions and recharge itself. The data can then be transferred from the buoy system to a satellite link and beamed back to a surface ship or shore station. Some AUVs store their data internally to be kept until the AUV is recovered (fig. 5).

AUVs may be programmed to make measurements while running specific courses, or they may relay their data acoustically to their controller; their course can be changed to improve their sampling. AUV power requirements depend on speed and distance traveled as well as the power required for their sensors and data transmission. Small size, low weight, and low power requirements mean the AUV can spend more time submerged and so can extend its range.

AUVs use internal software and sonar to survey terrain and avoid obstacles. They can use their proximity to the sea floor and its topography as a course guide, or they can navigate using preset acoustical seafloor beacons or acoustical buoys.

Satellites can monitor vast areas, but they track mainly surface features. A fleet of AUVs working in an area can monitor changing conditions and obtain more data in more detail over different depths than a single large research vessel. They are able to operate in any weather, and as the technology for standardizing their components develops, AUVs will cost less, more will be made, and the use of these robotic devices will increase. They are becoming the survey tool of choice for work in the Earth's last, largest, and least known frontier.

Readings

Bradley, A., D. Yoerger, and B. Walden. 1995. An AB(L)E Bodied Vehicle. *Oceanus* 38 (1):18–20.

Kunzig, R. 1995. A Thousand Diving Robots. *Discovery* 17 (4):60–71.

Sea Technology. 1996. 37 (12). (Issue devoted to submersibles, ROVs and AUVs.)

Internet Resources

Deep Sea Robots
http://www.physical.com/deep6/index.htm

WHOI Marine Operations—National Deep Submergence Facility—Go to ALVIN, ROVs, AUVs, Future Directions—images & information
http://www.marine.whoi.edu/ships/ships_vehicles.htm

Autonomous Underwater Vehicles Lab—images and information
http://seagrant.mit.edu/~auvlab

WHOI Deep Submergence Lab
http://www.dsl.whoi.edu/

Japanese Exploration Systems—information, images, and animation
http://www.jamstec.go.jp/jamstec/exp.html

Figure 5

Retrieving a programmed torpedo (AUV) beneath the Arctic Ocean ice.

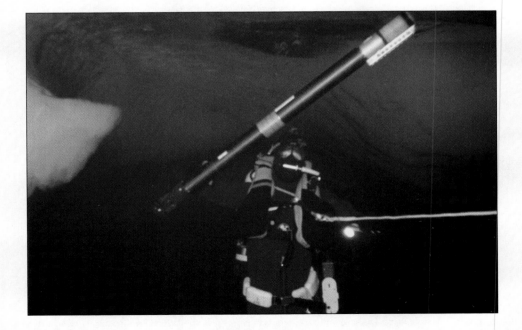

Water

Rain over the ocean.

Learning Objectives

After reading this chapter, you should be able to

- Describe the structure of the water molecule and how this structure affects the behavior of water.

- Describe the changes of state of water.

- Define density and explain how it is affected by changes in salinity, temperature, and pressure.

- Describe how water transmits energy in the form of heat, light, and sound.

- Discuss the practical use of sound in seawater.

- Understand the presence of salts as ions in seawater and the constant composition of seawater.

- Explain the concept of salinity.

- Explain variations in salinity along the coasts and in the surface waters of the deep sea.

- Discuss the processes that add and remove salt from seawater.

- Explain the importance of nutrients in seawater.

- Compare the distribution of dissolved oxygen and carbon dioxide in the oceans.

- Understand the importance of carbon dioxide as a buffer.

- Describe ways of extracting fresh water from seawater.

Acoustic Thermometry of Ocean Climate

*n*ot only is water the most common of substances on the Earth's surface, but it is also uncommon in many of its properties. It makes life possible, and its properties largely determine the characteristics of the oceans, the atmosphere, and the land. Seawater is salt water, but it is much more than just salty water. Seawater contains dissolved gases, nutrient molecules, and organic substances as well as salt. To understand the oceans, it is necessary to learn something of the physical and chemical characteristics of seawater and to follow the processes that influence and regulate it.

The Water Molecule 5.1

The properties that make water such a special, useful, and essential substance result from its molecular structure. The water molecule is deceptively simple, made up of three atoms: two hydrogen atoms and one oxygen atom (H_2O). These combine to form a V-shaped molecule with the hydrogen atoms on one side and the oxygen atom at the other (fig. 5.1a). The negatively charged electrons within the molecule are distributed unequally, staying closer to the oxygen atom and giving this end a slightly negative charge. The hydrogen end of the molecule carries a slightly positive charge. Although the water molecule as a whole is neutral, its opposite sides have opposite charges. These opposing charges make water molecules attract each other, and bonds form between the positively charged side of one water molecule and the negatively charged side of another water molecule (fig. 5.1b). These bonds are known as **hydrogen bonds,** and each water molecule can establish hydrogen bonds with four other water molecules. Any single bond is relatively weak, and as one bond is broken, another is formed. As a result, water is not a typical liquid. Compared with other liquids, it has extraordinary properties that are the consequence of attractions among water molecules.

> Describe the relationship between the hydrogen and oxygen atoms in a water molecule.
>
> How does the arrangement affect the behavior of water molecules?

Changes of State 5.2

Water exists on the Earth in three physical states: as a solid, a liquid, and a gas. When it is a solid, we refer to it as ice, and when it is a gas, we call it water vapor. Because of the bonding between the water molecules, it takes a great deal of energy to separate water molecules from the water's surface to form vapor and to separate water molecules from ice, melting it to water.

Heat is a form of energy that can be measured in **calories.** One calorie is the amount of heat needed to raise the temperature of 1 gram of water 1°C. If you are unfamiliar with the Celsius temperature scale, see Appendix C. Do not confuse heat energy with temperature. If 2 cups of water and 2 quarts of water at the same temperature are placed on the same size stove burners at the same time, it will take longer to boil the larger amount because more energy is needed to bring the larger quantity to a boil. Both amounts of water boil at the same temperature, but the amount of heat required to achieve the boiling is different in each case.

Water changes its state, between liquid and gas or liquid and solid, by the addition or loss of heat and the breaking and forming of bonds between molecules. When enough heat is added to ice at 0°C, these bonds break and the ice melts. When heat is added to liquid water, the temperature of the water increases and some of the water molecules escape from the liquid or evaporate to form water vapor. When heat is removed from water vapor, the water vapor condenses and bonds form between molecules, producing a liquid, and when liquid water at 0°C loses sufficient heat, more bonds form and ice is the result.

The **heat capacity** of a material is the quantity of heat needed to produce a unit change of temperature in a unit mass of material. The heat capacity of liquid water is 1 calorie per gram per °C. The high heat capacity of water, compared to most substances (table 5.1), means it requires a comparatively large exchange of heat energy to produce a change in water's temperature. A large amount of heat energy must be added or subtracted to water to increase or decrease its temperature respectively. In other words, water tends to

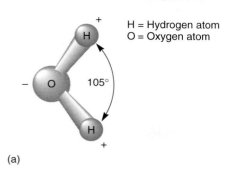

H = Hydrogen atom
O = Oxygen atom

105°

(a)

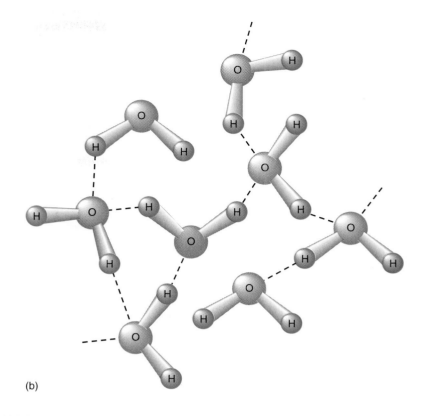

(b)

Figure 5.1

(a) Two hydrogen atoms bond to an oxygen atom and produce a water molecule with a positive side and a negative side. The angle between the hydrogen atoms and the oxygen atoms is about 105°. (b) The positive and negative charges allow each water molecule to form bonds with other water molecules.

table 5.1	Heat Capacity of Common Materials

Material	Heat Capacity (Calories/g/°C)
Acetone	0.51
Aluminum	0.22
Ammonia	1.13
Copper	0.09
Grain alcohol	0.23
Lead	0.03
Mercury	0.03
Silver	0.06
Water	1.00

change temperature much more slowly than other materials, such as the land.

To change pure fresh water from a solid at 0°C to a liquid at 0°C requires the addition of 80 calories for each gram of ice (fig. 5.2). This is called the latent heat of fusion; there is no change in temperature, only a change in the physical state of the water. This heat is released again in changing liquid water to ice. Ice is a very stable form of water, and it takes a lot of heat to melt it.

The change of state between liquid water and water vapor requires 540 calories of heat to convert 1 gram of water at 100°C to water vapor at 100°C. This is known as the latent heat of vaporization. When 1 gram of water vapor condenses and returns to the liquid state, 540 calories of heat are liberated. There is no change in temperature, only a change in the water's

physical state. This process may also be seen in figure 5.2. Water does convert from the liquid to the vapor state at temperatures other than 100°C; for example, rain puddles evaporate and clothes dry on the clothesline. This type of change requires slightly more heat to convert the liquid water to a gas at these lower temperatures.

Adding salt to water changes its boiling and freezing points: The boiling temperature is raised and the freezing temperature is lowered. The amount of change is controlled by the amount of salt added. The rise in the boiling temperature is of little consequence since seawater does not normally boil in nature, but the lowering of the freezing point is important in the formation of sea ice. Typical seawater freezes at about –2°C.

> What is heat and how is it measured?
>
> How does heat differ from temperature?
>
> Refer to figure 5.2 and compare the change of state between ice and water to that between water and water vapor. Consider heat requirements and bond formation.

Density 5.3

Density is defined as mass per unit volume of a substance and is usually measured in grams per cubic centimeter, or g/cm^3. Less dense substances will float in more dense liquids (for

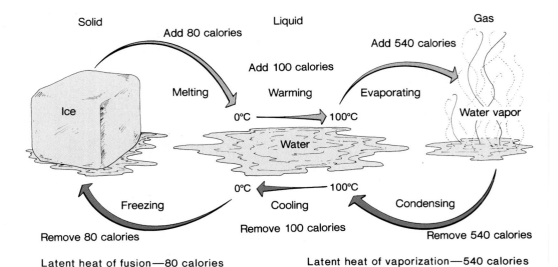

Figure 5.2

Heat energy must be added to convert a gram of ice to liquid water and to convert liquid water to water vapor. The same quantity of heat must be removed to reverse the process.

More Information: www.britannica.com/bcom/eb/article/2/0,5716,115012+10+108518,00.html

example, oil on water or dry pine wood on water). Ocean water is more dense than fresh water; therefore fresh water floats on salt water.

Water density is very sensitive to temperature changes. When water is heated, energy is added and water molecules move apart; therefore the mass per cubic centimeter becomes less because there are fewer water molecules per cubic centimeter. For this reason, the density of warm water is less than that of cold water. When water is cooled. it loses heat energy, and the water molecules slow down and come closer together; there are then more water molecules, or a greater mass, per cubic centimeter. Because cold water is more dense than warm water, it sinks below the warm water.

In pure fresh water, molecules move closer and closer together as water is cooled to 3.98°C, or about 4°C. At this temperature, fresh water reaches its greatest density, 1 g/cm³. As the temperature falls below 4°C, some water molecules start forming a crystal lattice (see fig. 5.1b), the water expands slightly, and the density decreases. Fresh water begins to freeze at 0°C when the water molecules no longer move enough to break the bonds with neighboring molecules. As the ice crystals form, the water molecules become widely spaced, resulting in fewer water molecules per cubic centimeter. Therefore, water at temperatures less than 4°C and ice are both less dense than water at 4°C (fig. 5.3).

When salts are dissolved in water, the density of the water increases because the salts have a greater density than water. The typical density of average ocean water at 4°C is 1.0278 g/cm³, in comparison to 1 g/cm³ for fresh water. Densities of water with and without dissolved salt are shown in table 5.2. Notice in this table that (1) if the salt content remains constant, density increases as temperature decreases and (2) if temperature remains constant, density increases with increasing salt content. High densities are associated with cold salty water. Fresh water is densest at 4°C, and freshwater ice is less dense than either fresh or salty water. Chapter 7 presents a discussion of changes in seawater density with dissolved salts and temperature; see figure 7.1.

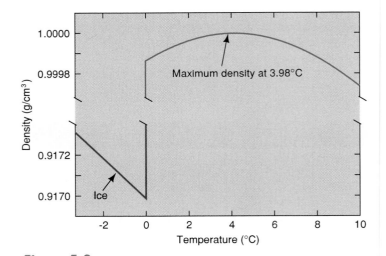

Figure 5.3

The density of pure water reaches its maximum at 3.98°C. Pure water is free of dissolved gases.

Density is also affected by pressure. Increasing the pressure increases the density of water by crowding the molecules together and reducing the volume occupied by a fixed mass of water. Pressure in the oceans increases with depth (fig. 5.4). For every 10 m of descent, the pressure increases by about 1 atmosphere, or 14.7 lbs/in². (See Appendix C for other units of pressure.) Pressure in the deepest ocean trench, 11,000 m (36,000 ft), is about 1100 atm. These great pressures have only a small effect on the volume of the oceans because of the low compressibility of water. The average pressure acting on the total world's ocean volume results in a reduction of ocean depth by about 37 m (121 ft). In other words, if ocean water were truly incompressible, sea level would stand about 37 m (121 ft) higher than it does at present. The effect of pressure on density is small enough that it can be ignored in most instances except where very accurate determination of seawater density is required.

table 5.2

Density of Water With and Without Dissolved Salt, g/cm³

°C	No Salt	Salt 20 g/kg	25 g/kg	30 g/kg	35 g/kg
−1	ice 0.917	1.01606	1.02010	1.02413	1.02817
0	0.99984	1.01607	1.02008	1.02410	1.02813
1	0.99990	1.01605	1.02005	1.02406	1.02807
2	0.99994	1.01603	1.02001	1.02400	1.02799
3	0.99996	1.01598	1.01995	1.02393	1.02791
4	0.99997	1.01593	1.01988	1.02384	1.02781
5	0.99996	1.01586	1.01980	1.02374	1.02770
10	0.99970	1.01532	1.01920	1.02308	1.02697
15	0.99910	1.01450	1.01832	1.02215	1.02599
20	0.99820	1.01342	1.01720	1.02098	1.02478
25	0.99704	1.01210	1.01585	1.01960	1.02336
30	0.99565	1.01057	1.01428	1.01801	1.02175

Figure 5.4

These oceanography students hold research cruise souvenirs — Styrofoam coffee cups that have been attached to a sampler and lowered 2000 m (6600 ft) into the sea. The water pressure has compressed the cups to the size of thimbles.

Define density.

Why are ice and water vapor less dense than water?

How are density, temperature, and the salt content of seawater related?

How does pressure cause changes in density?

Transmission of Heat 5.4

There are three ways in which heat energy moves through water: by **conduction,** by **convection,** and by **radiation.** Conduction is a molecular process. When heat is applied at one location, the molecules move faster due to the addition of energy; gradually, this molecular motion passes on to adjacent molecules, and the heat spreads by conduction. For example, if the bowl of a metal spoon is placed in a hot liquid, the handle soon becomes hot. Metals are excellent conductors; water is a poor conductor and transmits heat slowly in this way.

Convection is a density-driven process in which a heated fluid moves and carries its heat to a new location. Water rises when it is heated from below and its density is decreased; it sinks when it is cooled at its surface and its density is increased.

Radiation is the direct transmission of heat from its energy source. The Sun provides the Earth's surface with radiant energy, which warms the surface waters. Some of this solar radiation is reflected from the surface and adds no heat to the water, and some is absorbed and raises the temperature of the surface water. Since warming the surface water makes it less dense, this warm water remains floating at the surface and convection does not occur. Some heat gained at the surface is conducted downward, aided by the natural stirring action of the wind and currents. The ocean's surface waters may lose heat to the atmosphere by direct transfer of heat from warm water to colder air by conduction and convection in the air, and the transfer of water vapor to the atmosphere by evaporation and release of heat to the atmosphere when the water vapor condenses.

Distinguish among the three methods of heat transmission in water.

Transmission of Light 5.5

Seawater transmits only the visible wavelengths of sunlight (fig. 5.5). About 60% of the entering light energy is absorbed in the first meter (3.3 ft), and about 80% is gone after 10 m (33 ft). Only 1% of the total light available at the surface is left in the clearest water below 150 m (500 ft), and no sunlight penetrates

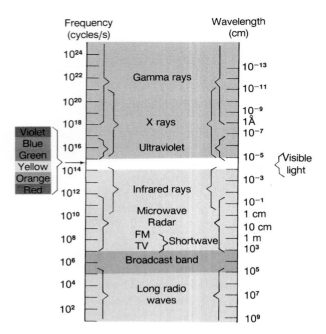

Figure 5.5

The electromagnetic spectrum. Visible light occupies only a small portion of the spectrum. The visible light is bounded on one side by longer infrared rays and on the other side by shorter ultraviolet rays. Water is relatively opaque to all wavelengths of this spectrum except visible light.

More Information:
imagine.gsfc.nasa.gov/docs/science/know_l1/emspectrum.html

below 1000 m (3300 ft). Not all wavelengths of visible light are transmitted equally (fig. 5.6). The long wavelengths at the red end of the spectrum are absorbed rapidly, while the short wavelengths of blue-green light are transmitted to the greater depths. The light undergoes **absorption** and **scattering** by suspended particles, including silt, single-celled organisms, and the water and salt molecules. This decrease in the intensity of light over distance is known as **attenuation.** The clearer the water, the greater the light penetration and the smaller the attenuation.

Color perception is due to the reflection back to our eyes of wavelengths of a particular color. The oceans usually appear blue-green because wavelengths of this color, being absorbed the least, are most available to be reflected and scattered. Coastal waters appear green or even brown or red, because this water usually contains silt from rivers as well as large numbers of microscopic organisms that add their color. Open-ocean water is often clear and blue, which indicates that it is nearly empty of both living and nonliving suspended materials. The clearer the water, the less suspended material matter present and the deeper the light penetrates.

When light passes from the air into the water, it is bent by **refraction** because its speed is faster in air than in water. Because the light rays bend as they enter the water from the air, objects seen through the water's surface are not where they appear to be (fig. 5.7). Refraction is affected slightly by changes in salinity, temperature, and pressure. Devices that measure refraction of seawater (refractometers) can be used for convenient but low-precision determination of salinity.

Oceanographers use a light meter to measure the intensity of light in water, but the simplest way of measuring light attenuation in surface water is to use a **Secchi disk** (fig. 5.8). This is a white disk about 30 cm (12 in) in diameter, which is lowered to the depth at which it just disappears from view. The measure of this depth can be used to determine the average attenuation of light. In waters rich with living organisms or suspended silt, the Secchi disk may disappear from view at depths of 1 to 3 m (3 to 10 ft) while in the open ocean it may be visible down to 20 to 30 m (65 to 100 ft).

What portion of the electromagnetic spectrum is transmitted in seawater?

What processes affect the attentuation of light in seawater?

Why does the sea generally appear blue?

Why are light waves refracted in water and what is the result?

Figure 5.6

Total available solar energy in the sea (shown by the areas under the curves) decreases as depth increases. The longer, red wavelengths are absorbed first. The color peak shifts to the shorter, blue wavelengths as the depth increases.

More Information:
www.oceansonline.com/light_in_the_sea.htm

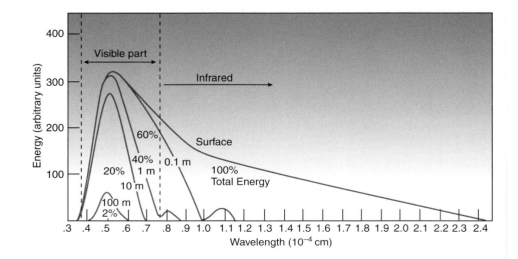

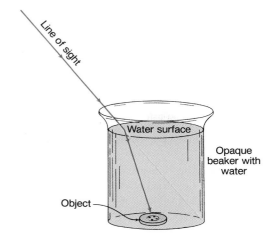

Figure 5.7

Objects that are not directly in one's line of sight can be seen in water due to the refraction of the light rays. The refraction is caused by the decreased speed of light in water.

More Information: wigner.byu.edu/LightRefract/LightRefract.html

Figure 5.8

A Secchi disk measures the transparency of water.

Transmission of Sound   5.6

Sound travels farther and faster in seawater than in air (average velocity in seawater is about 1500 m [5000 ft] per second as compared with 334 m [1100 ft] per second in dry air at 20°C). The speed of sound in seawater increases with increasing temperature, pressure, and salt content, and decreases with decreasing temperature, pressure, and salt content.

Because sound is reflected back after striking an object, sound can be used to find objects, sense their shape, and determine their distance from the sound's source. If a sound signal is sent into the water and the time required for the return of the

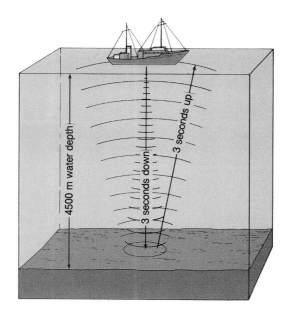

Figure 5.9

Traveling at an average speed of 1500 meters per second, a sound pulse leaves the ship, travels downward, strikes the bottom, and returns. In 4500 meters of water, the sound requires 3 seconds to reach the bottom and 3 seconds to return.

reflected sound, or echo, is measured accurately, the distance to an object may be determined. For example, if 6 seconds elapse between the outgoing sound pulse and its return after reflection, the sound has taken 3 seconds to travel to the object and 3 seconds to return. Since sound travels at an average speed of 1500 m per second in water, the object is 4500 m away, as shown in figure 5.9.

Oceanographic vessels use echo sounders or depth recorders to direct a narrow sound beam vertically to the sea bottom. The depth recorder chart produces a continuous reading of depth as the vessel moves along its course. If high-intensity sound pulses are transmitted, sound energy can penetrate the sea floor and reflect back from layers in the sediments, providing a display of the sea floor's structure on the depth recorder chart (fig. 5.10). Section 6 in chapter 4 discusses this topic as well.

Sonar (sound navigation and ranging) is a location system that directs a beam of sound pulses through the water and detects their echoes. Trained technicians are able to distinguish between the echoes produced by a whale, a school of fish, or a submarine. It is also possible to distinguish animals from vessels by listening to the sounds they generate, even to determine the type and class of a vessel from the sounds produced by its propellers and engines.

However, sound may be bent or refracted as it passes at an angle through water layers of differing density. When refraction occurs, targets appear displaced from their actual positions. Figure 5.11 illustrates the curvature of sound beams and how **sound shadow zones,** into which the sound does not penetrate, can form. Sound beams bend toward regions in which

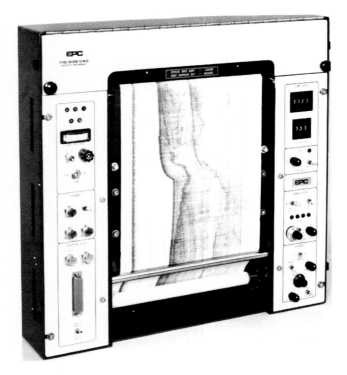

Figure 5.10

A precision depth recorder (PDR) displays bottom and subbottom profiles. A second image of the seafloor surface appears as the deepest layer at right. This image comes from the first seafloor echo bouncing off the sea surface to make a second round trip.

sound travels more slowly and away from regions in which the sound waves travel more rapidly. Understanding the behavior of sound in the oceans is crucial to naval vessels, both above and below the surface, and is the subject of much ongoing research.

At about 1000 m (3300 ft), the combination of salt content, temperature, and pressure creates a depth zone of minimum velocity for sound, the **sofar** (sound fixing and ranging) **channel** (fig. 5.12). Sound waves produced in the sofar channel do not escape from it unless they are directed outward at a sharp angle. Instead, the majority of the sound energy bounces back and forth along the channel for great distances.

This channel is being used by a project known as ATOC (Acoustic Thermometry of Ocean Climate) to send sound pulses over long distances to look for long-term changes in the temperature of ocean waters. (See the box "Acoustic Thermometry of Ocean Climate," in this chapter.)

How can sound be used to measure depth?

How does sound refraction affect sonar?

Explain the effect of the sofar channel on sound.

Figure 5.11

Sound waves change velocity and refract as they travel at an angle through water layers of different densities. The angle at which the sound beam leaves the ship indicates a target in the indicated, or ghost, position. To determine the actual position of the target the degree of refraction and change in velocity with depth must be known.

Source: Adapted from G. Neumann and W. J. Pierson, Principles of Physical Oceanography, page 51, Prentice-Hall, 1996.

More Information:
www.fas.org/irp/program/index.html

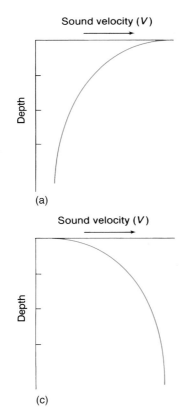

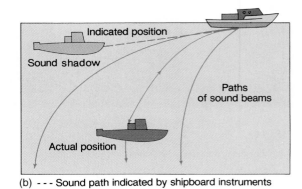

(a)

(b) - - - Sound path indicated by shipboard instruments

—— Actual path of refracted sound beam

(c)

(d)

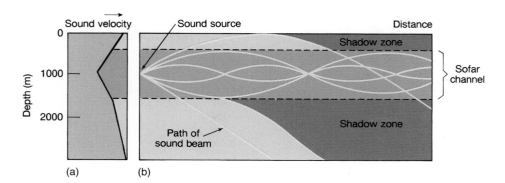

Figure 5.12

(a) The temperature, salinity, and pressure variation with depth combine to produce a minimum sound velocity at about 1000 meters. (b) Sound generated at this depth is trapped in a layer known as the sofar channel.

From Fundamentals of Acoustics, *by Lawrence E. Kinsler and Austin R. Frey. Copyright © 1962, John Wiley & Sons, Inc. Reprinted by permission of John Wiley & Sons, Inc.*

More Information: www.oal.whoi.edu/index.html, or www4.nas.edu/beyond/beyonddiscovery.nsf/web/ocean6?OpenDocument

Dissolving Ability of Water   5.7

Water is an extremely good solvent. More substances—solids, liquids, and gases—dissolve in water than in any other common liquid. The dissolving ability of water is related to its molecular chemistry. For example, common table salt dissolves in water. Common salt is chemically sodium chloride (NaCl); each atom of sodium has exactly one matching atom of chlorine. When the salt is in its natural crystal form, the positively charged sodium and the negatively charged chlorine are bonded in an alternating pattern because of the strong attractive force between positive and negative charges. When sodium chloride is placed in water, the bonds between the atoms break and the resulting charged atoms are called **ions.** This reaction can be written as:

Sodium chloride → Sodium ion⁺ + Chloride ion⁻

or it can be written with chemical abbreviations as:

$$NaCl \rightarrow Na^+ + Cl^-$$

The positive sodium ions are attracted to the negatively charged oxygen side of the water molecules, while the negative chloride ions are attracted to the positively charged hydrogen side. The ions thus are surrounded and separated by water molecules (fig. 5.13). The ability of water molecules to separate compounds into their ions makes water an excellent solvent.

This property of water, as well as those discussed earlier, appears in table 5.3, a summary of water's properties.

> What happens to soluble salts in seawater?

Salts in Seawater   5.8

Oceanographers measure the salt content of ocean water in grams of salt per kilogram of seawater (g/kg), or parts per thousand ($^0/_{00}$) or in Practical Salinity Units (PSU). For most purposes $^0/_{00}$ and PSU are numerically equal. The total quantity of dissolved salt in seawater is known as **salinity.** The dissolved salts are present as positively or negatively charged ions or groups of ions.

Six ions make up more than 99% of the salts dissolved in seawater: sodium (Na^+), magnesium (Mg^{2+}), calcium (Ca^{2+}),

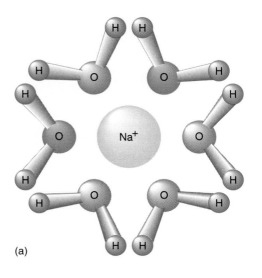

(a)

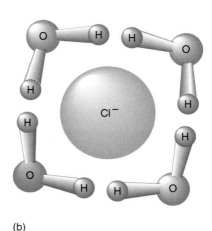

(b)

Figure 5.13

Salts dissolve in water because the polarity of the water molecule keeps positive salt ions separated from negative ions. (a) Sodium ions are surrounded by water molecules with their negatively charged portion attracted to the positive ion, cation. (b) Chloride ions are surrounded by water molecules with their positively charged portion attracted to the negative ion, anion.

table 5.3 Properties of Water

Definition	Comparisons	Effects
Physical States Gas, Liquid, Solid Addition or loss of heat breaks or forms bonds between molecules to change from one state to another.	The only substance that occurs naturally in three states on the Earth's surface.	Important for the hydrologic cycle and the transfer of heat between the oceans and atmosphere.
Heat Capacity One calorie per gram of water per °C.	Highest of all common solids and liquids.	Prevents large variations of surface temperature in the oceans and atmosphere.
Surface Tension Elastic property of water surface.	Highest of all common liquids.	Important in cell physiology, water-surface processes, and drop formation.
Latent Heat of Fusion Heat required to change a unit mass from a solid to a liquid without changing temperature.	Highest of all common liquids and most solids.	Results in the release of heat during freezing and the absorption of heat during melting. Moderates temperature of polar seas.
Latent Heat of Vaporization Heat required to change a unit mass from a liquid to a gas without changing temperature.	Highest of all common substances.	Results in the release of heat during condensation and the absorption of heat during vaporization important in controlling sea surface temperature and the transfer of heat to air.
Compressibility Average pressure on total ocean volume 200 atmospheres; ocean depth decreased by 37 m (121 ft).	Seawater is only slightly compressible, $4–4.6 \times 10^{-5}$ cm^3/g for an increase of 1 atmosphere of pressure.	Density changes only slightly with pressure. Sinking water can warm slightly due to its compressibility.
Density Mass per unit volume: grams per cubic centimeter, g/cm^3.	Density of seawater is controlled by temperature, salinity, and pressure.	Controls the ocean's verticle circulation and layering. Affects ocean temperature distribution.
Viscosity Liquid property that resists flow. Internal friction of a fluid.	Decreases with increasing temperature. Salt and pressure have little effect. Water has a low viscosity.	Some motions of water are considered friction free. Low friction dampens motion; retards sinking rate of single-celled organisms.
Dissolving Ability Dissolves solids, gases, and liquids.	Dissolves more substances than any other solvent.	Determines the physical and chemical properties of seawater and the biological processes of life-forms.
Heat Transmission Heat energy transmitted by conduction, convection, and radiation.	Molecular conduction slow; convection effective. Transparency to light allows radiant energy to penetrate seawater.	Affects density; related to vertical circulation and layering.
Light Transparency Transmits light energy.	Relatively transparent for visible wavelength light.	Allows plant life to grow in the upper layer of the sea.
Sound Transmission Transmits sound waves.	Transmits sound very well compared to other fluids and gases.	Used to determine water depth and to locate targets.
Refraction The bending of light and sound waves by density changes that affect the speed of light and sound.	Refraction increases with increasing salt content and decreases with increasing temperature.	Makes objects appear displaced when viewed by light or sound.

potassium (K^+), chloride (Cl^-), and sulfate (SO_4^{2-}). An ion with a positive charge is a **cation:** an ion with a negative charge is an **anion.** The salts in seawater are mostly present in their ionic forms as anions and cations. Table 5.4 lists these six ions and five more known as the **major constituents** of seawater.

Note that sodium and chloride ions together account for 86% of the salt ions present in seawater.

All the other elements dissolved in seawater are present in concentrations of less than 1 part per million and are called **trace elements.** Some of these elements are important to or-

table 5.4 Major Constituents of Seawater

Constituent	Symbol	g/kg in Seawater[1]	Percentage by Weight	
Chloride	Cl^-	19.35	55.07	
Sodium	Na^+	10.76	30.62	
Sulfate	SO_4^{2-}	2.71	7.72	
Magnesium	Mg^{2+}	1.29	3.68	99.36
Calcium	Ca^{2+}	0.41	1.17	
Potassium	K^+	0.39	1.10	
Bicarbonate	HCO_3^-	0.14	0.40	
Bromide	$Br-$	0.067	0.19	
Strontium	Sr^{2+}	0.008	0.02	
Boron	B^{3+}	0.004	0.01	
Fluoride	F^-	0.001	0.01	
Total		~ 35.00	99.99	

From Riley and Skirrow, Chemical Oceanography, *Vol. 1, 2d ed., p366. Copyright © 1975 Academic Press Ltd., London. Used by permission of Academic Press.*
[1]*Salinity = 35⁰/₀₀.*

ganisms that are able to concentrate the ions. For example, long before the presence of iodine could be determined chemically as a trace element of seawater, it was known that shellfish and seaweeds were rich sources for this element, and seaweed has been harvested commercially for iodine extraction.

Seawater is a well-mixed solution; currents at the surface and at depth, vertical mixing processes, and wave and tidal action have all helped to stir the ocean over geologic time. Because of this thorough mixing, the ionic composition of the major ions of open-ocean seawater (table 5.4, excluding bicarbonate and fluoride) is the same from place to place and from depth to depth. That is to say, the ratio of one major ion or seawater constituent to another remains the same. The total salinity may change as fresh water is removed or gained, but the major ions exist in the same proportions. This principle of constant proportions may not apply along the shores, where rivers may bring in large quantities of dissolved substances or reduce the salinity to very low values. The major ions are also called **conservative ions,** because they do not change their ratios to each other with changes in salinity, and because they are not generally removed or added by living organisms. Certain ions present in much smaller quantities, some dissolved gases, and assorted organic molecules do change their concentrations with biological and chemical processes; these are called **nonconservative ions.**

What is meant by salinity?

Distinguish between the major constituents and the trace elements found in seawater.

How does the ratio of major salt ions change when open-ocean salinity changes?

Determining Salinity 5.9

The great advantage of most of the major ions being in constant proportion is that it allows one to determine salinity by measuring the concentration of one type of ion. Historically, the quantity of chloride ions has been measured to establish a sample's salinity. To do so, silver nitrate is added to the sample so that the silver combines with the chloride ion, and when the amount of silver required to react with all of the chloride ion is known, the amount of chloride is known. However, the silver also combines with bromine, iodine, and some other trace elements; the chloride concentration measured in this way has the special name chlorinity. Chlorinity (Cl ⁰/₀₀) and salinity (S ⁰/₀₀) in parts per thousand are related by the following equation:

$$\text{Salinity } ^0/_{00} = 1.80655 \times \text{Chlorinity}$$

or

$$S\ ^0/_{00} = 1.80655 \times Cl\ ^0/_{00}.$$

Because of the ions it contains, seawater conducts electricity; the more ions there are, the greater the conductivity. This relationship makes it possible to determine salinity by using an electrical conductivity meter. The great advantage of the salinometer is that the measurements are made quickly and directly on the water sample with an electrical probe. It is therefore not necessary to analyze a water sample chemically except to test and calibrate the conductivity instruments. Because the conductivity of seawater is affected by both salinity and temperature, a conductivity instrument must correct for the temperature of the sample if it is calibrated to read directly in salinity units.

Why can salinity be calculated from the measurement of chloride ions?

What other method can be used to measure salinity?

Ocean Salinities 5.10

In the major ocean basins, a typical kilogram of seawater is made up of 965 grams of water and 35 grams of salt. The average ocean salinity is approximately 35⁰/₀₀. The salinity of ocean surface water is associated with latitude. The relationship between evaporation, precipitation, and surface salinity with latitude is shown in figure 5.14. Notice the low surface salinities in the cool and rainy belts at 40–50°N and S latitude, high evaporation rates, and high surface salinities in the desert belts of the world centered on 25°N and S, and low surface salinities again in the warm, rainy tropics centered at 5°N. Sea surface salinities during the Northern Hemisphere summer are shown in figure 5.15. Refer to section 5.3 and table 5.2 for a discussion of the effect of salts on seawater density, and section 7.1 and 7.2

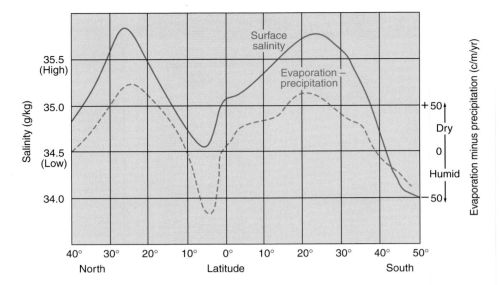

Figure 5.14

Mid-ocean average surface salinity values match the average changes in evaporation minus precipitation values that occur with latitude.

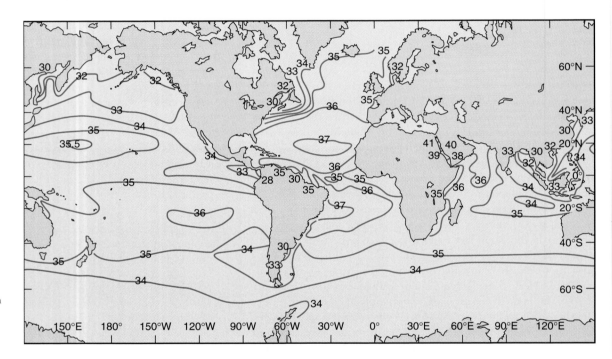

Figure 5.15

Average sea surface salinities in the Northern Hemisphere summer, given in parts per thousand ($^0/_{00}$).

for information on the effects of temperature and salinity on ocean density.

In coastal areas of high precipitation and river inflow, surface salinities fall below the average. For example, during periods of high flow, the water of the Columbia River lowers the Pacific Ocean's surface salinity to less than $24^0/_{00}$ as far as 30 km (20 mi) out to sea. In subtropic regions of high evaporation and low freshwater input, the surface salinities of nearly landlocked seas are well above the average: 40 to $42^0/_{00}$ in the Red Sea and the Persian Gulf, and 38 to $39^0/_{00}$ in the Mediterranean Sea. In the open ocean at these same latitudes, the surface salinity is closer to $36.5^0/_{00}$. Surface salinities change seasonally in polar areas, where the surface water forms sea ice

in winter, leaving behind the salt and raising the salinity of the water under the ice. In summer, the water returns as a freshwater layer when the sea ice melts.

Relate changes in open-ocean salinity to latitude and explain these changes.

Why do coastal surface salinities differ from open-ocean salinities?

What happens to surface salinities in the polar seas during the year?

The Salt Balance   5.11

The salts of the oceans come from the land through rivers, from the chemical reactions of seawater with sediments, from the gases produced by volcanoes, and from the spreading centers of the mid-ocean ridge and rise systems. During volcanic eruptions sulfide and chloride gases are released, dissolved in rainwater, and carried to the oceans as Cl^- (chloride) and SO_4^{2-} (sulfate). River water carrying chloride and sulfate ions is acidic and erodes and dissolves the rock over which it flows, helping to liberate the positive ions, such as Na^+ (sodium), Ca^{2+} (calcium), and Mg^{2+} (magnesium), from the Earth's crust. At the mid-ocean ridge systems where molten rock rises from the mantle, magma chambers are formed, and seawater becomes heated as it flows through the fractured crust. The heated water reacts chemically with the rocks of the crust, and copper, iron, manganese, zinc, potassium, and calcium are dissolved into the water. Follow the input of ions into the oceans from these sources in figure 5.16.

Although salt ions continually flow into the oceans, chemical and geologic evidence leads researchers to believe that the salt composition of the oceans has not changed for about the last 1.5 billion years. Therefore, the addition of salts must be balanced by the removal of salts if ocean salinity is to remain the same.

Salts are removed from seawater in a number of ways; some are depicted in figure 5.16. Sea spray from waves is blown ashore, depositing a film of salt on the land. Over geologic time, shallow arms of the sea may become isolated, the water evaporates, and the salts are left behind to become land deposits. Salt ions can also react with each other to form insoluble products that settle to the ocean floor. At the mid-ocean ridges, magnesium and sulfate ions are transferred from the water to form mineral deposits. Biological processes concentrate salts, which are removed if the organisms are harvested. Organisms' excretion products trap ions, which are transferred to the sediments or returned to the seawater. Other biological processes remove calcium by incorporating it into shells, and silica is used to form the hard parts of certain plantlike organisms and animals. These hard parts accumulate to form sediments when the organisms die. Chapter 4 discussed the sediments produced by biological processes.

Tiny clay mineral particles, weathered from rock and brought to the oceans by the rivers or the winds, bind ions and molecules to their surfaces by a process known as **adsorption.** Ions and trace metals sink with the clay particles and are eventually incorporated into sediments. The fecal pellets and skeletal remains of small organisms also act as adsorption surfaces. Although many processes remove ions from seawater and transfer them to the sediments, adsorption is the most important single process of ion removal.

Ions deposited in the sediments are trapped there, but geologic processes elevate some sediments from the sea floor to positions above sea level. Erosion then works to dissolve and wash these deposits back to the sea.

The relative abundance of the salts in the sea is due in part to the ease with which they are introduced from the Earth's crust and in part to the rate at which they are removed from the seawater. Sodium is moderately abundant in fresh

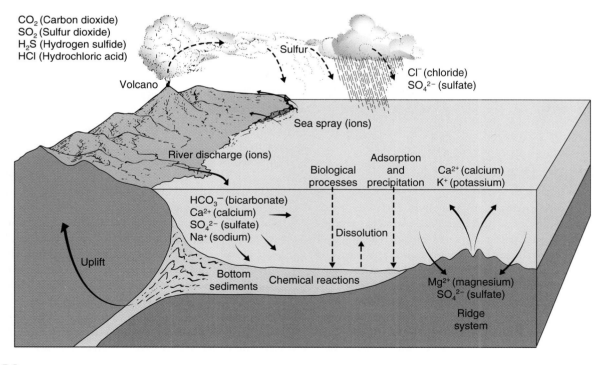

Figure 5.16
Processes that distribute and regulate the major constituents in seawater.

table 5.5   Approximate Residence Time of Ions in the Oceans

Ion	Time in Years
Chloride	100 million–∞
Sodium	210 million
Magnesium	22 million
Potassium	11 million
Sulfate	11 million
Calcium	1 million
Manganese	1.4 thousand
Iron	1.4 hundred
Aluminum	1 hundred

table 5.6   Nutrients in Seawater

Element	Concentration (µg/kg[1])
Nitrogen (N)	500
Phosphorus (P)	70
Silicon (Si)	3000

[1]Parts per billion.

water, but because it reacts slowly with other substances in seawater, it remains dissolved in the ocean. Calcium is removed rapidly from seawater, forming limestone and the shells of marine organisms.

The average time that a substance remains in solution in the oceans is called its **residence time** (table 5.5). Aluminum, iron, and chromium ions react with other substances quickly and form insoluble mineral solids in the sediments. They have short residence times, in the hundreds of years. Sodium, potassium, and magnesium are very soluble, combine with other substances slowly, and so have residence times in the millions of years.

Compare processes that add and remove salts from the oceans.

What is the role of adsorption in controlling ocean salinity?

How are the abundance, solubility, and residence time of ions related?

Nutrients and Organics   5.12

Ions required for plant growth are known as **nutrients;** these are the fertilizers of the oceans. As on land, plants require nitrogen and phosphorus in the forms of nitrate (NO_3^-) and phosphate (PO_4^{3-}) ions. A third nutrient required in the oceans is the silicate ion (SiO_4^-), which is needed to form the hard outer wall of the single-celled plantlike organisms called diatoms and the skeletal parts of some protozoans. These three nutrients are among the dissolved substances brought to the sea by the rivers and land runoff. Nitrates and phosphates are continually recycled from decomposing organic matter. Despite their importance, nitrates, phosphates, and silicates are present in very low concentrations (table 5.6).

The concentration ratios of some nutrient ions vary, because some of these ions are closely related to the life cycles of living organisms. Populations of organisms remove nutrients during periods of growth and reproduction, temporarily reducing the amounts in solution. Later, when the population declines, decay processes and animal wastes return the ions to the seawater. Therefore nutrients are nonconservative for they do not maintain constant ratios in the way that most major salt ions do. The biological cycling of nutrients is discussed in chapter 10.

A wide variety of organic substances is present in seawater. Proteins, carbohydrates, lipids (or fats), vitamins, hormones, and their breakdown products are all present. Some are eventually reduced to their inorganic components; others are used directly by organisms and are incorporated into their systems. Another portion of the organic matter accumulates in the sediments, where over geologic time it slowly forms deposits of oil and gas.

What are nutrients and why are they important?

The Gases in Seawater   5.13

The most abundant gases in the atmosphere and in the oceans are nitrogen (N_2), oxygen (O_2), and carbon dioxide (CO_2). The percentages of each of these gases in the atmosphere and in seawater are given in table 5.7. Oxygen and carbon dioxide play important roles in the ocean because they are necessary to life, and biological activities modify their concentrations at various depths. Nitrogen is not used directly by living organisms except for certain bacteria. Gases such as argon, helium, and neon are present in small amounts, but they neither interact with the ocean water nor are used by its inhabitants.

The amount of any gas that can be held in solution without causing the solution either to gain or to lose gas is the **saturation value.** The saturation value changes because it depends on the temperature, salinity, and pressure of the water. Colder water holds more dissolved gas than warmer water; less salty water holds more gas than more salty water, and water under more pressure holds more gas than water under less pressure.

table 5.7   Abundance of Gases in Air and Seawater

Gas	Symbol	Percentage by Volume in Atmosphere	Percentage by Volume in Surface Seawater[1]	Percentage by Volume in Total Oceans
Nitrogen	N_2	78.08	48	11
Oxygen	O_2	20.99	36	6
Carbon dioxide	CO_2	0.03	15	83
Argon, helium, neon, etc.	Ar, He, Ne	0.90	1	
Totals		100.00	100	100

[1]Salinity = $36\,^0/_{00}$, temperature = $20°C$.

In the process known as **photosynthesis,** plants and plantlike organisms use carbon dioxide to form organic molecules and produce oxygen as a by-product. Since plants require sunlight for photosynthesis, plantlike marine organisms are confined, on the average, to the top 100 m (330 ft) of the sea, where sufficient light is available. Therefore oxygen is produced in surface water, and carbon dioxide is consumed there. By contrast, **respiration,** which breaks down organic substances to provide energy, requires oxygen and produces carbon dioxide and energy. All living organisms respire in order to produce energy in living cells, and respiration occurs at all depths in the oceans. Decomposition, the bacterial breakdown of nonliving organic material, also requires oxygen and releases carbon dioxide. The processes of photosynthesis and respiration and their equations are discussed in more detail in chapter 10.

Oxygen can be added to the oceans only at the surface, from exchange with the atmosphere or as a waste product of photosynthesis. Carbon dioxide also enters from the atmosphere at the sea surface, but it is available at all depths from respiration and decomposition. Figure 5.17 shows typical oxygen and carbon dioxide concentrations with depth.

If the water is quiet, the nutrients and sunlight abundant, and a large population of plants is present, oxygen values at the surface can rise above the saturation value to 150% or more. This water is **supersaturated.** Wave action tends to liberate oxygen to the atmosphere and return the condition to the saturated state.

Below the surface, the **oxygen minimum** occurs at about 800 m (2600 ft) depth. At depths greater than this, the rates of removal of oxygen and production of carbon dioxide both fall because the population density of animals and the abundance of decaying organic matter have decreased. The slow supply of oxygen to depth by water sinking from the surface gradually increases the oxygen concentration above that found at the oxygen minimum depth.

Very low or zero concentrations of oxygen occur in the bottom waters of isolated deep basins, which have little or no exchange or replacement of water. These include the bottoms of trenches, deep basins behind a shallow entrance sill (the Black Sea, for example), and the bottoms of deep fjords. The water is trapped and becomes stagnant; respiration and decomposition use up the oxygen faster than the slow circulation of water to this depth can replace it. The bottom water becomes **anoxic,** or stripped of dissolved oxygen.

The average overall concentration of carbon dioxide throughout the oceans is not substantially affected by biological processes, but tends to remain almost constant, controlled by temperature, salinity, and pressure.

The carbon dioxide cycle and its impact on the Earth's atmosphere is discussed in section 6 of chapter 6.

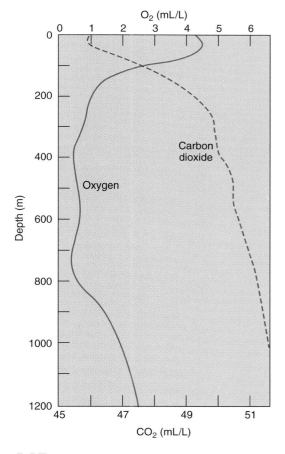

Figure 5.17

The distribution of O_2 and CO_2 with depth.

7.5 and 8.5; an average pH value for the world's oceans is approximately 7.8.

Carbon Dioxide as a Buffer 5.14

Carbon dioxide plays an additional and extremely important role in the ocean: in water it acts as a **buffer.** A buffer prevents sudden changes in the acidity or alkalinity, or **pH,** of a solution. The buffering capacity of seawater is important to organisms requiring a steady pH for their life processes and to the chemistry of seawater, which is controlled, in part, by pH. The pH scale indicates acidity and alkalinity by measuring the concentration of hydrogen ions in a solution (fig. 5.18). A pH of 7 indicates neutrality, neither acid nor alkaline. Values between 1 and 6 are acidic while the pH values between 8 and 14 are alkaline, or basic. The pH range of seawater is between

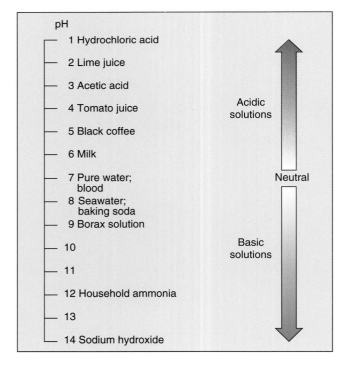

Figure 5.18

The pH scale.

More Information: www.chemtutor.com/acid.htm

Salt and Water 5.15

About 30% of the world's salt is extracted from seawater. In warm, dry climates, seawater is directed into shallow ponds and evaporated down to a concentrated brine solution. The process is repeated several times, until a dense brine is produced. Then the evaporation is allowed to continue until a thick, white salt deposit is left on the bottom of the pond. The salt is collected and refined to produce table salt. This technique has been recently used in southern France, Puerto Rico, and California (fig. 5.19).

In cold climates, salt has been recovered by freezing seawater in shallow ponds. The ice that forms is nearly fresh; the salts are concentrated in the brine beneath the ice. The brine is heated to remove the last of the water.

Sixty percent of the world's magnesium comes from the sea, and so does 70% of the bromine. There are vast amounts of dissolved minerals in the world's seawater, including 10 million tons of gold and 4 billion tons of uranium, but the concentration is very low (one part per billion or less), and the cost of extraction is too high for economic production at the present time. Dense, hot salt brines at the bottom of the Red Sea are estimated to contain minerals with values in the billions of dollars.

Desalination is the process of obtaining fresh water from salt water. The greatest drawback to desalination is the cost, which is linked to the energy required. There are several possible desalination methods; two processes use a change of state of the water (liquid to solid or liquid to vapor), and another process uses a membrane permeable to water molecules but not to salt ions.

The simplest and least energy-expensive process involving a change of state is the solar still. In this process a pond of seawater is capped by a low plastic dome. Solar radiation penetrates the dome, causing evaporation of the seawater. The evaporated water condenses on the undersurface of the dome and trickles down the curved surface to be caught in a trough, where it accumulates and flows to a freshwater reservoir. The rate of production is slow, and a very large system is needed to supply the water requirements of even a small community. Yet the costs for this type of system are low and are principally confined to construction of the facility, since the solar energy is free. This system is illustrated in figure 5.20.

When salt water is distilled by heating it to boiling, evaporation proceeds at a rapid rate and large quantities of fresh water are produced, but considerable energy is required. If water is introduced into a chamber with a reduced air pressure,

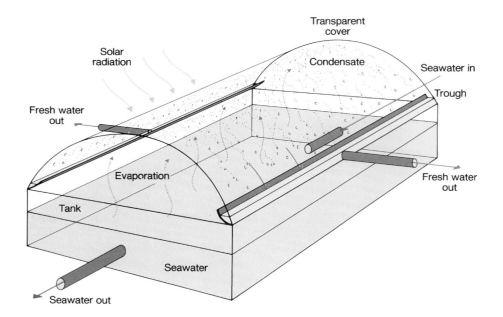

Solar radiation

Transparent cover

Condensate

Seawater in

Trough

Fresh water out

Evaporation

Fresh water out

Tank

Seawater

Seawater out

Figure 5.20

Solar energy is used to evaporate fresh water from seawater. Solar radiation penetrates the transparent cover of the still and causes evaporation of the seawater contained in the tank.

the boiling occurs at a much lower temperature, with less expenditure of energy.

In areas like Kuwait, Saudi Arabia, Morocco, Malta, Israel, the West Indies, California, and the Florida Keys, water is a limiting factor for population and industrial growth. Thousands of cubic meters of fresh water are produced daily in these areas by desalination. Around the world, nearly 4000 desalination plants produce a total of about 13 billion liters (3.4 billion gal) of fresh water from seawater each day. About 60% of the world's desalination capacity is located on the Arabian Peninsula. Water costs there are low, because fuel costs are low. Desalination plants in California and Texas produce fresh water at more than twice the cost of water from other sources; it is cheap enough to use for drinking water at approximately $4.00 per 1000 gallons and for certain industrial applications, but far too expensive for agriculture. Small evaporation-type desalination plants are used on military and commercial vessels.

It is also possible to produce fresh water from seawater by applying pressure to seawater. If the pressure applied to the seawater exceeds 24.5 atmospheres, water molecules will move through a membrane that is permeable only to water in a process called **reverse osmosis** (fig. 5.21). Fresh water is squeezed out of the salt water, leaving behind the salt ions and other impurities. The theoretical energy requirement is about one-half that needed for the evaporative process.

Reverse osmosis is the most rapidly growing form of desalination technology. Older evaporative desalination plants are being replaced with reverse osmosis plants. Ninety-nine percent of Florida's 110 desalination plants use reverse osmosis. Santa Barbara, California has built the nation's largest reverse osmosis plant, producing 26 million liters (7 million gal) per day. Twenty-six miles off the coast from Los Angeles, Santa Catalina Island is producing 5×10^5 liters (132,000 gal) per day, about one-third of

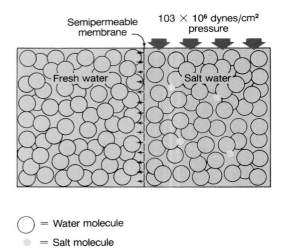

Figure 5.21

Reverse osmosis. Water moves from the saltwater side to the freshwater side when pressure is applied.

More Information: www.watermakers.com/how.htm

its potable water needs. Small commercial units using this process produce 400 liters (105 gal) per day; these are available for ship and summer home use. Units powered by hand pumps are available for emergency use in lifeboats and life rafts. The military uses portable reverse osmosis plants under emergency conditions.

What products are presently extracted from seawater?

Explain the operation of a solar still.

Where are most of the world's desalination plants located and why?

What is the process of reverse osmosis?

Summary

The water molecule has a specific shape with oppositely charged sides. Because of the distribution of the charges, water molecules interact with each other, forming bonds between molecules. The molecule's structure is responsible for the properties of the water.

Water exists as a solid, a liquid, and a gas. Changes from one state to another require the addition or extraction of heat energy. Water has a high heat capacity; it is able to take in or give up large quantities of heat with a small change in temperature.

The density of water increases with a decrease in temperature and an increase in salt content. The effect of pressure on density is small. Less dense water floats on more dense water. In the oceans, pressure increases 1 atm for every 10 m of depth, but the water is nearly incompressible.

Seawater transmits energy as light, heat, and sound. Heat is transmitted from the sea surface slowly downward by conduction. Convection transfers heat effectively from the sea surface to the atmosphere. Radiant energy penetrates seawater because seawater transmits visible light. The long red wavelengths of light are lost in the first 10 m (33 ft); only the shorter wavelengths of blue-green light penetrate to depths of 150 m (500 ft) or more. The intensity of light is decreased over distance by scattering and absorption, leading to attenuation. Light attenuation is measured either by a light meter or with a Secchi disk. Light is also refracted by water.

Sound transmission is affected by temperature, pressure, and salt content of the seawater. Echo sounders are used to measure the depth of water, and sonar is used to locate objects. Sound is refracted as it passes at an angle through the density layers of the ocean, and sound shadows are formed. The sofar channel, in which sound travels for long distances, is being used in experiments to monitor ocean temperatures and global warming.

The ability of water to dissolve substances is exceptionally good. Soluble salts are present as ions in seawater. Six major constituent ions make up 99% of the salt in seawater;

elements present in small quantities are known as trace elements. The proportion of most major ions to each other remains the same for all waters of the open oceans. The average salinity of ocean water is $35^0/_{00}$. The salinity of the surface water changes with latitude and is affected by evaporation, precipitation, and the freezing and thawing of sea ice.

Ions from weathering and erosion of the Earth's crust are added to the sea by rivers; other ions from the gases of volcanic eruptions are dissolved in river water and rain, and others come from seawater circulating through magma chambers at the spreading centers. Since the average salinity of the oceans remains constant, there must be an equal loss of salts. Salts are removed from seawater by sea spray, evaporite formation, and adsorption. Other mechanisms that remove salts include production of hydrogenous and biogenous sediments as well as geological uplift and activity at spreading centers. The residence time for salts in solution depends on their reactivity.

Nutrients include the nitrates, phosphates, and silicates required for plant growth. A wide variety of organic products is also present.

The saturation value of gases dissolved in seawater varies with salinity, temperature, and pressure. Carbon dioxide is added to seawater from the atmosphere at the sea surface and by respiration and decay processes at all depths; it is removed at the surface by photosynthesis. Oxygen is added only at the surface from the atmosphere and the photosynthetic process; it is depleted at all depths by respiration and decay. Seawater may be supersaturated with oxygen or it may become anoxic. Carbon dioxide levels tend to remain constant over depth. Carbon dioxide is the seawater's buffer, keeping the pH range of ocean water between 7.5 and 8.5.

Salt, magnesium, and bromine are extracted commercially from seawater. Desalination methods include change-of-state processes and reverse osmosis. The practicality of desalination is determined by cost and need.

Key Terms

absorption, 106	nonconservative ions, 111
adsorption, 113	oxygen minimum, 115
anion, 110	pH, 116
anoxic, 115	photosynthesis, 115
attenuation, 106	radiation, 105
buffer, 116	refraction, 106
calorie, 102	residence time, 114
cation, 110	respiration, 115
conduction, 105	reverse osmosis, 118
conservative ions, 111	salinity, 109
convection, 105	saturation value, 114
density, 103	scattering, 106
desalination, 116	Secchi disk, 106
heat capacity, 102	sofar channel, 108
hydrogen bonds, 102	sonar, 107
ion, 109	sound shadow zone, 107
major constituents, 110	supersaturated, 115
nutrients, 114	trace element, 110

Critical Thinking

1. Why might you expect to find that traces of every known naturally-occurring substance are dissolved in seawater?
2. If the substances in table 5.1 are exposed to the same quantity of heat energy, which liquid and which solid undergo the greatest temperature change?
3. Why does the addition of rain and river water to the oceans not decrease the ocean's overall average salinity?
4. How many kilograms of seawater would have to be processed to obtain 1 kg of boron?
5. If fresh water is added to seawater (salinity $35^0/_{00}$) and the salinity of the seawater is reduced to $30^0/_{00}$, what is the sodium ion (Na^+) concentration?

Suggested Readings

Alper, J. 1991. Munk's Hypothesis. *Sea Frontiers* 37(3):38–41. (The Heard Island acoustic experiment.)

Berner, R., and A. Lasaga. 1989. Modeling the Geochemical Carbon Cycle. *Scientific American* 260(3):74–81.

Bowditch, N. 1984. *American Practical Navigator*, vol. 1. U.S. Defense Mapping Agency Hydrographic Center, Washington D.C. 1414 pp. (Soundings are covered in chapter 28 and sound in chapter 35.)

Brown, N. 1991. The History of Salinometers and CTD Sensor Systems. *Oceanus* 34(1):61–66.

Forbes, A. 1994. Acoustic Monitoring of Global Ocean Climate. *Sea Technology* 35(5):65–67.

Friedman, R. 1990. Salt-Free Water from the Sea. *Sea Frontiers* 36(3):48–54. (Desalination methods.)

Gabianelli, V. J. 1970. Water: The Fluid of Life. *Sea Frontiers* 16(5):258–270.

MacIntyre, F. 1970. Why the Sea Is Salt. In *Ocean Science, Readings from Scientific American* (1977) 223(5):104–15.

Mikhalevsky, P., A. Braggeroer, A. Gavrilov, and M. Slavinsky. 1995. Experiment Tests Use of Acoustics to Monitor Temperature and Ice in Arctic Ocean. *EOS* 75(27):265, 268–69.

Open University. 1989. *Seawater: Its Composition, Properties and Behavior*. Pergamon Press, Oxford, England.

Stewart, W. K. 1991. High Resolution Optical and Acoustic Remote Sensing for Underwater Exploration. *Oceanus* 34(1):10–22. (Concerns sonar, optical imaging, cameras, and video.)

internet references

worldwide websites

Density
Seawater Density Calculator
http://www.es.flinders.edu.au/~mattom/Utilities/density.html

Light
Electromagnetic Spectrum
http://imagine.gsfc.nasa.gov/docs/science/know_l1/emspectrum.html

Light in the Sea
http://www.oceansonline.com/light_in_the_sea.htm

Light Refraction Demonstration
http://wigner.byu.edu/LightRefract/LightRefract.html

Secchi Disk
http://www.mlswa.org/secchi.htm

Sound
Woods Hole Oceanographic Institute Acoustics Lab
http://www.oal.whoi.edu/index.html

SOFAR Channel
http://www4.nas.edu/beyond/beyonddiscovery.nsf/web/ocean6?OpenDocument

Seawater Sound Speed and Density
http://www.spawar.navy.mil/sti/publications/pubs/tr/1669/nradtr1669apg.html

World Wide Web Virtual Library: SONAR: Gallery of SONAR
http://vision.dai.ed.ac.uk/ashley/Sonar/gal.html#ocean

Chemistry
Seawater Properties
http://www-class.unl.edu/geol109/seawater.htm,
and
http://www.britannica.com/bcom/eb/article/2/0,5716,115012+10+108518,00.html

Acids and Bases Tutorial
http://www.chemtutor.com/acid.htm

Carbon Cycle
Global Carbon Cycle Home Page
http://www.pmel.noaa.gov/co2/co2-home.html

Desalination
Reverse Osmosis and Desalination
http://www.watermakers.com/how.htm

Latest News on Desalination
http://www2.hawaii.edu/~nabil/latest.htm

Water Desalination Act of 1996
http://www2.hawaii.edu/~nabil/desalact.htm

World Wide Water
http://www.world-wide-water.com/index.html

Twenty years ago, based on the principle that sound's speed in seawater is determined primarily by the temperature of the water, oceanographers Carl Wunsch of the Massachusetts Institute of Technology and Walter Munk of the Scripps Institution of Oceanography in California envisioned a transoceanic experiment to follow ocean temperature response to global warming. If the oceans are warming,

ATOC, acoustic thermometry of ocean climate, began in the Pacific Ocean in 1995.

ATOC broadcasts sound from underwater sources, near San Francisco and Hawaii. The sound is picked up by arrays of sensitive hydrophones as far away as Christmas Island and New Zealand (fig. 1). After eighteen months of testing, temperature readings of the

Acoustic Thermometry of Ocean Climate

detectable worldwide decreases in the travel time of sound will occur, because sound waves travel faster in warm water than in cold water.

In 1991 the travel time of low-frequency sound pulses transmitted through the sofar channel were repeatedly measured as they moved from a site near Heard Island in the southern Indian Ocean to special listening stations around the world. The precision of sound travel-time measurements from source to receiver was about 1 millisecond over a path 1000 km (660 mi) long, so temperature changes of a few thousandths of a degree could be detected over long distances. Test results were promising, and a new series of tests called

Pacific Ocean were found to be even more precise than had been projected. Scientists can detect variations as small as 20 milliseconds in the hour-long travel time of the pulses. This precision enables researchers to calculate the average ocean temperature along the sound pulse path to within 0.006°C. Repeating the measurements will allow long-term temperature changes at mid-ocean depths to be measured before they can be deduced by any other method. At least ten years of observations will be required to find any human-induced climate effect, which is expected to be only a few thousandths of a degree per year in the deep Pacific.

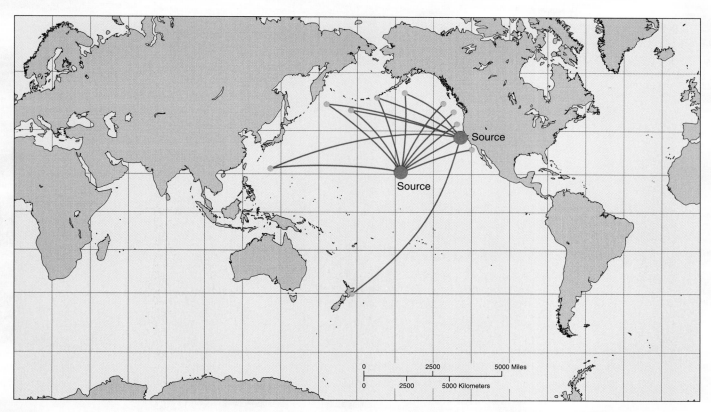

Figure 1

Acoustic thermometry of ocean climate (ATOC) takes the temperature of the Pacific Ocean using two sound sources, California and Hawaii, and twelve receivers.

More information: atocdb.ucsd.edu/

Questions concerning the effect of sound signals on marine mammals began with the Heard Island tests and delayed the start of the ATOC experiment. Marine mammalogists have monitored the behavior of whale and elephant seal populations near the sound source off California. They report having seen no changes in the animals' swimming activity or distribution. Mammal researchers have also released deep-diving elephant seals farther out to sea and used satellite tags to track the paths of the animals as they returned to shore. The seals made no attempt to avoid the sound source. Additional experiments are being conducted to see whether whale vocalizations are affected.

A similar experiment that measured water temperatures in the Arctic Ocean was performed in the spring of 1994 by a joint U.S.-Russian-Canadian project, Transarctic Acoustic Propagation Experiment (TAP). Sound signals were sent from an ice camp north of Spitsbergen to a camp 900 km (540 mi) away in the Lincoln Sea and to another camp 2600 km (1600 mi) away in the Beaufort Sea. Travel times were predicted by using water temperatures from earlier research, but the measured travel times were shorter, implying that the mid-depth Atlantic water that penetrates the Arctic Ocean had warmed by 0.2°–0.4°C since the mid-1980s. Scientists emphasize that it is too early to say whether the change is due to global warming or whether it is a part of some other natural cycle.

Readings

Forbes, A. 1994. Acoustic Monitoring of Global Ocean Climate. *Sea Technology* 35 (5): 65–67.

Georges, T.M. 1992. Taking the Ocean's Temperature with Sound. *The World & I* (July): 282–89.

The Heard Island Experiment. 1991. *Oceanus* 34 (1): 6–8.

Mikhalevsky, P., A. Braggeroer, A. Gavrilov, and M. Slavinsky. 1995. Experiment Tests Use of Acoustics to Monitor Temperature and Ice in Arctic Ocean *EOS* 76 (27): 265, 268–69.

Internet Resources

Acoustic Thermometry of Ocean Climate (ATOC) Home Page

http://atocdb.ucsd.edu/

Woods Hole Oceanographic Institute Acoustics Lab

http://www.oal.whoi.edu/index.html

chapter six

The Atmosphere and the Oceans

Cumulus clouds over the Caribbean.

Learning Objectives

After reading this chapter, you should be able to

- Explain the variation of solar radiation with latitude.
- Define the Earth's heat budget.
- Define heat capacity and compare the heat capacity of the land and oceans.
- Understand sea surface temperature changes over the year.
- Describe the formation of sea ice and icebergs.
- Explain how the atmosphere moves with changes in density.
- Explain the greenhouse effect.
- Understand the formation of winds and the role of the Coriolis effect.
- Show the latitude and direction of the major wind bands and name them.
- Explain the jet stream and its relationship to surface winds.
- Describe how the wind systems change over the year.
- Explain wind changes along coasts and those caused by monsoon effects.
- Distinguish the kinds of sea fog.
- Understand how El Niño is related to changes in oceanic and atmospheric conditions.
- Explain what causes a storm surge.

*t*he Sun's energy reaches the Earth through a thin envelope of air—the atmosphere. The processes that control the atmosphere are closely related to ocean processes, and together they form much of what people call weather and climate. Clouds, winds, storms, rain, and fog are all the result of the interaction of the Sun's energy, the atmosphere, and the Earth's water covering. Some of these processes are more predictable than others; some are better understood than others. The complex of interactions between Sun, atmosphere, and water provides the Earth's daily weather, sometimes pleasant and stable, at other times severe and turbulent. This chapter presents an overview of these self-adjusting relationships as well as specific examples of their complex interactions.

Distribution of Solar Radiation   6.1

The amount of solar radiation per unit area that reaches the Earth's surface at a specific instant in time is the instantaneous solar radiation per unit area at that point. Instantaneous solar radiation per unit surface area of the Earth has its greatest intensity at the equator, moderate intensity in the middle latitudes, and least intensity at the poles. Radiation intensity is greatest between $23^1/_2°$N (Tropic of Cancer) and $23^1/_2°$S (Tropic of Capricorn), because only between these latitudes does sunlight strike the surface of the Earth at a right angle. All other latitudes receive less instantaneous solar energy because of the decreasing angle at which the Sun's rays strike the Earth's surface (fig. 6.1). The intensity of solar radiation is also reduced because radiation is absorbed by the atmosphere on its way to the Earth's surface. The greater the latitude, the longer the distance through the atmosphere the Sun's rays must travel (fig. 6.1). As the Earth turns on its axis, the intensity of solar radiation along each latitude line changes with time, from a maximum at high noon to a minimum of zero between sunset and sunrise. In addition, a semiannual variation in the average intensity of solar radiation appears at tropical latitudes, because the Sun crosses the equator twice during the year, as it moves between the Tropics of Cancer and Capricorn.

Although the intensity of instantaneous solar radiation changes with daily and annual cycles, the hours of daylight and darkness are also important in determining the amount of radiation received over a 24-hour period. At the equator both the intensity of radiation and the length of daylight change little over the year. At the poles, the high daily levels of average radiant energy reflect the long periods of daylight in summer rather than high levels of instantaneous solar radiation. The maximum daily radiation levels at the South Pole are slightly larger than those at the North Pole, because the Earth is closer to the Sun during the Southern Hemisphere summer. The Earth's surface responds to this annual pattern of incoming solar radiation by warming the tropical areas and keeping the polar areas cold.

> Why does the intensity of solar radiation vary with latitude?
>
> Why are daily solar radiation values at the poles high in summer?
>
> Why do Southern Hemisphere summer radiation values exceed the summer values in the Northern Hemisphere?

The Heat Budget   6.2

In order to maintain its long-term mean temperature of 16°C, the Earth must reradiate as much heat back to space as it receives from the Sun. The gains and losses in heat are represented in a **heat budget.** Just as incoming funds must equal outgoing funds in a balanced monetary budget, in the heat budget incoming radiation must equal outgoing radiation. If incoming funds (or deposits) are insufficient, or if outgoing funds (or withdrawals) become too great, serious financial problems arise. In the same way, if less heat were returned to space than is gained, the Earth would become hotter, and if more heat were lost to space than is gained, the Earth would become colder.

To follow the long-term deposits and withdrawals in the Earth's heat budget, assume that the solar energy available to the

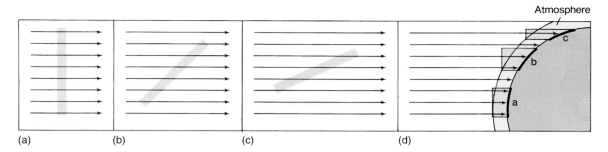

Figure 6.1

Areas of the Earth's surface that are equal in size receive different levels of solar radiation as they become more oblique to the Sun's rays (see a, b, and c). As latitude increases, the angle between the Sun's rays and the Earth decreases and the solar radiation received on the surface decreases (see d). Note also that the Sun's rays must travel an increasing distance through the atmosphere as latitude increases.

Earth and its atmosphere is 100 units or 100% of available solar energy. Incoming solar radiation from space contains short to long wavelengths of radiation, ultraviolet, visible, and infrared. The energy reradiated to space from the Earth's surface and the heat reradiated to space from the atmosphere are assumed to be infrared radiation. Refer to figure 6.2 as you read this discussion. Thirty-one of these units are reflected directly back to space from the atmosphere; no heating of the Earth or atmosphere occurs. Four more units are reflected back by the Earth's surface through the atmosphere to outer space and also take no part in heating the Earth or atmosphere. Of the remaining 65 units, 47.5 units are absorbed by the Earth's surface, and 17.5 units are absorbed by the atmosphere. To balance the budget, 65 units must be returned to space, 5.5 units from the surface and 59.5 units from the atmosphere. The budget is balanced; incoming radiation balances outgoing radiation.

Closer inspection shows that although the Earth's surface absorbs 47.5 units of solar radiation, it loses only 5.5 units to space, for a gain of 42 heat units, while the atmosphere absorbs 17.5 units and loses 59.5 units to space, for a loss of 42 units. The 42 units gained by the Earth must be transferred to the atmosphere. First, 29.5 of the units are transferred by evaporation processes, which cool the Earth's surface and liberate heat when condensation of water vapor occurs in the atmosphere. Second, the remaining 12.5 units are transferred to the atmosphere by conduction or reradiation and are absorbed as heat, raising the temperature of the air. On the world average, the atmosphere is primarily heated from below by heat given off from the Earth. These radiation averages do not take into account variation of heat exchange with latitude, time, or properties of the Earth's surface. Changes in the heat budget related to changes in the atmosphere are discussed in section 6.6.

On the average, at latitudes less than 45°N or S each unit of surface area absorbs more radiant energy annually than it loses to space, and at latitudes greater than 45° each unit area

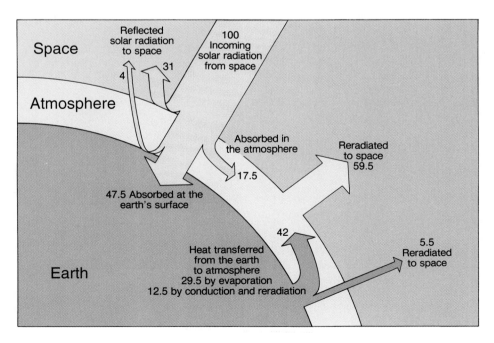

Figure 6.2

The Earth's heat budget. Incoming solar energy is balanced by reflected and reradiated energy. The atmosphere's loss of heat is balanced by heat transferred from the Earth to the atmosphere by evaporation, conduction, and reradiation.

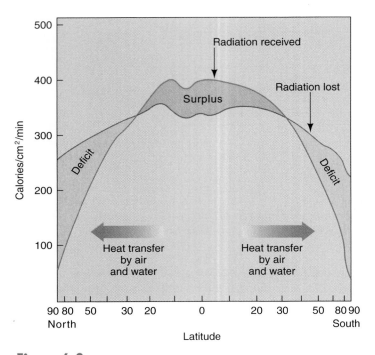

Figure 6.3

Comparison of incoming and outgoing radiation with latitude. A transfer of energy is required to maintain a balance.

Source: NOAA Meteorological Satellite Laboratory.

loses more radiant energy than it absorbs. See figure 6.3. However, the surplus of energy at low latitudes sets the atmosphere and oceans in motion, and heat is transferred by winds and ocean currents from lower to higher latitudes, maintaining a more even temperature distribution than might be expected.

> Why must incoming radiation be balanced by outgoing radiation?
>
> Compare the effect of reflected radiation to the effect of absorbed radiation.
>
> How is the Earth's atmosphere heated?

Earth Surface Temperatures and Heat Transfer   6.3

Land and sea respond differently to solar radiation. The low heat capacity of the land results in large temperature changes as heat is gained or lost between day and night or summer and winter. The high heat capacity of the oceans allows them to absorb and release large amounts of heat with little change in temperature. Review the concept of heat capacity and refer to table 5.1. The average temperature of the sea surface is shown for the Northern Hemisphere summer in figure 6.4. At each lat-

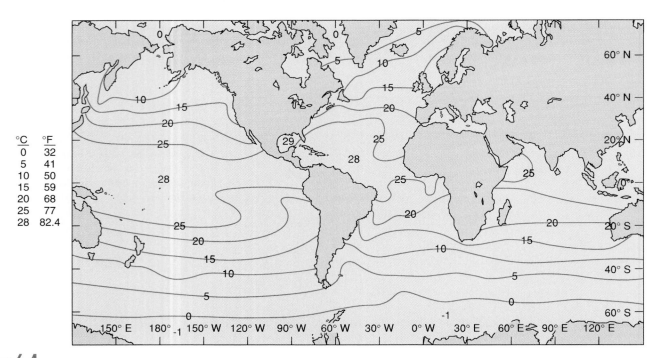

Figure 6.4

Typical average sea surface temperatures (in degrees Celsius) during the Northern Hemisphere summer months.

More Information: www.websites.noaa.gov/guide/sciences/ocean/seasur.html

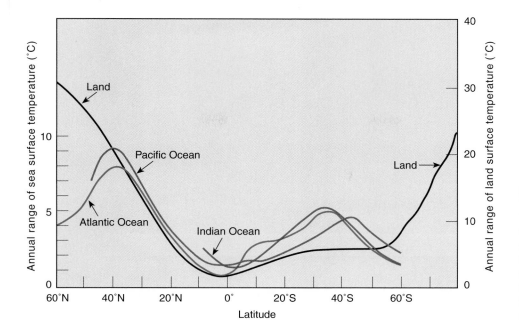

Figure 6.5

The annual range of mid-ocean sea surface temperatures is considerably less than the annual range of land surface temperatures at the highest latitudes. The maximum annual range of sea surface temperatures occurs at the middle latitudes. Pay particular attention to the different temperature scales used for the sea surface and land surface.

itude, the annual temperature changes are controlled by changes in available solar radiation and heat loss combined with the heat capacity of the surface material. The average annual temperature change for both land and mid-ocean surface water (fig. 6.5) shows how heat capacity affects sea and land differently. The maximum seasonal change in sea surface temperature, about 8 to 9°C, occurs at temperate latitudes in the Northern Hemisphere. At higher and lower latitudes, sea surface temperatures change less over the annual cycle. At polar latitudes, seasonal energy gains and losses are associated with latent heat of fusion as sea ice melts and freezes, rather than changes in the temperature of the water. At low latitudes, changes in sea surface temperatures are small because solar energy is nearly constant over the annual cycle.

The continents undergo large annual changes in surface temperatures at high latitudes. Because of their low heat capacity the continents respond to the large annual changes of solar energy, rising in temperature during the summer and cooling in winter. In addition there is no freezing and melting of sea ice to absorb and release energy without a change in temperature. The differences between land and ocean annual surface temperature changes can be seen by looking at figure 6.6. Compare the summer (fig. 6.6a) and winter (fig. 6.6b) temperatures of landmasses and water areas at about 60°N. The contrast in seasonal temperature patterns is caused by the presence of large landmasses in the Northern Hemisphere. At 60°S, figures 6.6a and 6.6b show little seasonal difference due to the lack of land at this latitude in the Southern Hemisphere. These very different annual temperature cycles cause distinctly different wind and weather patterns in the two hemispheres. Seasonal wind changes in the two hemispheres are discussed in section 6.11.

Where evaporation at the sea surface is high, at tropic and subtropic latitudes, there is a high rate of removal of heat energy, and large quantities of water vapor enter the atmosphere. Each gram of water evaporated and condensed in the air transfers the latent heat of vaporization—540 calories—from the Earth surface to the atmosphere. The moisture-laden atmosphere moves to higher latitudes along the Earth's surface as it rises and cools. The clouds that result from condensing water vapor produce high-precipitation zones at about 50 to 60°N and S and at 5°N. Therefore there is a large net transfer of heat to the atmosphere at low latitudes; the heat is carried to higher latitudes by the moving air (return to figure 6.3).

The oceans play a significant role in stabilizing the surface temperature of the Earth. Their ability to store and release large quantities of heat without large changes in temperature moderates surface temperatures both seasonally and over the day and night cycles. Evaporation, conduction, and reradiation from the Earth's surface all add heat to the atmosphere at the lower latitudes, which is transferred by winds and currents to the higher cooler latitudes.

Compare the heat capacity of land and water.

Explain the significance of heat capacity to the Earth's surface temperatures.

Why do sea surface temperatures undergo only small annual changes at both polar and equatorial latitudes?

What is the result of evaporation at subtropic latitudes?

Sea Ice and Icebergs 6.4

As seawater begins to freeze in the polar winters, a layer of slush forms, covering the ocean with a thin sheet of ice. Sheets of new **sea ice** may be broken into "pancakes" by waves and

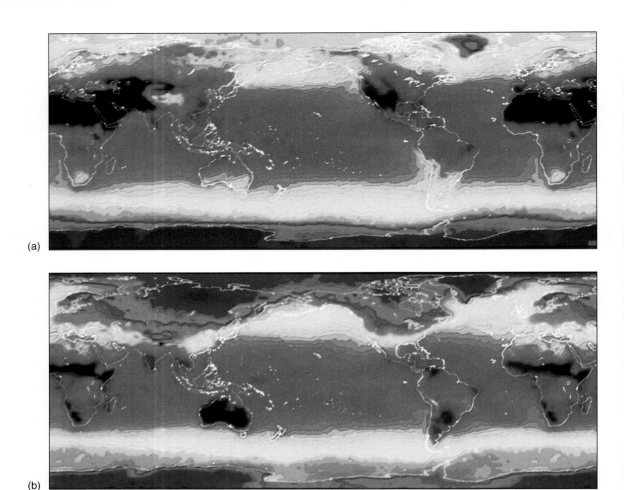

(a)

(b)

Figure 6.6

Meteorological satellites, such as NOAA's *TIROS*, carry high- resolution infrared sensors that measure long-wave radiation emitted from the Earth's surface and atmosphere. This radiation is related to the Earth's surface temperatures. Green and blue indicate temperatures below 0°C; warmer temperatures are shown in red and brown. (a) In July, the Northern Hemisphere landmasses have considerably warmer temperatures, but the ocean waters do not change dramatically from winter to summer. The oceans' warm surface water moves north and south with the change in the seasons. North-south currents along the coasts of continents are also visible. (b) In January, Siberia and Canada show surface temperatures near –30°C; at the same time, latitudes between 30° and 50° south show warm summer temperatures.

wind; then as the freezing continues the pancakes move about, unite, and form floes. Ice floes move with the currents and the wind, collide with each other, and form ridges and hummocks. Some floes shift constantly, breaking apart and freezing together; others remain anchored to a landmass. Continuous, or nearly continuous sea ice is also called **pack ice.** As sea ice forms, the heat of fusion is transferred to the cold atmosphere, and the sea water temperature remains at the freezing point.

Ice about 2 m (6 ft) thick may be formed in one season. The thickness of the ice is limited, because the latent heat of fusion from the underlying water must be extracted through the ice by conduction, a slow process even at polar temperatures. Snow that falls on the ice surface acts as an insulator, further retarding extraction of heat from the water below.

As sea ice is formed, some seawater is trapped in the voids between the ice crystals. If the ice forms slowly, most of

the trapped seawater drains out and escapes; if the ice forms quickly, more salt water is trapped. As time passes, the salt water slowly escapes through the ice, and eventually the ice becomes fresh enough to drink when melted. The formation of the fresh sea ice concentrates salt in the underlying seawater, increasing its salinity and density and causing it to sink.

Sea ice is seasonal around the bays and shores of parts of the northeast United States, Canada, Russia, Scandinavia, and Alaska. The ice exists year-round in the central Arctic and around Antarctica. Sea ice covers the entire Arctic Ocean in the northern winter and pushes far out to sea around Antarctica during the Southern Hemisphere's winter, and melts back along the edges in both oceans during summer (fig. 6.7). In areas where the ice does not melt entirely during the year, new ice is added each winter, and a thickness of 3.5 to 5 m (10 to 15 ft) can accumulate.

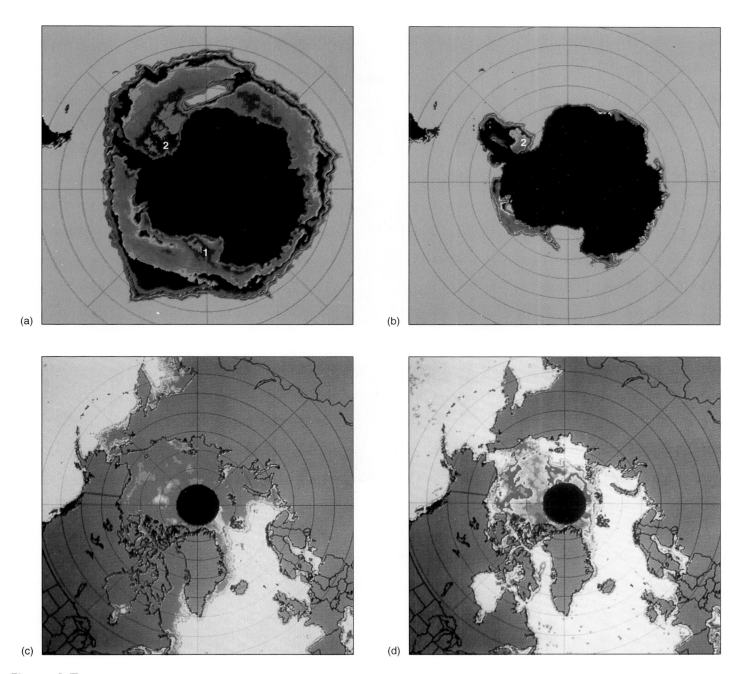

(a)

(b)

(c)

(d)

Figure 6.7

Sea ice around the Arctic and Antarctica. Data from NASA's *NIMBUS-5* satellite was used to construct seasonal maps of sea ice coverage at high latitudes. The Antarctic ice extends more than 100 kilometers from the land into the Ross (1) and Weddel (2) seas during the winter (a) and is much reduced during the summer (b). The Arctic Ocean is filled with ice in the winter (c) but by the next fall the ice cover is much less (d). Short-term motions of the ice boundaries can be calculated from repeated satellite images. These show that the ice floes move up to 50 km (30 mi) per day. Purple indicates high ice concentration; and blue indicates open water in (a) and (b); yellow indicates open water in (c) and (d).

More Information: polar.wwb.noaa.gov/seaice/Analyses.html

Icebergs are massive, irregular in shape, and float with about 12% of their mass above the sea surface. They are formed by glaciers, large masses of fresh-water ice, that begin inland in the snows of central Greenland, Antarctica, and Alaska, and inch their way toward the sea. The forward movement, the melting at the base of the glacier where it meets the ocean, and wave and wind action cause blocks of ice to break off and float out to sea.

The castle bergs produced in the Arctic (fig. 6.8a) drift south with the currents as far as New England and the busy shipping lanes of the North Atlantic. It was one of these icebergs, probably from a glacier in Greenland, that sank the

Figure 6.8

(a) An irregular castle berg drifting in the Atlantic Ocean north of the Grand Banks.
(b) A flat, tabular berg in the South Atlantic Ocean. Its approximate length is 500 m (1600 ft).

More Information: www.uscg.mil/lantarea/iip/home.html

(a)

(b)

Titanic with a loss of 1517 lives on its maiden voyage in 1912. Alaskan icebergs are usually released in narrow channels and bays; they do not readily escape into the open ocean. Flat, tabular icebergs (fig. 6.8b) produced from the broad continental ice sheets of Antarctica tend to stay close to the polar continent, caught by the circling currents, although they have been known to reach latitudes of 40°S. In late 1987, a large tabular berg, 155 km (96 mi) long and 230 m (755 ft) thick, with a surface area about the size of Delaware, broke from Antarctica and drifted 2000 km (1250 mi) along the Antarctic coast. Two years later, it grounded and broke into three pieces. The volume of ice in this berg was estimated to have been enough to provide everyone on Earth with two glasses of water daily for about 2000 years, or enough water for the city of Los Angeles for 100 years.

How is sea ice formed?

What limits the thickness of seasonal sea ice?

How are icebergs formed, and where are they most commonly found?

The Atmosphere 6.5

The atmosphere extends 90 km (54 mi) above the Earth and is made up of a series of layers; see figure 6.9. In the first layer, or **troposphere,** where the air is warmed by heat reradiated and conducted from the Earth's surface and by evaporation of water vapor and its condensation, the temperature decreases with altitude. Precipitation, evaporation, wind systems, and clouds are all found in the troposphere. In the next layer, or **stratosphere, ozone,** an unstable form of oxygen, absorbs ultraviolet radiation, raising the temperature with increasing elevation. Above the stratosphere are the mesosphere and the thermosphere.

The atmosphere is composed of a mixture of gases, suspended microscopic particles, and water droplets. This mixture is commonly simply called air. Atmospheric gases are typically categorized as being permanent or variable. The permanent gases are present in a constant relative percentage of the atmosphere's total volume while the concentration of the variable gases changes in both time and location (table 6.1).

The density of air is controlled by three variables; temperature, the amount of water vapor in the air, and atmospheric pressure. Air becomes less dense when it is warmed and more

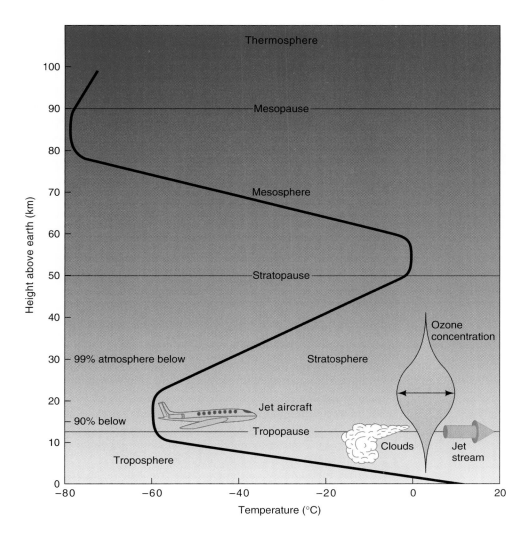

Figure 6.9

The structure of the atmosphere.

dense when it is cooled. The density of air will decrease if it's humidity increases, or the concentration of water vapor increases. When water vapor is added to the atmosphere, the relatively low molecular weight water molecules replace higher molecular weight permanent gases (compare the molecular weight of water with the molecular weights of the permanent gases listed in table 6.1). In general, cold, dry air will be more dense than warm, moist air. Finally, the density of air decreases with decreasing atmospheric pressure, or increasing elevation.

Atmospheric pressure is the force with which a column of overlying air presses on an area of the Earth's surface. The average atmospheric pressure at sea level is 1013.25 millibars (1 bar = 1×10^6 dynes/cm^2), or 14.7 lbs/in^2. Regions of air with a density greater than average are known as **high-pressure zones,** and regions with a density less than average are **low-pressure zones.**

What is air?

What processes regulate the density of air?

How are density and pressure zones related?

table 6.1   Composition of the Atmosphere

Permanent Gases

Gas	Formula	Percent by Volume	Molecular Weight
Nitrogen	N_2	78.08	28.01
Oxygen	O_2	20.95	32.00
Argon	Ar	0.93	39.95
Neon	Ne	1.8×10^{-3}	20.18
Helium	He	5.0×10^{-4}	4.00
Hydrogen	H_2	5.0×10^{-5}	2.02
Xenon	Xe	9.0×10^{-6}	131.3

Variable Gases

Gas	Formula	Percent by Volume	Molecular Weight
Water Vapor	H_2O	0 to 4	18.02
Carbon Dioxide	CO_2	3.5×10^{-2}	44.01
Methane	CH_4	1.7×10^{-4}	16.04
Nitrous Oxide	N_2O	3.0×10^{-5}	44.01
Ozone	O_3	4.0×10^{-6}	48.00

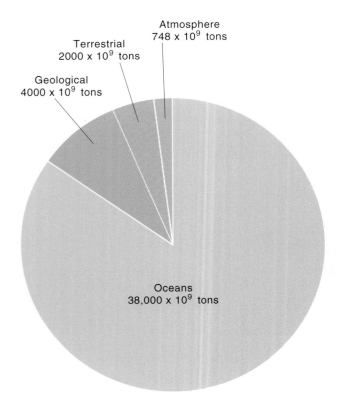

Figure 6.10

World carbon dioxide distribution. Values are in billions of tons.

More Information: www.pmel.noaa.gov/co2/co2-home.html

The atmosphere has changed continuously over the last 4.6 billion years, but today it is changing measurably and rapidly as a result of human activities. The rising level of carbon dioxide (CO_2) is of increasing concern. There are three active reservoirs for carbon dioxide (CO_2): the atmosphere, the oceans, and the terrestrial system; in addition there is the inactive reservoir of the Earth's crust. The oceans store the largest amount of CO_2, and the atmosphere has the smallest amount; see figure 6.10. The atmosphere is the link between the other reservoirs, and the oceans play a major part in determining the atmosphere's CO_2 concentration by physical (mixing and circulation), chemical, and biological means (fig. 6.11).

Carbon dioxide is one of several gases that are transparent to incoming short-wave radiation but absorb outgoing, long-wave radiation from the Earth's surface. An increase in the atmosphere's carbon dioxide level also increases the absorption of incoming infrared radiation by the atmosphere and decreases the infrared reaching the Earth's surface. This increase in the atmosphere's CO_2 content allows less transfer of long-wave energy from the Earth's surface to space and increases the transfer of this energy from the Earth's surface to the atmosphere. The net result of increased absorption of infrared energy in the atmosphere due to increasing CO_2 is the warming of the atmosphere. The retention of the outgoing radiation keeps the Earth warm by what is commonly known as the **greenhouse effect.** Human activities have recently added an excess of CO_2 from fossil fuels and deforestation to the atmosphere, an excess currently estimated at approximately *7 billion tons per year.*

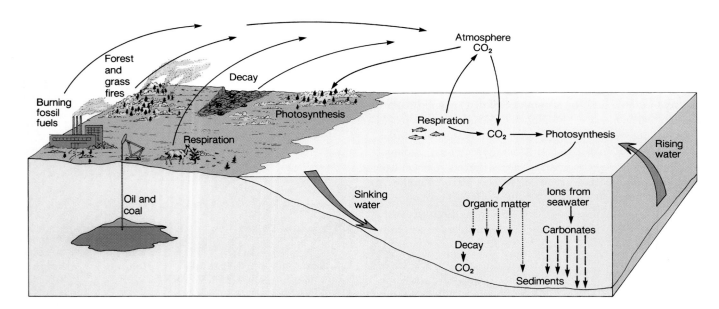

Figure 6.11

Major carbon dioxide pathways through the Earth's environment. Weathering and erosion of land deposits and volcanism also liberate CO_2.

In 1997 representatives of over 150 nations met in Kyoto, Japan and agreed on an international climate treaty called the Kyoto Protocol. The treaty calls for thirty-eight industrial nations to cut their emissions of greenhouse gases by an average of 5.2% from 1990 levels by the year 2012. The United States, the world's leader in greenhouse gas emissions at 25% of the total, agreed to a 7% reduction, the nations of the European Union committed to 8%, and Japan agreed to a cut of 6%. If achieved, these levels will reduce projected 2012 emissions by two-thirds, but they will not prevent total greenhouse emissions from rising because developing nations that are not bound by the treaty, such as China, India, and Russia, are expected to account for about 58% of the world's total greenhouse emissions by the year 2012. Controversy over the details of the treaty has prevented its implementation and the United States has not yet signed it. The United States supports a plan that would allow developed nations to count carbon dioxide absorbed by forests, so-called carbon sinks, against emissions reduction targets. The European Union is concerned that countries may claim that all their greenhouse gases are being absorbed by sinks and that they therefore do not need to make any actual reductions in the pollution they emit from vehicle exhausts and other sources. Representatives of the countries involved in the treaty failed to reach agreement on this issue once again at a meeting at The Hague, Netherlands in the Fall of 2000. There are tentative plans for the conference to resume sometime in 2001.

As a part of the effort to compare current CO_2 levels with those of the recent past, scientists have sampled the gases trapped in the polar ice and glaciers during the past 160,000 years. Bubbles in the ice contain samples of the air present at the time the ice was formed and allow comparisons of its composition before and after the Industrial Revolution. Since 1850, the concentration of carbon dioxide in the atmosphere has in-creased from 280 parts per million to 360 parts per million. Recently, the average rate of increase has been 1.5–2 parts per million per year.

In addition, for more than forty years, scientists have been recording CO_2's annual cycle and the steady increase in its concentration (fig. 6.12). In the Northern Hemisphere, CO_2 decreases in the spring and summer when plants increase photosynthesis and remove CO_2 faster than it is contributed by respiration and decay. In the autumn and winter, plants lose their leaves, photosynthetic activity is reduced, decay processes release CO_2, and atmospheric CO_2 increases. The effects of deforestation, conversion of forest land to agriculture (often accompanied by burning), the burning of fossil fuels in homes and industries, and the growth of human populations are responsible for the increasing values that have been superimposed on the natural cycle. If this trend were to continue unstopped, the concentration of carbon dioxide would double over historic levels some time before the end of this century.

Global warming of 2 to 4°C has been predicted by some in response to such an increase in CO_2 levels. A change in temperature of this magnitude in a few decades is comparable to that which has occurred since the last ice age, over 10,000 years ago. It is difficult to predict the extent of the effects brought about by such a change, because many other uncertain factors are involved. However, if warming does occur, it is expected to affect the high latitudes more than the low latitudes, causing melting of polar land ice. That, along with increasing seawater temperature and volume, might lead to a sea-level rise of about 1 meter. Another possibility is that the polar ice caps may expand rather than diminish because of increased evaporation caused by the warming, which causes increased precipitation at high latitudes.

In the stratosphere, ozone absorbs much of the ultraviolet radiation from the Sun, protecting life-forms on land and at

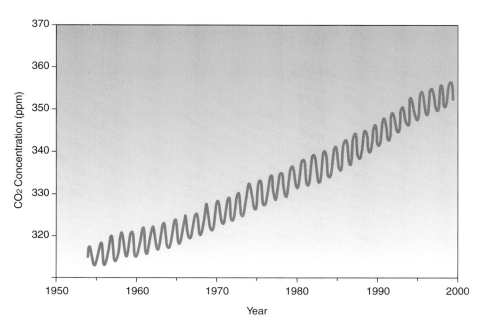

Figure 6.12

Concentration of atmospheric carbon dioxide in parts per million (ppm) of dry air versus time in years as observed with a continuously recording nondispersive infrared gas analyzer at Mauna Loa Observatory, Hawaii.

More Information:
cdiac.esd.ornl.gov/trends/co2/sio-mlo.htm

the sea surface. A decrease in the atmosphere's ozone allows more ultraviolet radiation to pass through the atmosphere and increases this radiation reaching the Earth's surface. Significant reduction of ozone in the stratosphere has been occurring over Antarctica since the late 1970s. The Arctic winters are not as cold as those of the Antarctic, and ozone loss over the Arctic is less. It is estimated that the average global loss of ozone since 1978 is about 3 percent; at polar latitudes the loss is estimated at about 8 percent per decade. Studies done in 1990 in the Southern Ocean around Antarctica indicated that springtime increases in ultraviolet radiation are reducing the annual production of single-celled plantlike organisms at the sea surface. The average reduction is estimated to be 2 to 4 percent of normal. These single-celled organisms are at the base of the food chains that support the Antarctic's marine life.

The most widely accepted theory of ozone destruction is related to the release of chlorine into the atmosphere. Chlorine is a product of vulcanism, but it is commonly released in in-

creased quantities as a component of chlorofluorocarbons (CFCs) used for refrigeration, air-conditioning, in solvents, and for production of insulating foams.

The microscopic plantlike organisms at the sea surface produce dimethyl sulfide (DMS) gas. Current estimates are that the oceans emit about 60 million tons of DMS, or 30 million tons of sulfur, to the atmosphere each year. During their time in the atmosphere, sulfur compounds influence the amount of incoming radiation by reflecting radiation back into space before it can reach the Earth's surface. These sulfur compounds are also thought to control the density of marine clouds and their reflective properties. Denser, more reflective clouds reduce incoming radiation and decrease the heating of the ocean surface and the growth of the plantlike organisms that produce the DMS. In this way DMS may act as a feedback mechanism to help control ocean surface temperatures, another link in the complexity of interactions between the atmosphere and the oceans (fig. 6.13). Industrially produced sulfates

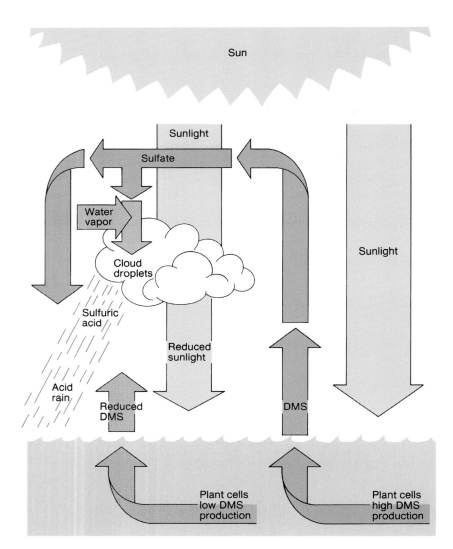

Figure 6.13

A proposed self-regulating thermal control system links dimethyl sulfide (DMS) to the sea surface and the atmosphere, a cycle among plantlike cells, the Sun, and clouds. Acid rain, pH ~ 3.5, has little impact on seawater because of seawater's carbon dioxide buffer.

also contribute to acid rain; they form clouds that block solar radiation and cool the Earth's surface.

In 1991 Mount Pinatubo erupted in the Philippines. The sulfur-based gases and the particulate materials from this eruption created a globe-circling haze that decreased incoming radiation and caused measurable global cooling in 1992.

The complexity of these interconnected systems makes it very difficult to accurately predict the rate and extent of anticipated changes. However, despite the complexity, there is a consensus that the increasing CO_2 content of the atmosphere is producing a warming trend.

What is the greenhouse effect?

How is carbon dioxide related to the greenhouse effect?

Explain the Northern Hemisphere's annual carbon dioxide cycle.

How have carbon dioxide levels in the atmosphere changed since the Industrial Revolution, and how do we know?

What is the significance of the loss of ozone from the stratosphere?

Why is it difficult to predict the rate of global warming and its effects? Give examples.

Explain the feedback mechanism of DMS.

Winds on a Non-rotating Earth   6.7

The air above the Earth moves because less-dense air rises away from the Earth and more-dense air sinks toward the Earth. A horizontal airflow or wind above the Earth's surface moves toward the area of sinking air and another horizontal airflow or wind at the Earth's surface moves toward the area of rising air. These air motions form convection cells driven by changes in the air's density (fig. 6.14). The result is two wind levels, one at the Earth's surface and the other aloft. The average density of the air column combined with Earth's gravity produces air pressure; in regions of low density the air pressure is less than in regions of high density. The air is forced along the suface of the Earth from high- to low-pressure regions, and the greater the air pressure change per distance along the Earth's surface the greater the speed of the air flow.

Imagine an Earth heated and cooled as described in sections 6.2 and 6.3, but with no continents and no rotation. Due to the unequal distribution of the Sun's heat over the Earth's surface, large amounts of heat and water vapor are transferred to the atmosphere around the equator. The warm air rises and cools, so that the water vapor condenses, producing precipitation. Equatorial regions are known for their warm wet climate.

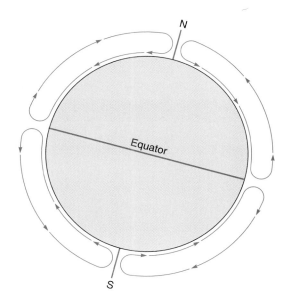

Figure 6.14

A convection cell is formed in the atmosphere when air is warmed at one location and cooled at another. In this figure, heating at the equator and cooling at the poles produces a single, large convection cell in each hemisphere on a nonrotating water-covered Earth model. Actual atmospheric circulation is more complicated (see fig. 6.16).

The dry air aloft flows toward the poles where its density causes it to sink, producing a cold, dry climate. The air then moves over the water back toward the equator, increasing its water vapor content and gaining heat. Review figure 6.14.

In this system, the winds at the Earth's surface in the Northern Hemisphere blow from north to south, and the upper winds blow from south to north; in the Southern Hemisphere the reverse is true. It is important to remember that winds are named for the direction *from which they blow*. A north wind blows from north to south, while a south wind blows from south to north.

How are winds formed?

Explain how the air would circulate in the Northern and Southern Hemispheres if there was no land and the Earth did not rotate.

How are the winds named?

The Effect of Rotation   6.8

Adding rotation to the system just described changes the atmospheric circulation. Each of us is moving with the Earth in its daily eastward motion. The speed of this motion varies with latitude because the size of circles of latitude decreases with

table 6.2	Eastward Speed of Earth's Surface with Latitude	
Latitude	Speed (km/hr)	Speed (mi/hr)
90°N&S	0	0
75°N&S	433	269
60°N&S	837	520
45°N&S	1184	735
30°N&S	1450	900
15°N&S	1617	1004
0°	1674	1040

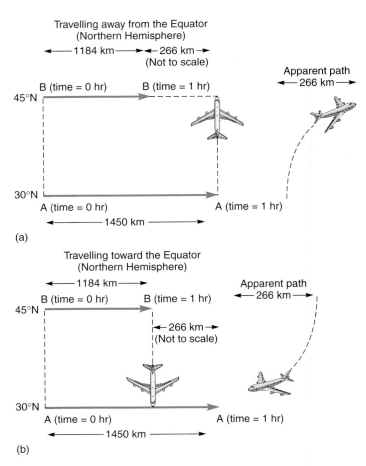

(a)

(b)

Figure 6.15

Moving parcels of air in the atmosphere, or water in the oceans, appear to be deflected from straight-line movement as a result of the Coriolis effect. The Coriolis effect is a consequence of the variation in eastward speed of the surface of the rotating Earth with latitude. The apparent deflection is to the right in the Northern Hemsiphere. This can be seen by imagining the path of a plane traveling north or south. (a) A plane traveling from point A at 30°N to point B at 45°N will arrive at 45°N east of point B because the plane's eastward velocity is greater than the eastward speed of the surface of the rotating Earth at 45°N. Hence, its path will appear to veer to the right. (b) A plane traveling in the opposite direction from point B at 45°N to point A at 30°N will arrive at 30°N west of point A because the plane's eastward velocity is less than the eastward speed of the surface of the rotating Earth at 30°N. Once again its path will appear to veer to the right. If you perform this same thought experiment with the plane in the Southern Hemisphere you will see that the Coriolis effect causes an apparent deflection to the left.

More Information: www.windpower.dk/tour/wres/coriolis.htm

distance from the equator (table 6.2). Standing on the Earth's surface we are unaware of this motion, but moving air masses are affected by this change in rotational speed since they are not attached to the surface. One way of visualizing the way they are affected is to consider a plane traveling along a north-south line. Imagine a plane traveling north from point A at latitude 30°N to point B at latitude 45°N (fig. 6.15a). The plane takes one hour to fly from 30°N to 45°N. When the plane begins the flight it is moving eastward with the Earth's rotation at 1450 km/hr and it will continue to have this eastward speed once airborne. As it flies true north, the surface of the Earth is moving to the east at a progressively slower rate beneath it. After one hour of flight the plane arrives at 45°N. During the hour-long flight it traveled 1450 km east while point B only traveled 1184 km east. The plane arrives at 45°N 266 km east of point B. As viewed from the ground the plane's flight path appears to have veered to the right (fig. 6.15a). Now imagine a plane flying from point B at latitude 45°N to point A at latitude 30°N (fig. 6.15b). The plane takes one hour to fly from B to A. When the plane begins the flight it is moving eastward with the Earth's rotation at 1184 km/hr and it will continue to have this eastward speed once airborne. As it flies south, the surface of the Earth is moving to the east at a progressively faster rate beneath it. After one hour of flight the plane arrives at 30°N. During the hour-long flight it traveled 1184 km east while pont A traveled 1450 km east. The plane arrives at 30°N 266 km west of point A. As viewed from the ground the plane's flight path appears to have veered to the right again. Moving parcels of air, or the wind, will be affected in the same way. This deflection is caused by an apparent force, not a true force since no true force is exerted on the wind. The deflection of moving air relative to the Earth's surface is called the **Coriolis effect,** after Gaspard Gustave de Coriolis (1792–1843), who mathematically solved the problem of deflection in frictionless motion when the motion occurs relative to a rotating body. If the same experiment were performed in the Southern Hemisphere the deflection would be to the left of the intended direction of motion. At the equator no deflection occurs; the Coriolis effect is zero.

The Coriolis effect is equally important in determining the relative motion of ocean currents. *Both moving air and moving water are deflected relative to their direction of motion, to the right in the Northern Hemisphere and to the left in the Southern Hemisphere.*

Why is the Coriolis effect not a true force?

How does the Earth's rotation affect wind direction?

The Wind Bands   6.9

When the Earth rotates, air continues to rise at the equator and flows toward the north and south poles. As it moves, it is deflected to the right in the Northern Hemisphere and to the left in the Southern Hemisphere. This deflection short-circuits the single hemispheric convection cells of the earlier stationary model (fig. 6.14). The deflected air sinks at 30°N and 30°S; it then moves over the water, either back toward the equator or toward 60°N and 60°S. The air that cools and sinks over the poles then moves to lower latitudes, warming and picking up water vapor; at 60°N and 60°S it rises again. The result is three convection cells (fig. 6.16) in each hemisphere for the rotating Earth.

The surface winds between 30°N and 30°S are deflected relative to the Earth, blowing from the north and east in the Northern Hemisphere and from the south and east in the Southern Hemisphere. These are the **trade winds:** the northeast trade winds north of the equator, and the southeast trade winds south of the equator. Between 30°N and 60°N, the de-

flected surface flow produces winds that blow from the south and west, while between 30°S and 60°S, they blow from the north and west. In both hemispheres these winds are called the **westerlies,** or **prevailing westerlies.** Between 60°N and the North Pole, the winds blow from the north and east, while between 60°S and the South Pole, they blow from the south and east. In both cases they are called the **polar easterlies.** The six surface wind bands are shown in figure 6.16.

At 0° and at 60°N and S, moist low-density air rises; these are areas of low atmospheric pressure, zones of clouds and rain. Zones of high-density descending air at 30° and 90°N and S are areas of high atmospheric pressure, zones of low precipitation and clear skies. Air flows over the Earth's surface from atmospheric high pressure to atmospheric low pressure. In the zones of vertical motion, between the wind belts, the surface winds are unsteady. Such areas were troublesome to the early sailors, who depended on steady winds for propulsion. The area of rising air at the equator is known as the **doldrums,** or **intertropical convergence,** and the high-pressure areas at 30°N and S are known as the **horse latitudes.** In all these areas, sailing ships could find themselves becalmed for days. The

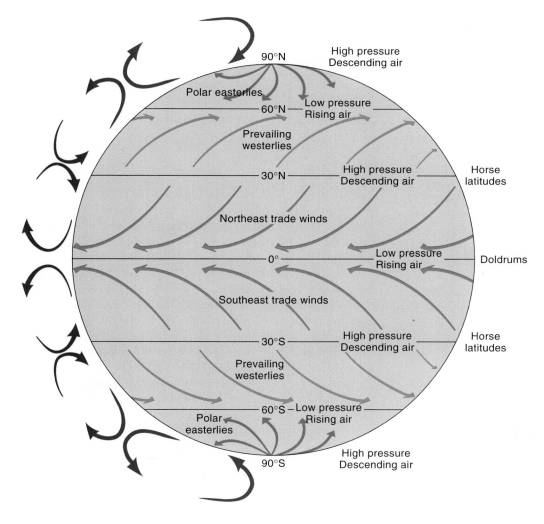

Figure 6.16

The circulation of the Earth's atmosphere results in a six-band surface wind system on a rotating water-covered Earth.

origin of the word "doldrums" is obscure, but the horse latitudes are said to have gotten their name from the stories of ships carrying horses that were thrown overboard when the ships were becalmed and the freshwater supply became insufficient for both the sailors and the animals.

> Locate the trade winds, the westerlies, and the polar easterlies with latitude.
>
> Describe the properties of the zones between the wind bands.
>
> What and where are the doldrums and the horse latitudes?

The Jet Streams   6.10

The weather is generated by changes in the atmosphere, but annual heating cycles, wind patterns, and seasonal changes in the Earth's surface temperature do not paint a complete picture of day-to-day weather. The winds of the upper troposphere have considerable effect on short-term weather patterns. The polar **jet stream** is the best known of the strong winds that lie above the boundary between the polar easterlies and the westerlies. This wind flows rapidly (in excess of 300 km (180 mi) per hour) eastward and can oscillate more than 2000 km (1250 mi) north to south during the winter as pressure systems change (fig. 6.17). The oscillations in the jet stream flow are caused by strong high-pressure systems to the south and intense low-pressure systems to the north. The westerlies move the wave form of the stream and its associated high- and low-pressure systems eastward around the Earth. In winter, jet stream velocities are higher and displacements are greater. The slow eastward migration of this system constantly alters weather patterns as they move over the Earth's surface.

Above the zone between the trades and westerlies, there are also upper-atmosphere winds, the subtropical jets. In the equatorial regions, the zone where the northeast and southeast trade winds come together is the intertropical convergence zone (or doldrums). This zone oscillates north and south with the seasonal cycle of solar heating. Thunderstorms and great cumulus cloud systems formed in the rising air of the intertropical convergence generally migrate westward.

> Why is the flight time of high-altitude jet aircraft usually shorter when flying eastward rather than westward across the United States?
>
> How is the jet stream related to the Earth's surface winds?

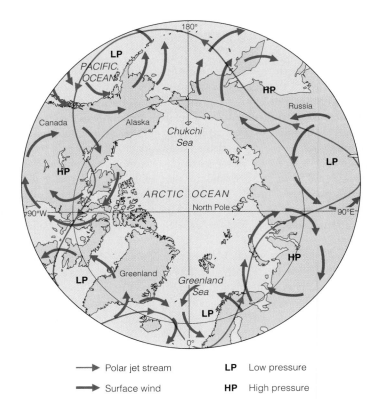

| → Polar jet stream | **LP** Low pressure |
| → Surface wind | **HP** High pressure |

Figure 6.17

The polar jet stream circles the Earth in the Northern Hemisphere above the boundary between the polar easterlies and the westerly winds. It is deflected north and south by the alternating air-pressure cells of the northern temperate zone.

More Information: www.usatoday.org/weather/tg/wjstream/wjstream.htm

Seasonal Wind Changes   6.11

Unlike our model Earth that was one worldwide ocean, on the real Earth large blocks of continents interrupt the oceans, and the land surfaces change temperature differently than the oceans. Throughout the year ocean and land surfaces remain warm at the equatorial latitudes and the oceans remain cold at polar latitudes, but the land goes through large seasonal temperature changes at higher latitudes. In the middle latitudes, the land is warmer than the ocean during the warm summer months (refer to fig. 6.6). The density of air over the warm land is reduced, creating a low-pressure area, while the air density over the cold sea increases producing a high-pressure area (fig. 6.18). Low-pressure zones at 60°N and 0° tend to combine over the land, breaking the high-pressure belt at 30°N into several discrete high-pressure cells located over the oceans during the Northern Hemisphere's summer. In the winter the land becomes much colder than the water, and the air pressure is low over the water and high over the land, creating a low-pressure zone over the ocean. Over the land, the polar high-pressure

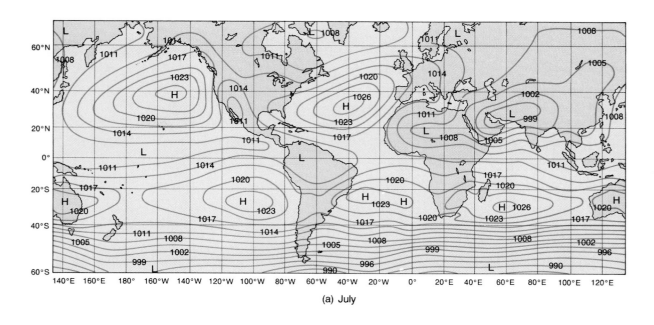

(a) July

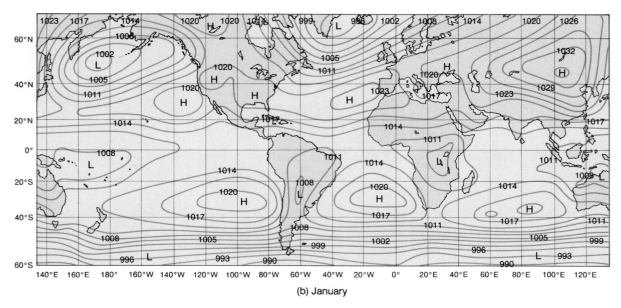

(b) January

Figure 6.18

Average sea-level atmospheric pressures expressed in millibars for (a) July and (b) January. Large landmasses at mid-latitudes in the Northern Hemisphere cause high and low atmospheric pressure cells to change their positions with the seasons. In the Southern Hemisphere, large landmasses do not exist at mid-latitudes, and average air pressure distribution changes little with the seasons.

zone spreads toward the high-pressure zone at 30°N, breaking the low-pressure belt centered on 60°N into separate low-pressure cells over the warmer ocean water.

During the Northern Hemisphere's summer, the cool air in the high-pressure cells over the central part of the North Atlantic and North Pacific Oceans descends and flows outward toward the low-pressure areas. As the descending air moves outward from the center of a high-pressure system, it is deflected to the right, producing winds that spiral in a clockwise direction around the high (fig. 6.19). On the northern side of these high-pressure cells are the westerlies, and on the southern side are the northeasterlies; the eastern side of the cell has northerly winds, and the western side has southerly ones. In the winter, the air circulates counterclockwise about the low-pressure cell over the northern oceans, and the prevailing wind directions are those of the polar easterlies and the westerlies. Figure 6.20 shows the first global measurements of ocean wind data obtained by satellite.

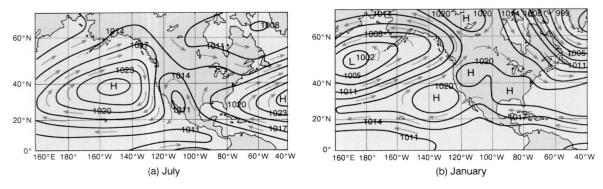

(a) July   (b) January

Figure 6.19

Atmospheric pressure cells control the direction of the prevailing winds. Atmospheric pressures are expressed in millibars. In the Northern Hemisphere, air flows clockwise around a region of high pressure, and counterclockwise around a low-pressure area. (a) Average wind conditions for summer. (b) Average wind conditions for winter.

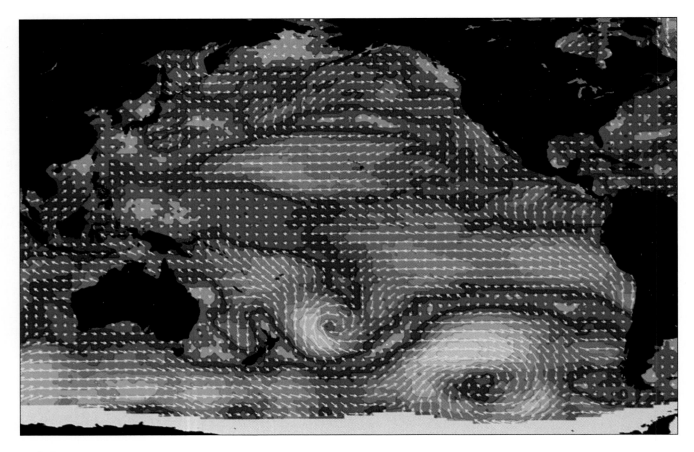

Figure 6.20

Pacific Ocean wind distributions. The SEASAT satellite obtained the first global measurements of ocean wind data during three days in 1978. Arrows point in the direction the winds blow; longer arrows indicate greater wind speed. Light winds (less than 4 m/s or 9 mph) are colored blue, and strong winds (greater than 14 m/s or 31 mph) are colored yellow. Wind speeds and directions were calculated from radar reflections from the small wind-driven waves that roughened the sea surface. The approximate accuracy is ± 2m/s and ± 20°. Winds blowing from the west in the Southern Hemisphere blow almost continuously around the entire Southern Ocean except near the tip of South America. Intense storms are shown in the South Pacific. The strong and constant southeast trade winds are shown to their north. Summer in the Northern Hemisphere is characterized by a large high-pressure cell in the North Pacific. Circulation about this cell produces the winds along the west coast of the U.S. that induce upwelling. Winds and squalls are formed at the boundary between the northeast and southeast trade winds, the doldrums.

The strength of the wind is governed by the change in pressure with distance, or the pressure gradient. Along the Pacific coast of the United States, the northerly winds cool the coastal areas in the summer, and the southerly winds warm them in the winter. The New England coast receives warm, moist air from the low latitudes in the summer and cold air from the high latitudes in the winter. Although these seasonal changes modify the wind and pressure belts of the water-covered model of the Earth, the wind and pressure belts of the model are still identifiable in the northern latitudes when the atmospheric pressures are averaged over long time periods.

Rotation of the airflow about high- and low-pressure air cells is reversed in the Southern Hemisphere due to the Coriolis effect. At the middle latitudes in the Southern Hemisphere there is little land to separate the water; therefore, the water temperature predominates and there is little seasonal effect (refer to figures 6.5 and 6.6). The atmospheric pressure and wind pattern created by surface temperatures in the Southern Hemisphere changes little over the annual cycle and are very similar to that developed for the water-covered model (see fig. 6.16).

> Why do high- and low-pressure systems alternate through the temperature latitudes of the Northern Hemisphere?
>
> Why do the winds circulate clockwise around a high-pressure zone in the Northern Hemisphere?
>
> What happens to winds around a low-pressure zone in the same hemisphere?
>
> How does the lack of land in the Southern Hemisphere affect the seasonal wind pattern?

Land Effects on Winds 6.12

The difference in temperature between land and water produces large-scale and small-scale effects in coastal areas. A particularly good example of this is found in the Indian Ocean. In India and Asia during the summer, the air rises over the hot land, creating a low-pressure system. The rising air is replaced by warm, moist air carried on the southwest winds from the Indian Ocean. As this onshore airflow rises over the elevated land, the moisture condenses, producing a steady, heavy rainfall. This is the wet or summer **monsoon** (from the Arabic word meaning "season") (fig. 6.21a). In the winter, a high-pressure cell forms over the land, and northeast winds carry the dry, cool air southward from the Asian mainland and out over the Indian Ocean. This movement produces cool, dry weather over India known as the dry or winter monsoon (fig. 6.21b).

The same effect is seen on a smaller scale along a coastal area or along the shore of a large lake. During the day, the land is warmed faster than the water and the air rises over the land. The air over the water moves in to replace it, creating an **onshore** breeze. At night, the land cools rapidly, and the water becomes warmer than the land. Air rises over the water, and the air from the land replaces it, this time creating an **offshore** breeze (fig. 6.22). The onshore breeze reaches its peak in the afternoon, when the temperature difference between land and water is at its maximum. The offshore breeze is strongest in the late night and early morning hours. Fishing boats relying only on their sails used this daily wind cycle, leaving harbor early in the morning and returning in the late afternoon or early evening.

As the winds sweep across the oceans, they are forced to rise to continue moving across the land, as shown in figure 6.23. The upward deflection cools the air, causing rain on the windward side of islands and mountains; on the leeward (or sheltered)

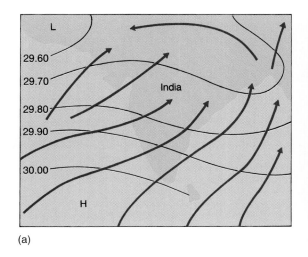
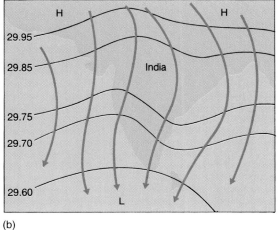

(a)   (b)

Figure 6.21

The seasonal reversal in wind patterns associated with (a) the summer (wet) monsoon and (b) the winter (dry) monsoon. The isobars of pressure are given in inches of mercury. Land topography increases friction between winds and land, thereby decreasing the Coriolis effect. The winds blow more directly from high- to low-pressure areas across the isobars.

More Information: nasadaacs.eos.nasa.gov/yearbooks/96/reckoning_winds.html and www.usda.gov/oce/waob/jawf/profiles/specials/monsoon/monsoon.htm

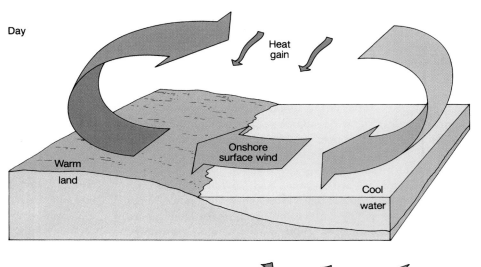

Day

Heat gain

Onshore surface wind

Warm land

Cool water

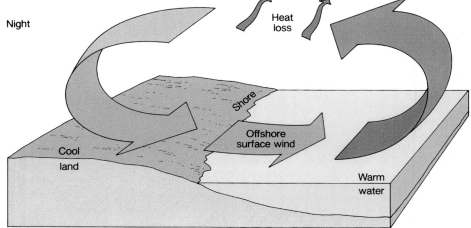

Night

Heat loss

Shore

Offshore surface wind

Cool land

Warm water

Figure 6.22

Differences between day and night land-sea temperatures produce an onshore breeze during the day and an offshore breeze at night.

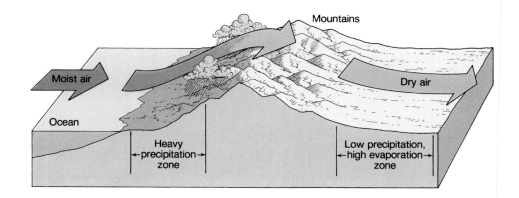

Mountains

Moist air

Dry air

Ocean

Heavy ←precipitation→ zone

Low precipitation, ←high evaporation→ zone

Figure 6.23

Moist air rising over the land loses its moisture on the mountains' windward side. The descending air on the leeward side is dry.

More Information:
www.weatherpages.com/rainshadow/

side, there is a low- precipitation zone, sometimes called a **rain shadow.** For example, the windward sides of the Hawaiian mountains have high precipitation and lush vegetation, while the leeward sides are much drier. On the west coast of Washington state, the westerlies rise to clear the Olympic Mountains and produce the Olympic rain forest, with a rainfall of as much as 5 m (200 in) per year; 100 km (60 mi) away, on the other side of the mountains, the rainfall is 40–50 cm (16–20 in) per year.

What produces the monsoon?

Why do local coastal winds frequently change direction from day to night?

Explain the effect of land elevation on precipitation.

Fog 6.13

When clouds form close to the ground, they are called **fog.** There are three basic types of fog, each formed differently. (1) The most common fog at sea is **advective fog,** formed when warm air saturated with water vapor moves over colder water. In northern California, Oregon, and Washington, when offshore warm, moist air moves over the cold coastal water, a coastal fog results, cooling the city of San Francisco (fig. 6.24a) and providing moisture to the redwood forests and grassy coastal cliffs. (2) Streamers of **sea smoke** rise from the sea surface on a cold winter day (fig. 6.24b). Dry, cold air from the land or from the polar ice pack moves out over the warmer water, which warms the cold air from below. This warmed air picks up water vapor from the sea surface, rises and cools rapidly, forming ribbons of fog. (3) **Radiative fog** is the result of warm days and cold nights. If the air holds enough moisture during the day, it condenses as the Earth cools at night, forming the low-lying thick, white fogs that are often found in river valleys and occasionally in bays and inlets along the coast. Such fogs usually disappear during the morning as the Sun gradually warms the air, changing the water droplets back to water vapor.

> Compare the direction of heat transfer for advective fog and sea smoke.

El Niño 6.14

El Niño begins with the breaking down of the normal systems that support the trade winds, the high-pressure system over the area of Easter Island, and the low-pressure area over Indonesia. The trade winds first strengthen, causing an accumulation of warm surface water on the west side of the Pacific near Indonesia; then they weaken and the warm surface water surges eastward across the Pacific carrying its overlying low-pressure storm systems with it. The result is warm surface water and associated wet weather along the west coast of South America in a region that normally has cold surface water and a dry climate (fig. 6.25). The increase in coastal surface temperatures usually ends in a few months but about every four to seven years large quantities of warm water spread north and south along the coast of the Americas, and surface temperatures remain elevated for more than a year.

El Niño is named for the Christ Child due to its frequent coincidence with the Christmas season in the waters off Peru and Ecuador. The intrusion of warm surface water along this coast causes the death of the usual populations of cold-water organisms and upsets the food chains on which fish, marine mammals, and seabirds depend. Intense rains often cause flooding and landslides along normally dry coast lands.

The climatic effects of El Niño are highly variable and appear to depend on the size of the warm pool of water and its temperature. The same region may experience higher-than-normal rainfall and flooding during one event and drought conditions

(a)

Figure 6.24

(a) Advective fog obscures the Golden Gate Bridge at the entrance to San Francisco Bay. (b) Sea smoke rising off ocean's surface as dry, cold air moves over warmer water.

More Information: http://www.usatoday.com/weather/wfog.htm

(b)

Figure 6.25

Ocean surface temperature data collected by satellite allows a global view of an El Niño event. Color is keyed to surface temperatures in °C.
(a) Annual mean sea surface temperatures. (b) During an El Niño event the pool of west Pacific warm surface water enlarges and expands across the Pacific and also into the Indian Ocean. Increased sea surface temperatures modify weather worldwide. (c) The El Niño ceases when the pool of warm surface water dissipates and heat is transported to higher latitudes.

Images processed by Dr. Xiao-Han Yan, University of Delaware.

More Information: www.ogp.noaa.gov/enso/ and www.elnino.noaa.gov/

during the next event. But, there do appear to be some very regular consequences of El Niño events. The northern United States and Canada generally experience warmer-than-normal winters. The eastern United States and normally dry regions of Peru and Ecuador typically have high rainfall while Indonesia, Australia, and the Philippines experience drought. In addition, El Niño years are associated with less intense hurricane seasons in the Atlantic Ocean.

During the severe El Niño of 1982–83, ocean surface temperatures off Peru rose 7°C above normal, and tropical species were displaced as far north as the Gulf of Alaska. In 1982–83 the polar jet stream was displaced far southward over the Pacific Ocean, bringing unusually dry conditions to Hawaii but high winds and high precipitation to areas along the west coast of the United States while the eastern United States had its mildest winter in 25 years. Heavy rains occurred in Ecuador, Peru, and Polynesia, but drought conditions were experienced in Central America, Africa, Indonesia, Australia, India, and China. There was an estimated $8 billion in damages worldwide ($2 billion in the United States) from this event. The severity of this El Niño and its worldwide effects led to the initiation in 1985 of a decade long study called the Tropical Ocean Global Atmosphere (TOGA) program. TOGA was designed to monitor ocean and atmospheric conditions in the equatorial South Pacific Ocean for the purpose of predicting future El Niño events. The ability to forecast these El Niño events and so regulate fisheries, predict agricultural droughts, and prepare for severe weather conditions has great economic and commercial importance around the world. Continued monitoring is now being done using 70 stationary buoys deployed in the international Tropical Atmosphere and Ocean (TAO) project. Data collected during the TOGA and TAO projects has led to the creation of improved ocean-atmosphere models that have been used with some success to predict the onset of El Niño conditions, but further improvement is needed to accurately predict the severity and length of future El Niños.

The strongest El Niño on record occurred in 1997–98. The greatest warming occurred in the eastern topical Pacific where surface water temperatures were as much as 8°C above normal near the Galapagos Islands and off the coast of Peru. Normally dry regions in Ecuador and Peru that usually receive only 10 to 13 cm (4–5 in) of rain annually had as much as 350 cm (138 in or 11.5 ft). Severe drought struck areas of the west Pacific Ocean in the Philippines, Indonesia, and Australia. A series of strong storms caused severe beach erosion, flooding, and landslides along the California coast. The jet stream was diverted far to the south over North America, inhibiting the growth of hurricanes in the Atlantic Ocean.

Between El Niño episodes surface temperatures off Peru may drop below normal to begin **La Niña** (the girl) or **El Viejo** (the old one). This event occurs when stronger than normal trade winds lower the surface temperatures in the eastern tropical Pacific while it is warmer than normal to the west. This helps establish dry conditions over the west coast areas of South America, while rainfall and flooding increase in India, Burma, and Thailand.

The exact cause of El Niño is still in doubt, but certain processes have been identified with its appearance. When atmospheric pressure increases over Easter Island in the eastern Pacific, decreases over Indonesia in the western Pacific, and then reverses, it is known as the **Southern Oscillation.** It causes west Pacific surface water to surge across the Pacific and raise surface temperatures along South America. Because the relationship of El Niño and the Southern Oscillation, the two events together are known as **ENSO.**

How does El Niño begin?

What is the result of El Niño along the west coast of South America?

How does La Niña differ from El Niño?

Why is the prediction of El Niño important?

Intense Ocean Storms   6.15

Intense atmospheric storms known as **hurricanes** are born over tropical oceans when surface water temperature exceeds 27°C. On either side of the equator very strong low-pressure cells pump large amounts of energy from the sea surface into the atmosphere as warm moist marine air rises rapidly and condenses to form clouds and precipitation. The large amounts of heat energy liberated from the condensing water vapor fuel the storm winds, raising them to destructive levels, up to 300 km per hour (160 knots). A major hurricane contains energy exceeding that of a large nuclear explosion; fortunately, the energy is released much more slowly. The energy generated by a hurricane is about 3×10^{12} watt-hours per day, the equivalent of one million tons of TNT.

Hurricanes can form on either side of the equator but not at the equator, because the Coriolis effect is necessary to create their spiraling winds. When a storm of this type is formed in the western Pacific Ocean it is called a **typhoon** or **cyclone** instead of a hurricane. Areas that give rise to hurricanes and typhoons and their storm tracks are shown in figure 6.26.

There are often periods of excessive high water along a coast associated with these intense ocean storms and the changes in atmospheric pressure and wind that accompany them. These episodes are known as **storm surges** or **storm tides.** The low atmospheric pressure in the hurricane's or typhoon's center allows the sea surface to rise up into a dome or hill, while the sea surface is depressed farther away from the center where the atmospheric pressure is greater.

The strong winds circulating about the center of the storm plus the speed of the storm over the ocean create a wind-driven surface current. When the storm comes ashore, the wind pushes the water toward the coast; the elevated sea level, the large waves, and the continued strong onshore winds retard the runoff

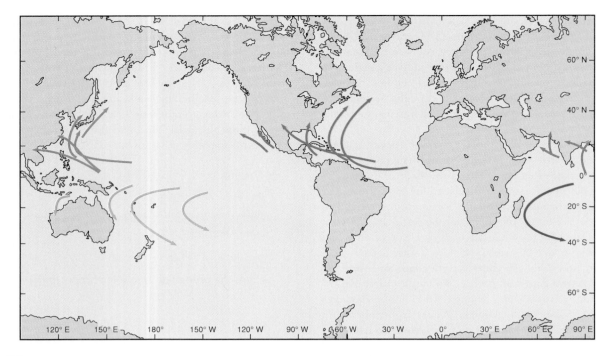

Figure 6.26

Hurricanes, typhoons, and cyclones form on either side of the equator in tropical seas. These storms follow preferred paths in different areas of the oceans.

More Information: www.nhc.noaa.gov/ and www.earthwatch.com/hurricane.html

of water from the land and increase the severity of the flooding. The effect of the high water is intensified when the coastal area is shallow and the adjacent land is low. If the storm tide arrives at the same time as the high tide, the disaster is magnified.

Along shallow areas of the East and Gulf coasts of the United States, storm surges have caused great damage. Hurricane Hugo came ashore with a storm tide of 5 m (16.5 ft) in South Carolina in 1989, destroying much property along the coast (fig. 6.27) and in the city of Charleston. In August 1992 Hurricane Andrew (fig. 6.28) struck Florida, but the coastal high-water damage was much less than that along the Carolina coast. Hurricane Andrew, traveling a short distance between the Bahamas and Florida, did not have enough time to produce a large storm tide and high waves or drive water onshore; in ad-

Figure 6.27

Storm tide damage caused by Hurricane Hugo, Pawley's Island, South Carolina, September 1989.

More Information: www.nhc.noaa.gov/

Figure 6.28

A satellite image of Hurricane Andrew approaching land south of Miami, August 24, 1992. Maximum sustained winds were 220 km per hour (138 mi/hr).

More Information: www.earthwatch.com/hurricane.html

dition Florida's extremely flat coast reduced the formation of high waves, and some of Andrew's wave energy was absorbed by the offshore reefs and coastal mangrove swamps. However, the damage caused by its severe onshore winds made Andrew the most destructive hurricane on record in the United States. An estimated 300,000 people lost their lives when storm surges struck the coast of Bangladesh in 1970; another 10,000 were killed in 1985, and 139,000 in 1991. The expanding populations of this area live and farm in the low-lying delta of the Ganges River, and there is no way to evacuate thousands of people when the storm tides cover hundreds of square miles. These storm tides or storm surges occur not only in the Indian Ocean and the Gulf of Mexico, but also in the North Sea and along any shallow coast under intense storm conditions.

What is the source of energy for intense tropical storms?

Where do hurricanes begin, and in what direction are they most likely to travel?

How is sea level affected by these storms?

Why are storm tides (storm surges) so destructive?

Summary

The intensity of solar radiation over the Earth's surface varies with the latitude and the season. Incoming radiation and outgoing radiation are equal when averaged over the whole Earth. Reflection, reradiation, evaporation, conduction, and absorption keep the heat budget in balance.

Earth surface temperatures change with latitude and seasonal changes in solar radiation as well as with the heat capacity of the Earth's surface materials, sea ice formation, and evaporation. The oceans gain and lose large quantities of heat, but their temperature changes very little because of the oceans' high heat capacity. Heat that is mixed downward reduces the temperature change at the surface.

Sea ice is formed in the extreme cold of polar latitudes. The process is self-insulating, so that large thicknesses of ice do not form each winter. Icebergs are formed from glaciers that break off into the sea.

The atmosphere is made up of layers. Weather occurs in the layer closest to the Earth's surface, the troposphere. Air is a mixture of gases. The density of air is controlled by the air's temperature, pressure, and water vapor content. High-pressure and low-pressure zones are related to air density.

Concern that the Earth's climate may be changing is related to changes in the Earth's atmosphere. Increasing carbon dioxide concentration, due to the burning of fossil fuels, is leading to predictions of global warming. The ozone layer is being depleted, increasing the ultraviolet radiation reaching the Earth's surface, and reducing populations of single-celled plantlike organisms in Antarctic seas.

Winds are the horizontal motion of air in the atmospheric convection cells produced by heating at the Earth's surface. Winds are named for the direction from which they blow.

Because of the Coriolis effect, winds are deflected to their right in the Northern Hemisphere and to their left in the Southern Hemisphere. This action produces a three-celled wind system in each hemisphere that results in the surface wind bands of the trade winds, the westerlies, and the polar easterlies. Zones of rising air occur at 0° and 60°N and S; these are low-pressure areas of clouds and rain. Zones of descending air at 30° and 90°N and S are high-pressure areas of clear skies and low precipitation. Surface winds are unsteady and unreliable in the zones of the doldrums and the horse latitudes. The polar jet stream is a high-speed upper-atmosphere wind that marks the boundary between the westerlies and the polar easterlies in the Northern Hemisphere.

Seasonal atmospheric pressure changes modify these wind bands and cause coastal winds to change direction seasonally in the Northern Hemisphere.

Differences in temperature between land and water produce the monsoon effect. This seasonal reversal in wind pattern causes the wet and dry monsoons of the Indian Ocean; a similar daily reversal causes the onshore and offshore winds of coastal areas. Winds from the ocean rising to cross the land also produce heavy rainfall.

Fog occurs when water vapor condenses to form liquid droplets. Advective fog occurs when warm, water-saturated air passes over cold water; sea smoke occurs when dry, cold air moves over warm water; and radiative fog occurs when warm, moist air is cooled at night.

Warm, tropical surface water moving eastward across the Pacific and accumulating along the west coast of the Americas is known as El Niño. Changes in atmospheric pressure across the Pacific Ocean are identified with El Niño's appearance and are known as Southern Oscillation; together they are called ENSO. ENSO models attempt to forecast El Niño and its associated severe storms and droughts in different parts of the world. Colder than normal surface water events are known as La Niña and appear to alternate with El Niño events.

In the tropics, rapid heat transfer to the atmosphere from the sea surface produces intense ocean storms. A rising sea surface under a storm's low-pressure center, combined with storm winds and waves, produces severe coastal flooding known as a storm tide or storm surge.

Key Terms

advective fog, 143	offshore, 141
atmospheric pressure, 131	onshore, 141
Coriolis effect, 136	ozone, 130
cyclone, 145	pack ice, 128
doldrums, 137	polar easterlies, 137
El Niño, 143	radiative fog, 143
fog, 143	rain shadow, 142
greenhouse effect, 132	sea ice, 127
heat budget, 124	sea smoke, 143
high-pressure zone, 131	Southern Oscillation, 145
horse latitudes, 137	storm surge/storm tide, 145
hurricane, 145	stratosphere, 130
iceberg, 129	trade winds, 137
intertropical convergence, 137	troposphere, 130
jet stream, 138	typhoon, 145
La Niña (El Viejo), 145	westerlies (prevailing
low-pressure zone, 131	westerlies), 137
monsoon, 141	

Critical Thinking

1. Why does the circulation of the atmosphere depend on its transparency to solar radiation and what would happen to atmospheric circulation if the upper atmosphere absorbed most of the solar radiation?

2. Why do Earth surface temperatures around 60°N change dramatically summer to winter while those around 60°S remain much the same?

3. Why is the severity of a storm tide different along each side of a hurricane's storm track?

4. A frictionless projectile is fired from the North Pole and is aimed along the prime meridian. It takes three hours to reach its landing point, halfway to the equator. Where does it land (latitude and longitude)? If the same projectile is fired from the South Pole under the same circumstances, where does it land (latitude and longitude)?

5. Sea ice fills the Arctic Ocean in winter but is much less in summer; at the same time the sea surface temperature of the Arctic Ocean changes little between summer and winter. Why?

Suggested Readings

Bell, M. 1994. Is Our Climate Unstable. *Earth* 3(1):24–31.

Berner, R. A., and A. C. Lasaga. 1989. Modeling the Geochemical Carbon Cycle. *Scientific American* 260(3):74–81.

Charlson, R., and T. Wigley. 1994. Sulfate Aerosol and Climatic Change. *Scientific American* 270(2):48–57.

Cobb, C. E., Jr. 1993. Bangladesh: When the Water Comes. *National Geographic* 183(6):118–34.

Collins, G. 1998. Springtime Arctic Ozone Levels Fall Further in 1997. *Physics Today* 51(1):18–19.

Davidson, K. 1995. El Niño Strikes Again. *Earth* 4(3):24–33.

Flanagan, R., and T. Yulsman. 1996. On Thin Ice. *Earth* 5(2):44–51.

Gorman, J. 1998. They Saw it Coming. *Discovery* 8(1):82–83.

Heidorn, K. C. 1975. Land and Sea Breezes. *Sea Frontiers* 21(6):340–343.

Karl, R., N. Nicholls, and J. Gregory. 1997. The Coming Climate. *Scientific American* 276(5):78–91.

Kerr, R. A. 1993. El Niño Metamorphosis Throws Forecasters. *Science* 262(5134):656–57.

Knox, P. N. 1992. El Niño — A Current Catastrophe. *Earth* 1(5):30–37.

Kretschmer, J. 1990. When the Winds Blow: A Mythology of Gust and Squall. *Sea Frontiers* 36(3):40–43.

Leetma, A. 1989. The Interplay of El Niño and La Niña. *Oceanus* 32(2):30–34.

Olson, D. B. 1990. Monsoons and the Arabian Sea. *Sea Frontiers* 36(1):34–41.

Smith, F. G. W. 1992. Hurricane Special: An Inside Look at Planet's Powerhouse. *Sea Frontiers* 38(6):28–31.

Stevens, J. 1996. Exploring Antarctic Ice. *National Geographic* 189(5):428–439.

Suplee, C. 1998. Unlocking the Climate Puzzle. *National Geographic* 193(5):38–71.

Valiela, I., et al. 1996. Hurricane Bob on Cape Cod. *American Scientist* 84(2):154–165.

Webster, P. J. 1981. Monsoons. *Scientific American* 245(5):108–119.

internet references

 ## worldwide websites

Sea Surface Temperature

University of Rhode Island Sea Surface Temperature Images
http://dcz.gso.uri.edu/avhrr-archive/archive.html

National Environmental Satellite, Data, and Information Service
http://psbsgi1.nesdis.noaa.gov:8080/PSB/EPS/SST/contour.html

National Oceanic and Atmospheric Administration Sea Surface Temperature
http://www.websites.noaa.gov/guide/sciences/ocean/seasur.html

Sea Ice

NOAA Ocean Modelling Branch Sea Ice Analysis Page
http://polar.wwb.noaa.gov/seaice/Analyses.html

USGS Monthly Average Polar Sea Ice Concentration
http://geochange.er.usgs.gov/pub/sea_ice/README.html

United States Coast Guard International Ice Patrol
http://www.uscg.mil/lantarea/iip/home.html

Carbon Dioxide

Pacific Marine Environmental Lab: Global Carbon Cycle
http://www.pmel.noaa.gov/co2/co2-home.html

Canada: National Environmental Indicator
http://www.ec.gc.ca/Ind/English/Climate/Tech_Sup/ccsup06_e.cfm

National Academy Press: Climate Change and Carbon Dioxide
http://www.nap.edu/books/0309058767/html/1.html

Trends: Atmospheric Carbon Dioxide
http://cdiac.esd.ornl.gov/trends/co2/sio-mlo.htm

The Kyoto Protocol
http://www.greenhouse.gov.au/pubs/factsheets/fs_sinks.html

EPA: Ozone Depletion
http://www.epa.gov/docs/ozone/

Carbon Dioxide Information Analysis Center
http://cdiac.esd.ornl.gov/

Atmospheric Circulation and Effects

Jet Stream
http://www.usatoday.org/weather/tg/wjstream/wjstream.htm

Satellite Wind Measurements
http://nasadaacs.eos.nasa.gov/yearbooks/96/reckoning_winds.html

Indian Monsoon
http://www.usda.gov/oce/waob/jawf/profiles/specials/monsoon/monsoon.htm

Rain Shadow
http://www.weatherpages.com/rainshadow/

Radiative Fog
http://www.usatoday.com/weather/wrfog.htm

Understanding Fog
http://www.usatoday.com/weather/wfog.htm

Satellite Images and Weather
http://www.dkrz.de/sat/sat-eng.html

Coriolis Effect
http://www.windpower.dk/tour/wres/coriolis.htm

ENSO

NOAA: ENSO Home Page
http://www.ogp.noaa.gov/enso/

NOAA: El Niño Page
http://www.elnino.noaa.gov/

NOAA: El Niño Theme Page
    http://www.pmel.noaa.gov/toga-tao/el-nino/
    nino-home.html
Additional El Niño sites and information
    http://www.elnino.noaa.gov/sites.html, and
    http://www.coaps.fsu.edu/lib/enso_sites.html
Tropical Atmosphere Ocean project
    http://www.pmel.noaa.gov/tao/
Tropical Ocean Global Atmosphere project
    http://www.ncdc.noaa.gov/coare/toga.html

Hurricanes
National Hurricane Center: Tropical Prediction Center
    http://www.nhc.noaa.gov/
Hurricanes
    http://www.hurricanes98.com/

Earthwatch: Hurricanes
    http://www.earthwatch.com/hurricane.html
Storm Surge Risk Analysis
    http://www.csc.noaa.gov/products/alabama/htm/
    hssra.htm
USA Today Hurricane Information
    http://www.usatoday.com/weather/huricane/
    whur0.htm
Hurricane Hazards
    http://hurricanes.noaa.gov/prepare/title_hazards.
    htm
Storm Surge Information
    http://www.ncstormsurge.com/srginf.html

Clouds in their different shapes, ever changing and always moving have been watched and studied by those at sea and those on land from the beginnings of human history to the present. We see clouds as huge and fluffy building up into great towers, or high and wispy spreading across the sky. We see them white in the sunshine, rosy and red at dawn and sunset, black and threatening as a storm ap-

Clouds

proaches. Different kinds of clouds are placed in categories based on the work of Luke Howard, an English pharmacist. In 1803 Howard proposed the first useful way to classify clouds: stratus or layered, cumulus or puffy, cirrus or wispy, and nimbus or clouds releasing snow or rain that travels all the way to the ground. This system is the basis for the cloud types recognized by today's World Meteorological Organization (WMO); see figure 1.

Cloud Formation

Clouds form as moist air rises and cools. When the rising air is cooled below the dew point, the temperature at which water vapor begins to condense, the cloud begins to form. Water vapor condenses around condensation nuclei, forming droplets. Near the Earth, in the lower troposphere, condensation nuclei may be small particles of dust, salt, or other matter; these particles are generally abundant, and condensation takes place readily. In the higher troposphere the nuclei are composed of crystallized water vapor or ice crystals and snow. If nuclei are sparse, the droplets are fewer and larger, increasing the chance of precipitation and

reducing further the amount of nuclei. Many small nuclei produce abundant small droplets, increasing cloud reflectivity and absorption and decreasing incoming solar energy. In oceanic areas where precipitation is very frequent, particles are removed by the rain, and a low supply of condensation nuclei may hinder cloud formation. In these areas satellite photos show passing ships leaving cloud trails as condensation occurs on particles from the ships' exhausts.

Air temperature in the troposphere decreases with height, therefore when air with a high water content and a high dew point temperature rises and cools, the water vapor condenses at a low elevation above the Earth's surface. When condensation occurs, the latent heat of vaporization is released to the air, warming it, and increasing its ability to rise and therefore cool further. If the air carries little water vapor with a low dew point temperature, it must rise higher to cool to the dew point. High altitude cirrus clouds are wispy because the air at high altitudes contains very little water vapor for forming ice crystals.

When temperatures are above freezing, clouds form from liquid water droplets; super-cooled water droplets form clouds when temperatures are below freezing. Clouds below 6 km (19,000 ft) elevation are usually formed of water droplets, while those above 8 km (26,000 ft) elevation are formed from ice crystals.

Clouds are part of the Earth's hydrologic cycle; refer to figure 2.13. In this figure the clouds are the visual manifestations of water droplets and ice crystals carried in the air, moving from a

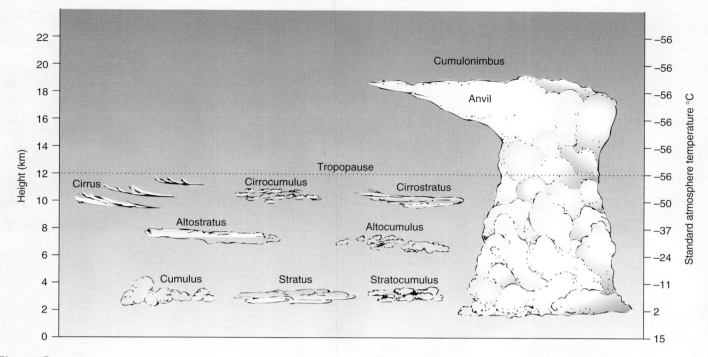

Figure 1

Cloud types used by the World Meteorological Organization.

source of evaporation and rising air to a region of precipitation where condensed droplets become large enough to fall to the Earth's surface.

Climate Effects

Clouds simultaneously heat and cool the Earth. Clouds reflect the Sun's energy, and they also absorb some of its incoming energy. Both processes reduce incoming energy to the Earth's surface and so help to cool the Earth. Clouds also intercept the Sun's energy reflected and reradiated from the Earth's surface, and because they are generally cooler than the Earth, the clouds reradiate only part of this absorbed energy to space. This tends to trap heat and warm the Earth and its atmosphere. The net effect is that clouds play a significant part in the Earth's heat budget (see fig. 6.2), helping to maintain the Earth's average temperature.

Recent research indicates that, on a global scale, the Earth's present cloud cover provides more shielding from incoming energy than trapping of reflected and reradiated energy. At present, cloud shielding results in a total net reduction in solar energy to the Earth of about 14 to 21 watts per square meter per month. It is estimated that, if there were no clouds, average Earth surface temperature would increase by about 10°C.

Clouds not only affect the Earth's climate, they are affected by it. If the Earth's climate changed, the feedback mechanism described previously could impose a new cloud-controlled climate and heat balance on the Earth. If the present reduction in solar energy due to clouds became less negative, the Earth's surface would warm. If cloud feedback produced a greater reduction in available solar energy, the Earth's surface would become cooler than it is at present. If the Earth

warmed due to the greenhouse effect (refer to section 6.6), an increase in evaporation would increase opaque, low-altitude clouds over the ocean; these clouds would then absorb and reflect incoming solar energy back into space and cool the Earth. Still another possibility is that increased water vapor could be carried by the winds over land at the higher latitudes, resulting in a greater accumulation of snow and hastening the beginning of a new ice age. But an increase in high-altitude ice clouds would warm the Earth, because ice clouds pass the Sun's incoming energy and reduce reflected energy loss, while absorbing heat energy reradiated from the Earth.

Figure 2

Storm clouds over western Scotland.

Figure 3

Subtropical cumulus clouds over the Florida Keys.

Figure 4

Sunset clouds at sea in the tropics.

Computers and Cloud Models

Although clouds are among the most common of atmospheric phenomena, they are believed to be the principal source of uncertainty in predicting the Earth's future climate. Because our understanding of cloud formation and distribution, and of climate change and feedback is presently incomplete, computer models of the atmosphere and its circulation have often led to uncertain and conflicting results. Presently the large computer programs or General Circulation Models (GCMs), used to predict climate, are helping researchers understand the complexities of cloud-climate systems. Since the formation and movement of clouds depends on everything from local winds to jet stream movement, more data is required before atmospheric circulation programs and heat-budget climate models can become reliable. At present, thirty different atmospheric science groups are attempting to model the past climate of the 1980s, in order to test the accuracy of the GCMs and learn more about the role of clouds. Predictions of future climate trends are not yet sufficiently accurate to use in policy planning, but scientists believe that their work will lead to this ability—in about ten years.

Readings

Baskin, Y. 1995. Under the Influence of Clouds. *Discover* 16 (9):62–69.

Glanz, J. 1993. Climate Researchers Look to the Clouds. *R and D Magazine* February: 56–62.

LaBreque, M. 1990. Clouds and Climate. A Critical Unknown. *Mosaic* 21 (2):2–11.

Internet Resources

Cloud Types—images and information
http://covis1.atmos.uiuc.edu/guide/clouds/html/

Clouds and Precipitation—cloud development and cloud types
http://ww2010.atmos/uiuc.edu/(Gh)/guides/mtr/cld/home.rxml

Planning and Executing a Successful Oceanographic Expedition

Oceanographic expeditions are complex events that require long and detailed planning. There are two teams involved in every research cruise; the science team led by the Chief Scientist, and the ship's crew, led by the Captain. The following are two reflections on the responsibilities of a Chief Scientist, Dr. Marcia McNutt, and a ship Captain, Captain Ian Young for planning and executing a successful oceanographic expedition. Dr. McNutt served on the faculty at the Massachusetts Institute of Technology for 15 years in the 1980s and 1990s, where she was the Griswold Professor of Geophysics. Captain Young sailed as Master of the R/V *Maurice Ewing* for many years during the 1990s. This ship is owned by the National Science Foundation and operated by Lamont-Doherty Earth Observatory. Both are currently employed at the Monterey Bay Aquarium Research Institute, where Captain Young sails on the R/V *Western Flyer* and Dr. McNutt is the President and CEO.

The Role of the Chief Scientist

All scientists are motivated by the thrill of discovery. For oceanographers such as myself, discovery does not lie in a test tube in the lab, but in the ocean itself. Most of the deep sea remains either completely unexplored or observed in only a cursory sense with modern technology, and thus there are ample opportunities to go where no one has gone before and to see what is there through new sets of technological "eyes." My own expeditions invariably return with wondrous new discoveries that were not predicted by any prevailing theories. For that reason, I live for going to sea.

Oceanographic expeditions are an expensive undertaking, however. The cost of the ship time alone is typically between $10,000 and $20,000 per day, and those costs do not include the salaries of the science party, travel to the vessel, use of very specialized equipment, or the eventual analysis of the data collected on the expedition. Therefore, it is imperative that each expedition address important scientific problems, use the best technology for the data to be collected, and be carefully planned and executed by the most capable scientific team.

I begin the process of conducting an expedition by formulating a testable hypothesis, which if verified or refuted, would lead to progress in answering an important scientific question. The next step is to determine what observations need to be made or what experiment needs to be performed to test the hypothesis. From this consideration naturally flows the equipment that must be assembled for the expedition and the expertise needed from my team members. I select other team members primarily based on their scientific ability and expertise, but personal considerations (such as do you want to be confined for 30–60 days on 250 ft of floating steel with this person?) weigh in as well.

The team then writes a proposal to a funding agency, commonly the National Science Foundation, to support the expedition, and waits. And waits. The proposal is reviewed by other scientists who make recommendations on which projects should be funded with the limited money available. Typically about six months later a decision is reached as to which expeditions will be supported. If my team is one of the lucky ones, we are then placed on the schedule for one of the ships in the national fleet of vessels operated by academic institutions. To which ship we are assigned depends on the type of equipment needed on the ship and where the research area is. Much of my own work is specialized enough that only one or two ships can conduct the mission. Therefore, we may have to wait another year or more for that ship to eventually sail to the right part of the world to conduct our experiment.

Once the ship is determined, I write up a summary of the proposed work for the benefit of the Captain and technical personnel on the designated ship. Unlike the science proposal, which is meant to excite other scientists about the importance of the proposed expedition, this summary is a detailed account of where we will go, what equipment we will use, and how we will use it. Several months before the expedition, I meet with the operators of the ship and its captain to discuss the mission and answer questions they may have about the summary. This is the chance to make sure that there are no unpleasant surprises on either side that only surface after we set sail (e.g., "We thought that you were bringing liners for the piston core." "No, I thought the ship already stocked liners for the core!"). This is also the time when we make sure that we have all of the necessary permits to work in the territorial waters of other nations.

If this is my first meeting with the Captain, I want very much to impress him with my knowledge of marine operations and thoroughness of preparation. Our teamwork must be built on a foundation of mutual respect: I must trust him not to place limitations on our plans for scientific operations at sea unless he truly feels that the safety of the vessel and its occupants are at risk. On the other hand, he must feel confident that I would not ask for anything extra from the crew or the ship unless the success of the science mission truly depended on it.

Once at sea we make the most of every minute. Operations go around the clock, with subsets of the crew and science party assigned to pairs of 4-hour watches. As chief scientist, it is my responsibility to make sure that the Captain and mates on the bridge know well in advance what the science plan is for the next watch. From a science perspective, the very best expeditions are those in which the cruise plan is constantly in flux as we follow up on new discoveries. But such expeditions are not always the favorite on the bridge, nor do they allow the chief scientist to get eight hours of uninterrupted sleep!

In the final analysis, the success of any expedition can often be attributed to the willingness of the Captain and his crew to put in the extra effort necessary to achieve what are invariably ambitious objectives in the face of bad weather, temperamental equipment, and Murphy's laws. That is why the last line in so many oceanographic research publications is some variant of the following: "Special thanks to the Captain and crew of the R/V _____ for making this expedition such a success."

The Role of the Ship's Captain

I've sailed professionally for more than twenty years on various types of vessels, with different levels of responsibility. Whether signed on a freighter as Third Mate bound for Africa, or wandering the world's oceans on a tramp research vessel, I've always loved being "Safely back to sea."

As Captain of a modern research vessel, I am primarily concerned with the readiness of the ship. Usually no two expeditions are alike, and each requires unique preparations. Information about each scientific voyage is provided by the Chief Scientist either through the ship's marine office, or directly to the Captain, 1–3 months in advance.

The Captain starts planning for the cruise months before it begins. The five Ps apply at all times: **Proper Planning Prevents Poor Performance.** This includes ordering charts for the research site, preparing equipment, and performing needed maintenance on the ship's machinery and systems (electrical and mechanical) that will be needed during the next cruise for around the clock operations.

The Captain relies on the crew; Science Technician, Chief Mate, Chief Engineer, and Steward, to make all preparations for each cruise. The ship's Science Technician is responsible for making sure that the ship's data collection systems are ready for, and compatible with, science computers. The Chief Mate will perform deck maintenance to have equipment ready and deck space clear for storing, staging, launching, and recovering equipment. All safety equipment has to be in good working order as required by the U.S. Coast Guard. The Chief Engineer plans ahead to order fuel, lubricating oil, hydraulic oil, and spare parts for all the systems under his care. These systems are all important for the success of the science mission. A failure in any one of these systems can delay, slow, or stop an expedition. The ship's Steward must choose and stock adequate provisions for forty persons for a 30–50 day voyage, a job that requires experience and attention to detail to assure quality meals.

Hopefully, clearances from the State Department will arrive prior to the vessel sailing. These clearances give the vessel legal permission to work in the territorial waters of foreign sovereign nations and are required to avoid conflict with foreign governments. Depending on the proposed location for the research, multiple countries may need to be contacted for the same expedition. I experienced an interesting example of this on a voyage that took the vessel into waters surrounding the Falkland Islands shortly after a conflict between Great Britain and Argentina. The ship's marine office typically applies for the necessary clearances 6–12 months in advance of the cruise. A delay in the receipt of a clearance can have serious consequences because it can force a cruise to be rescheduled for a later date. Rescheduling can result in the expedition missing the optimum weather window for the research site, which can degrade the quality and quantity of data collected.

The Captain has to work closely with the Chief Scientist to make sure the cruise objectives are achieved. This starts with the first science meeting held between them. Here the Captain gets the first full view of the cruise for which he has been preparing the vessel. The cruise can have many different aspects; station work, towed gear deployments, bottom instrument launches, buoy recoveries, and/or grid surveys. This meeting provides an opportunity for the Captain to contribute operations suggestions to help the scientists make better use of time and equipment because he knows the capabilities of the ship and personnel.

The Captain is always watching the weather. Sometimes work can be scheduled around certain weather patterns. Quite often there are "weather days" when work has to stop for the safety of the people on deck and/or the safety of the vessel. This may require the vessel heaving-to for a night or retrieving gear and leaving the area due to a hurricane or typhoon. This is a difficult decision for the Captain to make, because the Chief Scientist never wants to stop collecting data and lose time from the cruise. This is one of the many situations in which the Captain's decision has to be right, an error in judgment could jeopardize the safety of the vessel and the people on board, or cost him his job.

At the end of the voyage there is a time-honored ritual involving the Captain and Chief Scientist: negotiating the break-off time for the "end of science." The distance to port from the work area may be measured by a few hours or by a few weeks. The Captain and Chief Scientist will estimate the time required to return to port; frequently both will come up with different answers to this question. The Captain will allow for weather and currents while the Chief Scientist generally will not. At this point in the cruise the Chief Scientist likes to state: "If I can't have time to collect the last bit of data then the whole expedition will be for nothing." Of course this isn't true, but it contributes to the drama required to end a successful cruise. Usually the time requested to finish the scientific mission is within reason and the Captain is able to safely return to port on schedule. Even though there are days of transit to the next port, in the Captain's mind this trip is over, and he is already preparing for the next expedition to follow.

There is a river in the ocean. In the severest droughts it never fails, and in the mightiest floods it never overflows. Its banks and its bottom are of cold water, while its current is of warm. The Gulf of Mexico is its fountain, and its mouth is in the Arctic Seas. It is the Gulf Stream. There is in the world no other such majestic flow of waters. Its current is more rapid than the Mississippi or the Amazon...

chapter seven

Circulation Patterns and Ocean Currents

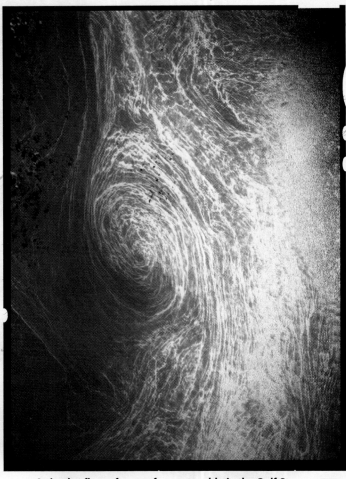

A circular flow of water forms an eddy in the Gulf Stream

Learning Objectives

After reading this chapter, you should be able to

- Understand the effects of changing surface temperatures and salinities.
- Know how temperature and salinity influence density with depth.
- Describe how surface water densities vary with latitude.
- Explain density-driven circulation and relate it to surface changes at different latitudes.
- Discuss the layered structure of the Atlantic Ocean and compare it with the Pacific and Indian Oceans.
- Understand how water properties are measured at sea.
- Relate the effects of surface winds to the movement of water and the ocean's surface currents.
- Explain the processes that produce the oceanic gyres.
- Describe the major ocean surface current systems.
- Compare the processes of divergence and convergence.
- Identify major areas of coastal upwelling and downwelling.
- Understand the role of eddies.
- Understand how ocean currents are measured.
- Describe how energy is produced from changes in seawater temperature and from energy flow.

i f we could cut vertically through the ocean and remove a slice of ocean water in the same way we might cut a slice of cake, we would find that, like a cake, the ocean is a layered system. The layers are invisible to us, but they can be detected by measuring the water's changing salt content and temperature and by calculating the density of water from the surface to the ocean floor. This layered structure is a dynamic response to processes that occur at the surface: the gain and loss of heat, the evaporation and addition of water, the freezing and thawing of sea ice, and the movement of water in response to winds. This chapter explores the structure of the oceans below the surface and follows the currents as they flow, merge, and move away from each other. It examines both horizontal and vertical circulation and the links between them.

Density-Driven Circulation 7.1

Variations in temperature, salinity, and pressure control the density of seawater. When a combination of these three properties produces water that is more dense than adjacent water, the more dense water will sink until it lies below the less dense water. Return to table 5.2 and notice that warm, high salinity water may have the same density value as cold low salinity water. Figure 7.1 shows these same relationships; here the constant density values are shown as curved lines. If a straight line, called a "mixing line," is drawn between any two points on the same constant density line, that line falls below the curved line. This indicates that the mixing of two bodies of water with different salinity and temperature values but the same density will produce a new type of water. The new water lies on the straight line and is more dense than either of the original waters. In this way density changes in seawater, although small, create the flows that readjust the seawater layers in the ocean.

In many places, the oceans have a well-mixed surface layer approximately 100 m (300 ft) deep. In this layer, the increasing pressure with depth causes the density to increase very slightly. Below this surface layer, density changes rapidly between 100 m and about 1000 m (300–3000 ft) because the temperature decreases rapidly. Also in this layer (between 100 and 1000 m), the salinity may increase, decrease, or remain unchanged; it has a smaller influence on the density over this depth. Below 1000 m, both temperature and salinity change slowly with depth, and the density increases slightly down to the sea floor because of the increasing pressure. Figures 7.2a & b show temperature, salinity, and density changes with depth.

At the ocean surface, density is controlled by the Earth's uneven heating and evaporation-precipitation patterns. Variations in surface densities control the vertical movement of the water. Use figure 7.3 with the following discussion. The warm, low-salinity surface water at the equator is

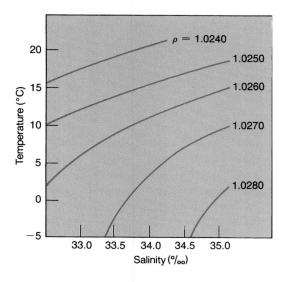

Figure 7.1

The density of seawater, measured in grams per cubic centimeter, is abbreviated as ρ (rho) and varies with temperature and salinity. The pressure factor is not included. Many combinations of salinity and temperature produce the same density. Low densities are at the upper left and high densities at the lower right.

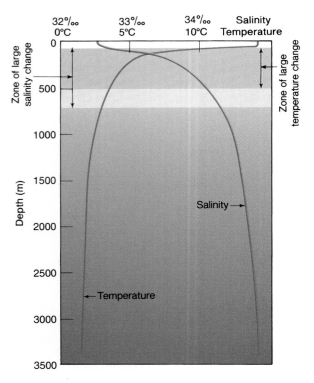

Figure 7.2a

Temperature and salinity values change with depth in seawater. Rapid changes in temperature and salinity with depth occur in the near-surface layer (50–700 m). (Based on data from the northeast Pacific Ocean.)

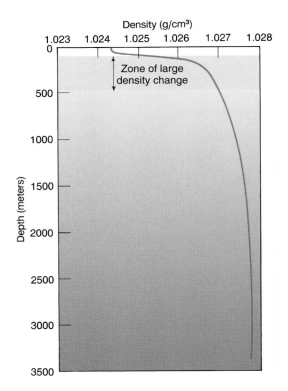

Figure 7.2b

Density increases with depth in seawater.

Figure 7.3

Water of different densities is distributed over depth. The density of each layer is determined by the salinity and temperature where each layer meets the surface.

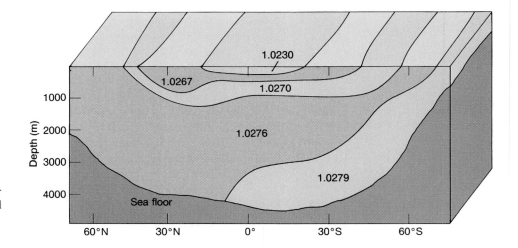

less dense than the surface water at latitudes 30°N and 30°S, which is slightly cooler and has a higher salinity because of evaporation and diminished rainfall. Therefore, surface water produced at 30° latitudes is denser than that produced at the equator, and it sinks below the equatorial water. The combined salinity and temperature of surface water at 50 to 60°N or S produce water that is denser than either the equatorial or the 30° latitude surface water, and so this water sinks below the equatorial and 30° waters. Winter conditions in the polar re-

gions lower the surface water temperature and, if sea ice forms, increase the surface salinity. This results in dense surface water that sinks toward the ocean floor at polar latitudes. The cold, dense water of the Arctic is confined by the seafloor topography to these high northern latitudes; similar water formed around Antarctica moves northward into the Southern Hemisphere oceans. The thickness and horizontal extent of each density layer is related to the rate at which the water is formed and the size of the surface region over which it is formed.

What are the effects of changing the salinity and temperature of surface water?

Explain how water samples with different salinities and temperatures have the same density.

How do temperature and salinity change with ocean depth?

Why does water density change more rapidly with depth in the upper 1000 m than near the sea floor?

Explain why surface water densities vary with latitude.

Thermohaline Circulation   7.2

Density driven vertical circulation controlled by surface changes in temperature and salinity is known as **thermohaline circulation.** If a surface process forms more-dense water on top of less-dense water, the water column is unstable, and as the denser water sinks or **downwells,** the less-dense water rises or **upwells.** This vertical displacement is called **overturn.** If the water column has the same density over depth, it has neutral stability, and the water is easily mixed by wind, waves, and currents. If the water column has an increasing density with depth, it does not overturn; it is stable, and is more difficult to mix vertically.

Large-scale thermohaline circulation ensures an eventual top-to-bottom exchange of water throughout the oceans. In the middle latitudes, the surface water temperature changes with the seasons (fig. 7.4). During the summer, the surface water

warms and the water column is stable, but in the fall the surface water cools, increases its density, and begins to overturn. Winter storms and winter cooling continue the mixing process. The shallow, warm, low-density surface water formed during the previous summer cools and sinks, and the upper portion of the water column becomes neutrally stable to greater depths. Spring brings surface warming, and the water column becomes stable and remains so through the summer.

Seasonal changes in temperature are more important than seasonal changes in salinity in altering density in the open ocean. For example, in the Atlantic Ocean the surface water at 30°N and S has a high salinity, but it is warm year-round so that there is no water sinking to great depths. On the other hand the surface water from 50° to 60° in the North Atlantic has a lower salinity but is cold, especially during the winter. This cold water downwells and flows below the saltier but warmer surface water.

Close to shore, however, the salinity of seawater may outweigh the temperature in controlling the density. This is particularly true along coasts that receive large amounts of fresh water runoff. Here even cold (0°–2°C), river water does not downwell when it meets the salt water but remains at the surface as a freshwater lid. In polar regions, when the sea ice melts, a layer of fresh water forms at the surface and slowly dilutes the surface seawater. When sea ice forms, the extraction of the water increases the salinity of the surface water and this denser water downwells. The extremely dense surface water produced by winter cooling and freezing in the Weddell Sea of Antarctica sinks all the way to the deep sea floor. In regions of high evaporation, the salinity of water in marginal seas, rises, increasing the density of the water. In the Red Sea and the Mediterranean Sea such water flows into the adjacent ocean and sinks to mid-depths; this is discussed in chapter 9, section 9.

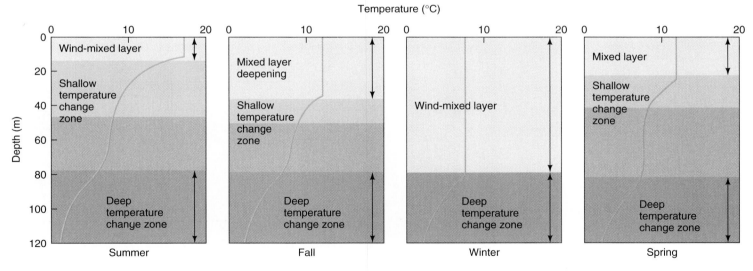

Figure 7.4

The surface layer temperature structure varies over the year. In the absence of strong winds and wave action in the summer, solar heating produces a shallow temperature change zone. During the fall and winter the surface cooling and storm conditions cause mixing and vertical overturn, which eliminate the shallow temperature change zone and produce a deep wind-mixed layer. In spring the shallow temperature change zone re-forms.

Upwellings and downwellings are measured at rates of 0.1 to 1.5 m (0.3–5 ft) per day, compared to fast-flowing ocean surface currents that reach speeds of 1.5 meters per second. Horizontal movement at depth due to thermohaline circulation is about 0.01 cm (0.004 in) per second. Water caught in this slow but relentlessly moving cycle may spend 100 years or more at depth before it again rises to the surface.

> How is overturn affected by stable, unstable, and neutral water column conditions?
>
> Why is oceanic vertical circulation called thermohaline circulation?
>
> How do seasonal climate changes control surface overturn at middle latitudes?
>
> Compare temperature and salinity control of surface density in the open ocean and along the coast.

The Layered Oceans 7.3

Thermohaline circulation and the resulting deep-water flows determine the distribution of temperature, salinity, and density with depth in the world's oceans. The subsurface density layering of the Atlantic Ocean is shown in figure 7.5, and figure 7.6 compares average salinity and temperature distribution for the Atlantic, Arctic, Pacific, and Indian Oceans.

The Atlantic Ocean

At the surface of the North Atlantic, water from high northern latitudes moves southward, while water from low latitudes moves northward along the coast of North America and then east across the Atlantic. These waters converge in areas of cool temperatures and high precipitation at approximately 60°N. The resulting mixed water has a salinity of about $34.9^0/_{00}$ and a temperature of 2° to 4°C. This water, known as **North Atlantic deep water,** sinks and moves southward (see fig. 7.5 and 7.6a and b). Above this water, at about 30°N, a shallow lens of very salty ($36.5^0/_{00}$), very warm (25°C) surface water remains trapped by the circular movement of the oceanic surface currents (to be discussed later in section 6). Between this surface water and the North Atlantic deep water lies water of intermediate temperature (10°–13°C) and salinity ($35–37^0/_{00}$). Very warm, high-salinity water from the Mediterranean Sea is added to the North Atlantic through the Strait of Gibraltar. This water finds its own density at approximately 1000 m (3000 ft) depth in the North Atlantic.

Near the equator, **Antarctic intermediate water,** warmer (5°C) and less salty ($34.4^0/_{00}$) than the North Atlantic deep water, remains above the denser water below, while less dense but high-salinity surface waters lie above it. Along the edge of Antarctica, very cold (−0.5°C), salty ($34.8^0/_{00}$), and dense water is produced at the surface by sea ice formation during the Southern Hemisphere's winter. It sinks to the ocean bottom, because this **Antarctic bottom water** is the densest water produced in the world's oceans. After it descends to the ocean floor, it begins to move northward. When it meets North Atlantic deep water, it creeps beneath it and continues northward along the east coast of South America. Antarctic bottom water does

Figure 7.5

A north-south midsection of the Atlantic Ocean shows the vertical and deep-water movements caused by thermohaline circulation. Wind-driven currents are indicated at the sea surface.

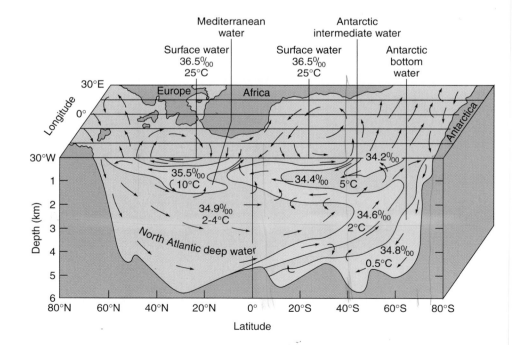

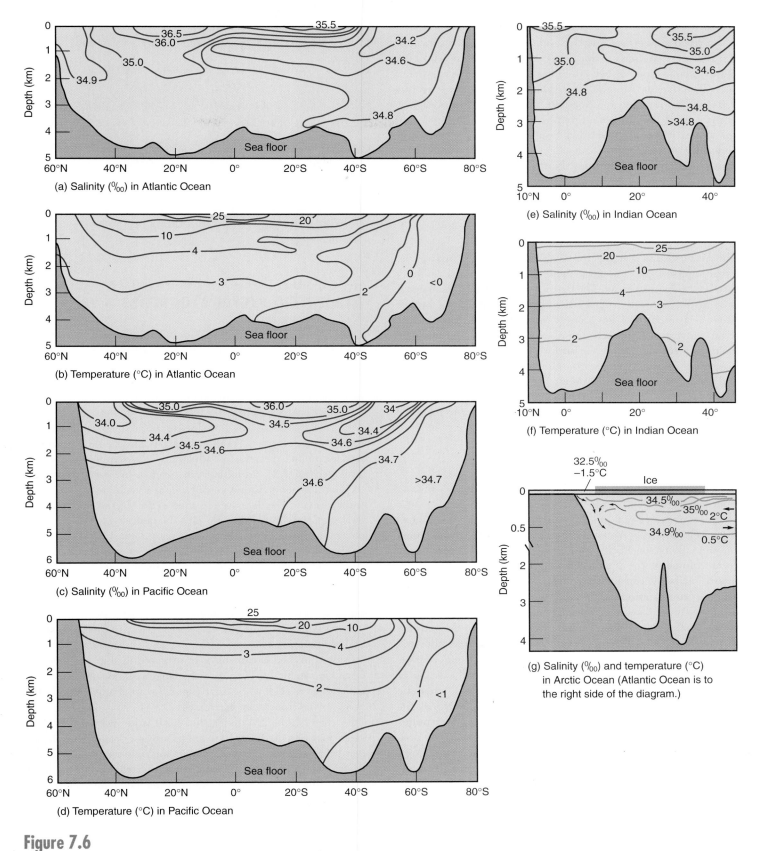

(a) Salinity (⁰/₀₀) in Atlantic Ocean

(b) Temperature (°C) in Atlantic Ocean

(c) Salinity (⁰/₀₀) in Pacific Ocean

(d) Temperature (°C) in Pacific Ocean

(e) Salinity (⁰/₀₀) in Indian Ocean

(f) Temperature (°C) in Indian Ocean

(g) Salinity (⁰/₀₀) and temperature (°C) in Arctic Ocean (Atlantic Ocean is to the right side of the diagram.)

Figure 7.6

Mid-ocean salinity and temperature profiles of the Atlantic, Pacific, and Indian Oceans (a–f). Temperature and salinity patterns caused by thermohaline circulation are shown as functions of depth and latitude. In the Arctic Ocean (g) wide continental shelves surround two central basins.

not accumulate enough height to be able to flow over the mid-ocean ridge system and into the basins along the African coast. It is therefore confined to the deep basins on the west side of the South Atlantic and has been found as far north as the equator.

The North Atlantic deep water trapped between the Antarctic bottom and intermediate waters rises to the surface in the area of the 60°S divergence. As it reaches the surface it splits; part moves northward as **South Atlantic surface water** and Antarctic intermediate water; part moves southward toward Antarctica. A mixture of North Atlantic deep water and Antarctic bottom water becomes the circumpolar water for the Southern Ocean around Antarctica.

The Arctic Ocean

The Arctic Ocean is connected directly to the North Atlantic between Spitsbergen and Greenland and to the Pacific Ocean through the Bering Strait. The density of the Arctic Ocean water is controlled more by salinity than by temperature. Its surface water is a mixture of low-salinity water from the Bering Sea, fresh water from Siberian and Canadian rivers, and seasonal melting of sea ice. Below the surface layer salinity increases, a product of cold, salty water from the annual formation of sea ice and water from the North Atlantic (fig. 7.6g). Water exiting the Arctic Ocean enters the North Atlantic between Greenland and Iceland where it combines with Gulf Stream water to form the North Atlantic deep water (fig. 7.5).

The Pacific Ocean

In the high latitudes of the vast Pacific ocean, waters sinking from relatively small areas of surface convergence lose their identities rapidly, making the layers difficult to distinguish (fig. 7.6c and d). Antarctic bottom water forms in small amounts along the Pacific rim of Antarctica but is lost in the great volume of the Pacific Ocean. The deeper water of the South Pacific Ocean has its source in the circumpolar flow of water around Antarctica. Because the North Pacific is isolated from the Arctic Ocean, only a small amount of water comparable to North Atlantic deep water can be formed. In the extreme western North Pacific, convergence of the southward-flowing cold water from the Bering Sea and the northward-moving water from the equatorial latitudes produces a small volume of water that sinks to mid-depths (fig. 7.6c). There is no dramatic source of deep water similar to that found in the North Atlantic. Warm, salty surface water is found at subtropic latitudes in each hemisphere, and Antarctic intermediate water is produced in small quantities, but its influence is small (fig. 7.6c). Deep-water movements in the Pacific are sluggish, and conditions are very uniform below 2000 m (6600 ft).

The Indian Ocean

Since the Indian Ocean is principally an ocean of the Southern Hemisphere, there is no counterpart of the North Atlantic deep water (fig. 7.6e and f). Small amounts of Antarctic bottom water are soon lost, and the deeper waters tend to be supplied from Antarctic circumpolar water. There is a small amount of Antarctic intermediate water and a lens of warm, salty water at the surface.

Identify the water layers of the Atlantic Ocean. What is the origin of each? In which direction does each flow?

Why is the layering of the Pacific Ocean less dramatic than the layering of the Atlantic Ocean?

In what ways is the Indian Ocean similar to the South Atlantic Ocean?

What is the origin of Arctic Ocean deep water?

Measuring Water Properties   7.4

Until recently all water samples for chemical analysis were collected in specially designed bottles hung from wire rope or cable. The open water bottles were fastened to the cable at predetermined intervals, and the water bottles lowered to the desired depths. The water samples were then recovered for later analysis. See figure 7.7.

Today deep sea oceanographers use **conductivity-temperature-depth (CTD)** sensors to record salinity by directly measuring the electrical conductivity and temperature of the seawater (fig. 7.8). Additional sensors measure variables such as oxygen, nutrients, sound velocity, light transmission, and water transparency. The data are returned to the ship as electronic signals that may be fed directly into the computer, recorded on a chart, or made available as numeric data. CTDs are often equipped with specialized water bottles that can be activated by electronic signal in order to collect water samples for additional measurements. CTDs acquire large amounts of data in this way, while water bottles carrying thermometers collect only one sample at each depth for later analysis. CTD sensors can be left in place, suspended from the vessel or a buoy, to monitor salinity and temperature changes with time, and the collected data can be stored in the instrument or transmitted to ship and land stations via satellite. CTDs are used from a stationary vessel, but CTD devices such as Seasoar (fig. 7.9) can change depth while being towed from a moving vessel.

The SLOCUM glider (fig. 7.10) is an independent, long range instrument, designed to monitor water properties independently for up to five years. It carries multiple sensors and is able to send data and update its navigation by satellite when it periodically returns to the surface. Another independent float or buoy known as a "profiler" (fig. 7.11a&b) sinks to a preset depth, drifts with the current, and then measures water properties as it rises. At the surface it transmits its data to a ship or satellite and then sinks to repeat the process. Refer back to section 1.9 for an overview of project Argo that uses this type of float.

Satellites monitor surface water properties as temperature, salinity, elevation, waves, currents, and some marine life.

Figure 7.7

Hanging a Nansen water bottle. A mechanical water bottle with three deep-sea thermometers attached is fastened to a cable. When the water bottle is lowered to the desired depth it will be signaled to close by a weight that slides down the cable. At the time the water sample is taken, the temperature registered by the thermometers will be recorded. Sample depth is determined by measuring the cable as it passes over the meter wheel at the top of the photograph.

Figure 7.8

Retrieving a conductivity-temperature-depth sensor, or CTD. The CTD is attached below a rosette of water bottles. Data are relayed to the ship's electronic processing and data center. Water samples are taken when an interesting water structure is found or when water samples are required to calibrate the CTD.

More Information: www.seabird.com/

They have greatly advanced our understanding of surface phenomena on a global scale but can only sense this data to the depth of a few meters.

> Why are water bottles mounted on the CTD cage in figure 7.8?
>
> Why do the conductivity and temperature measurements recorded by a CTD provide better information than the individual samples taken by mechanical water bottles in figure 7.7?
>
> Compare the distribution of data collected by a Seasoar or a SLOCUM glider to the data obtained by a conventional CTD.

Wind-Driven Water Motion   7.5

When the winds blow over the oceans, they set the surface water in motion, driving the large-scale surface currents in nearly constant patterns. Refer to chapter 6 for a discussion of the Earth's surface winds. Because the friction between the water and the Earth's surface is small, the moving water is deflected by the Coriolis effect, to the right of the wind direction in the Northern Hemisphere and to the left in the Southern Hemisphere. In the open sea, the deflection of the surface current is at a 45° angle to the wind direction, as shown in figure 7.12. The combined effect of the wind on the surface and the deflection of the water creates the large-scale pattern of the wind-driven surface currents in the open oceans. Once set in motion, these currents are further modified by the interaction that occurs among

Figure 7.9

Seasoar is a conductivity-temperature-depth sensor that is towed by a ship. The wings allow Seasoar to move vertically while it is towed horizontally.

More Information: www.soc.sotom.ac/OTD/seasys/otd2.html and matisse.whoi.edu/seasoarhome.html

Figure 7.10

The SLOCUM glider has multiple sensors designed to monitor water properties independently for up to five years. While at the surface it is able to send data and update its navigation by satellite.

More Information: www.webbresearch.com/slocum.html

Figure 7.11

This Argo subsurface drifter is equipped with conductivity-temperature-depth sensors. The instrument sinks to a pre-programmed depth where it drifts for a set time and then surfaces. The position and recorded data are then transmitted to the research vessel or to a satellite from which the data may be retrieved.

More Information: www.argo.ucsd.edu

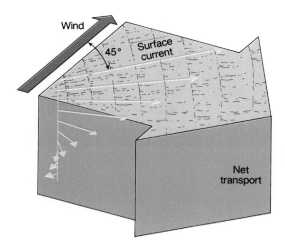

Figure 7.12

Water is set in motion by the wind. The direction and speed of flow change with depth to form the Ekman spiral. This change with depth is a result of the Earth's rotation and the inability of water, due to low friction, to transmit a driving force downward with 100% efficiency. The surface current is 45° to the right of the wind, while the net transport over the wind-driven column is 90° to the right of the wind in the Northern Hemisphere.

wind-driven currents, zones of converging and diverging water, and landmasses.

The water just below the surface receives less energy and moves more slowly than the water at the surface, and it is further deflected to the right (Northern Hemisphere) or left (Southern Hemisphere) of the surface water. The same is true for the water down to approximately 150 m (500 ft). The overall result is a spiral in which deeper water set in motion by the wind moves more slowly and with greater deflection than the water above, and the much-reduced current at the bottom moves exactly opposite to the surface (fig. 7.12). This spiral motion is called the **Ekman spiral,** after the Swedish physicist V. Walfrid Ekman, who developed its mathematical relationship in 1902. Over the depth of the spiral, the average flow of all the water set in motion by the wind, or the net flow known as **Ekman transport,** moves 90° to the right or left of the surface wind, depending on the hemisphere in which the motion takes place. Ekman transport, therefore, differs from the surface current, which moves at an angle of 45° to the wind direction.

If a north wind blows across the sea surface, what direction does the surface current flow in the Northern Hemisphere and in the Southern Hemisphere?

What direction is the net flow of all wind-driven water in the question above?

Current Flow and Gyres 7.6

In the open ocean between the westerlies and the trade winds (review section 6.9 and fig. 6.16), Ekman transport directs the surface water toward 30° latitudes N and S forming elevated convergences extending across the oceans in each hemisphere. As the sea level in each ocean is raised, pressure causes the water to slide downhill away from the convergences. When the inward Ekman transport and the outward pressure flow balance, wind-driven water moves parallel to elevation contours of the convergences.

Refer to figure 7.13 as you read the following description. The water under the westerlies moves away from an ocean's western shore or boundary and toward its eastern shore or boundary; the water under the trade winds moves the water away from an ocean's eastern boundary and toward its western boundary. North of the equator, when water moving with the trade winds accumulates at the land boundary on the west side of an ocean, the water flows north and parallel to the land toward the region from which the water moves eastward in the latitudes of the westerlies. Water moving with the westerlies and accumulating at the land boundary on the east side of a

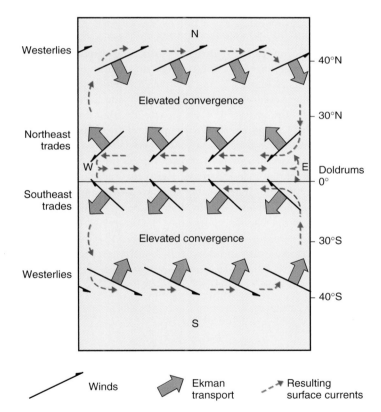

Figure 7.13

Wind-driven transport and resulting surface currents in an ocean bounded by land to the east and to the west. The currents form large oceanic gyres that rotate clockwise in the Northern Hemisphere and counterclockwise in the Southern Hemisphere.

Northern Hemisphere ocean flows south, parallel to the land, toward the region from which the water moves westward. This produces a series of interconnecting surface currents moving in a circular, clockwise path centered on 30°N latitude. In the Southern Hemisphere the water will move from east to west with the trade winds, then south to the latitude of the westerlies where it will move west to east with the westerlies, and then north back into the trade wind latitudes. South of the equator the path is circular and counterclockwise, centered on 30°S.

Such a wind-driven circular flow forms a mound of converging surface water at its center. This water is driven toward the center by the Coriolis effect and can result in an increase in the water's elevation by a meter (3 ft) or more. There is a direct relationship between the speed of the flow and the force of the Coriolis deflection. When the elevation of the mound is such that the outward, pressure driven flow equals the inward Coriolis effect, a balance is achieved. When this balance is in place the water flow is no longer deflected inward, but instead the water moves around the mound parallel to the mound's contours (fig. 7.14). These large, circular, wind-driven, oceanic flows are called **gyres.**

Wind-driven, open-ocean surface currents move at speeds that are about one one-hundredth of the average wind speed above the sea surface, between 0.1 and 0.5 meters per second (0.3–1.5 ft/s or 0.25–1 knot). The speed of the surface flow around a large ocean gyre is not necessarily the same on all sides. In the North Atlantic and North Pacific the currents flowing from low to high latitudes on the western sides of the gyres tend to be narrower and stronger than those on the eastern side. This is known as the **western intensification** of the currents. The increased flow on the western sides of the gyres is caused by (1) the increasing Coriolis effect with increasing latitude, (2) the change in wind strength and direction with increasing latitude, and (3) the friction between ocean currents and landmasses. The intensification of flow is most conspicuous in the Gulf Stream and Kuroshio currents of the Northern Hemisphere. In the Southern Hemisphere the intensification is obscured in the Atlantic by currents deflected by the coast of Brazil, and in the Pacific a portion of the westward-driven water escapes through the islands of Indonesia. Also the southern tips of South America and Africa deflect the circumpolar current flow around Antarctica into the eastern sides of the gyres in the southern oceans (see figure 7.15 in section 7.7).

Why is the sea surface elevated in the center of the major current gyres of the world's oceans?

Why do the current gyres rotate clockwise in the Northern Hemisphere and counterclockwise in the Southern Hemisphere?

Draw diagrams showing the direction of winds and currents in a Northern Hemisphere gyre and in a Southern Hemisphere gyre.

What factors obscure western intensification in the Southern Hemisphere?

Ocean Surface Currents   7.7

The major surface currents that result from the processes previously described are shown in figure 7.15. Use this figure to follow the current patterns described in this section.

The Pacific Ocean

In the North Pacific, the northeast trade winds push the water toward the west and northwest; this is the *North Equatorial Current.* The westerlies create the *North Pacific Current* or *North Pacific Drift* moving from west to east. A current's direction or set is the direction in which it flows. This is an easterly current. Note that the trade winds move the water away from Central and South America and pile it up against Asia, while the westerlies move the water away from Asia and push it against the west coast of North America. The water that accumulates in one area must flow toward areas from which the water has been removed. This movement forms two north-south currents: the *California Current,* moving south along the western coast of North America, and the *Kuroshio Current,* moving north along the east coast of Japan. The Kuroshio and California currents are part of a continuous flow or circular motion centered around 30°N latitude. This circular, clockwise flow of water is called the North Pacific gyre. Other major North Pacific currents include the *Oyashio Current,* driven by the polar

Figure 7.14

Currents flow *(V)* around a gyre when F_c, the inward Ekman transport due to the Coriolis effect, is balanced by F_g, the outward pressure force. The example is of a clockwise gyre in the Northern Hemisphere creating a convergence in the gyre's center. Diagram is not drawn to scale.

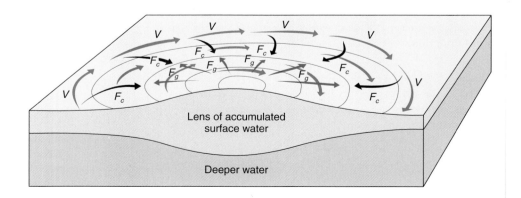

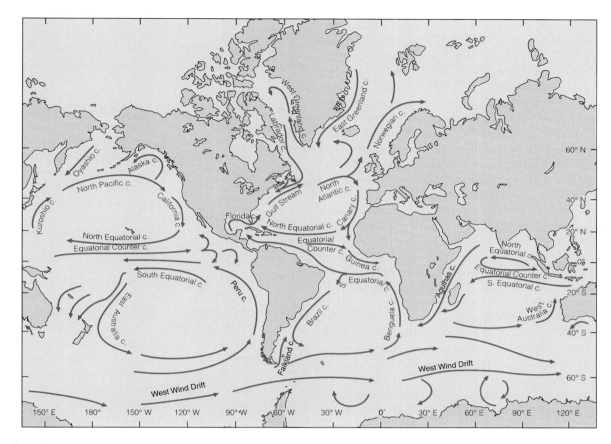

Figure 7.15

The major surface currents of the world.

More Information: www.athena.ivv.nasa.gov/curric/oceans/drifters/ocecur.htm and www.whoi.edu/coastal-briefs/Coastal-Brief-94-05.html

easterlies, and the *Alaska Current,* fed by water from the North Pacific Current and moving in a counterclockwise gyre in the Gulf of Alaska.

In the South Pacific Ocean, the southeast trade winds move the water to the left of the wind and westward, forming the *South Equatorial Current.* At a higher latitude, the westerly winds push the water to the east, where at these southern latitudes it moves almost continuously around the Earth as the *West Wind Drift.* The tips of South America and Africa deflect a portion of this flow northward on the east side of both the South Pacific and South Atlantic Oceans. As in the North Pacific, currents between the South Equatorial Current and the West Wind Drift complete the gyre. The *Peru Current* or *Humboldt Current* flows north along the coast of South America, and the *East Australia Current* moves south on the west side of the ocean. These four currents form the counterclockwise South Pacific gyre.

The North Pacific and South Pacific gyres are formed not on either side of 0° but on either side of 5°N, because the doldrum belt is displaced northward due to the unequal heating between the Northern and Southern Hemispheres. Also between the North and South Equatorial currents, under the

doldrums, there is a surface current moving down slope west to east, the *Equatorial Countercurrent.* This current helps to return surface water accumulated against the coast of Asia by the Equatorial Currents. See figures 7.13 and 7.15.

Current charts may show ocean surface flows measured over a specific period of time or they may show the average of currents over a much longer period of time. These charts do not indicate natural daily or seasonal variations, nor do they demonstrate long-term cyclic changes such as El Niño and La Niña (refer back to section 6.14). Recently oceanographers have begun the study of 20–30-year current and climate cycles in the Northeast Pacific, known as the *Pacific Decadal Oscillation* or PDO. These cycles are thought to affect environmental conditions along the northwest coast of the United States.

The Atlantic Ocean

The North Atlantic westerly winds move the water eastward as the *North Atlantic Current* or *North Atlantic Drift.* The northeast trade winds push the water to the west, forming the *North Equatorial Current.* The north-south currents are the *Gulf Stream,* moving north along the coast of North America, and the *Canary*

Current, flowing south on the eastern side of the North Atlantic. The Gulf Stream is fed by the *Florida Current* and the North Equatorial Current. The North Atlantic gyre rotates clockwise. The polar easterlies provide the force for the *Labrador* and *East Greenland Currents* that balance water flowing into the Arctic Ocean from the *Norwegian Current.*

In some areas of the Gulf Stream the current speed may reach 1.5 meters per second (5 ft/s or 3 knots); its volume transport rate is 55×10^6 m^3/s, more than 500 times the flow of the Amazon River. The Florida Current, exiting the Gulf of Mexico through the narrow channel between Florida and Cuba, may exceed 1.5 meters per second.

The gyre currents of the central North Atlantic isolate a lens of clear, warm surface water 1000 m (3300 ft) deep. This is the Sargasso Sea, bounded by the Gulf Stream on the west, the North Atlantic Current on the north, the Mid-Atlantic Ridge rising upward from the sea floor on the east, and the North Equatorial Current on the south. This region is famous for its floating mats of *Sargassum,* a brown seaweed collected by the water motion and held within the gyre. Except for the floating *Sargassum,* with its rich and specialized community of small plants and animals, the area is biologically a near desert because of the downwelling and nutrient-poor surface water associated with the gyre's center.

In the South Atlantic, the westerlies continue the West Wind Drift. The southeast trade winds move the water to the west, but the bulge of Brazil splits the *South Equatorial Current.* Much of this flow is deflected northward over the top of South America, into the Caribbean Sea, and eventually into the Gulf of Mexico, where it exits as the Florida Current to feed the Gulf Stream. A portion of the South Equatorial Current slides south of the Brazilian bulge along the western side of the South Atlantic to form the *Brazil Current.* The *Benguela Current* moves northward up the African coast, completing the South Atlantic gyre that rotates counterclockwise.

Because much of the South Equatorial Current is deflected across the equator, the Equatorial Countercurrent appears only weakly in the eastern tropical Atlantic. The northward movement of South Atlantic water across the equator means that there is a net flow of surface water from the Southern Hemisphere to the Northern Hemisphere. This flow is balanced by a flow of water at depth from the Northern Hemisphere to the Southern Hemisphere, the North Atlantic deep water discussed in section 3 of this chapter. Again, the equatorial currents are displaced north of the equator, although not as markedly as in the Pacific Ocean.

Studies of surface water changes in the North Atlantic provide evidence of a system similar to the *Pacific Decadal Oscillation.* Pools of warm and cool surface water with life spans of four to ten years circle the North Atlantic. Cold low salinity surface water affects the atmosphere and creates weather systems that are associated with cooler than average land temperatures, bringing hard winters to northern Europe and North America. These climate systems are referred to as the *North Atlantic Oscillation.* Cold, low salinity surface water also affects the rate of North Atlantic deep water formation.

The Arctic Ocean

The polar easterlies drive the water and ice of the Arctic Ocean in a large clockwise gyre centered over the Canadian Basin at 150°W and 80°N (fig. 7.16). A small amount of water enters the Arctic Ocean from the Pacific Ocean through the Bering Strait. Some water from the Atlantic enters the Arctic west of Spitsbergen, but most Atlantic water flows over the top of Norway and Finland as the *Norwegian Current* and then moves eastward along Siberia's Arctic coast. North of Greenland Arctic Ocean water exits as the *East Greenland Current* and flows south into the North Atlantic. The *West Greenland Current* carrying water from the Arctic Ocean joins the *Labrador Current* moving to the south along the Canadian coast.

The Indian Ocean

This ocean is mainly a Southern Hemisphere ocean. The southeast trade winds push the water to the west, creating the *South Equatorial Current.* The Southern Hemisphere westerlies still move the water eastward in the West Wind Drift. The gyre is completed by the *West Australia Current* moving northward and the *Agulhas Current* moving southward along the coast of Africa. In this Southern Hemisphere ocean the currents move to the left of the wind direction, and the gyre rotates counterclockwise. The northeast trade winds strengthen the *North Equatorial Current* during the dry monsoon season, and the

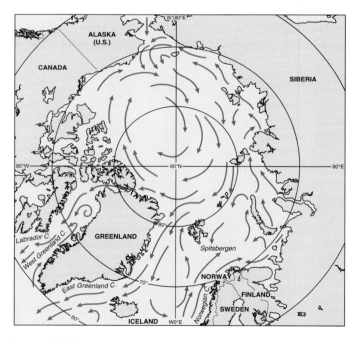

Figure 7.16

The circulation in the Arctic Ocean is driven by the polar easterlies producing a large, clockwise gyre. Water enters the Arctic Ocean from the North Atlantic by way of the Norwegian Current and exits to the Atlantic by the East Greenland Current and the Labrador Current.

Equatorial Countercurrent is reduced. The wet monsoon season winds strengthen the Equatorial Countercurrent and reduce the North Equatorial Current. The strong seasonal monsoon controls the surface flow of the Northern Hemisphere portion of the Indian Ocean. This strong seasonal shift is unlike anything found in either the Atlantic or the Pacific.

In which direction do the major ocean current gyres flow in the Northern Hemisphere and in the Southern Hemisphere?

Follow the major ocean gyres and identify the major currents in the North Atlantic, South Atlantic, North Pacific, South Pacific, and Indian oceans.

Where does water enter and exit the Arctic Ocean?

Why are the equatorial countercurrents located under the doldrums?

Which is the stronger current and why: (a) the Kuroshio or the California? (b) the Gulf Stream or the Canary?

How does deep-ocean circulation compensate for the surface flow across the equator in the Atlantic Ocean?

Explain the role of the monsoon on the currents of the northern Indian Ocean.

Open-Ocean Convergence and Divergence 7.8

Density-driven thermohaline circulation and wind-driven surface currents combine to create regions of open-ocean **convergence** and **divergence.** Surface waters converge or move into a region where downwelling is occurring. When the downwelled surface water reaches a level at which it is denser than the water above but less dense than the water below, it spreads horizontally as more water descends behind it. The downwelled water displaces less-dense water upward, creating upwelling and forcing the surface waters to diverge or move away from the area.

Open-ocean convergence and divergence zones are shown in figure 7.17. The tropical convergence is at the equator; it extends across the oceans centered under the doldrums. The two subtropical convergences are at approximately 30° to 40°N and S; the Arctic and Antarctic convergences occur about 50°N and S, and the coastal Antarctic convergence is at 60°S. There are three major oceanic divergence zones: two tropical divergences and the Antarctic divergence. Surface convergence zones produce downwelling that transports oxygen-rich surface water to depth, and upwellings at divergences deliver water rich in nutrients to the surface, where these fertilizers promote the production of plantlike organisms that supply the food chains supporting the remainder of the oceans' organisms.

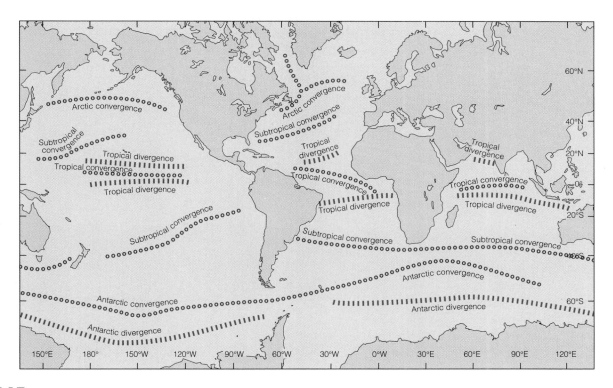

Figure 7.17

The principal zones of surface convergence and divergence associated with wind-driven and thermohaline circulation.

Convergences and divergences of surface currents also produce mixing between water from different geographic areas. Waters carried by the surface currents converge and form new water mixtures with intermediate values of temperature and salinity. These new water mixtures then sink to their density levels, move horizontally, blending and sharing properties with adjacent water, and eventually rise to the surface at a new location. Thermohaline circulation and wind-driven surface currents are closely related, so closely related that it is difficult to assign a priority of importance to one process over the other.

Over the history of the Earth interactions between thermohaline and wind-driven circulation have changed many times. Far back in time the changing sea levels, shapes of the ocean basins and locations of the continents influenced ocean circulation. Geological studies are providing evidence of changes in winds and sea surface processes during glacial and interglacial periods. Over long periods of time climate changes alter the locations and roles of downwellings and upwellings. Combinations of wind-driven surface currents and thermohaline circulation produce interconnected flows that redistribute heat, salt, and dissolved gases as the water moves from one ocean to another. Figure 7.18 presents a general view of the present-day flow of ocean water within and between oceans, at the surface and at depth.

Compare the processes of divergence and convergence.

If a convergence zone is present, why must there also be a divergence zone?

Refer to figures 7.5 and 7.16 and follow the water sinking at the Arctic convergence; identify the area in which it resurfaces.

Coastal Upwelling and Downwelling   7.9

Wind-driven Ekman transport can drive surface water toward or away from a coast. In coastal areas where the trade winds move the surface waters away from the western side of the continents, upwelling occurs nearly continuously throughout the

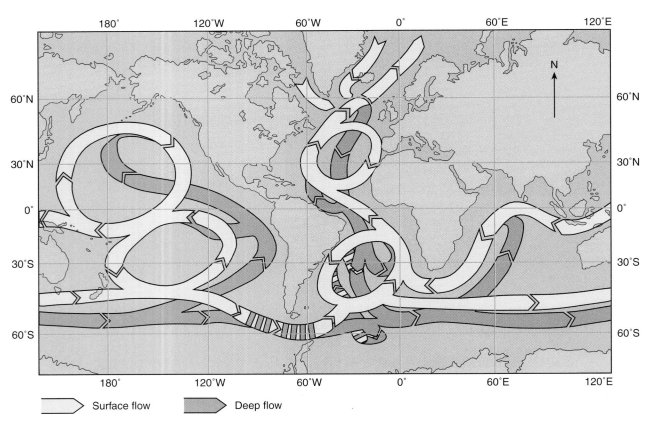

Surface flow   Deep flow

Figure 7.18

The present flow of ocean water within and between oceans circulates at the surface and at depth. Cold surface water sinks to the sea floor in the North Atlantic Ocean, then flows south to be further cooled by the Antarctic bottom water formed in the Weddell Sea at 40°W 65°S. This deep water moves eastward around Antarctica, feeding into the surface layer of the Indian Ocean and also into the deep basins of the Pacific Ocean. A return flow of surface water from the Pacific and Indian Oceans flows north to replace the surface water in the North Atlantic Ocean.

year. These zones of upwelling are the product of the winds, not of thermohaline circulation. The rich fishing grounds along the west coasts of Africa and South America are supported by this coastal upwelling of nutrient-rich water that stimulates the growth of marine plants. Refer back to the discussion of El Niño in section 6.14.

Seasonal wind-driven downwellings and upwellings also occur along ocean coasts, for example, North America from central California to Vancouver Island. The wind pattern in this region changes from southerly in winter to northerly in summer, causing changes in the coastal Ekman transport. Remember that open-ocean Ekman transport moves at an angle of 90° to the right or left of the wind direction depending on the hemisphere.

Along this coast the average wind blows from the north in the summer, and the net movement of the water is offshore, resulting in an upwelling along the coast (fig. 7.19a). In winter the winds blow from the south, and the net movement of the water is onshore against the coast, producing a downwelling in the same region (fig. 7.19b). The summer upwelling produces cold coastal water at San Francisco and helps cause its frequent summer fogs. Because of the lack of land to alter the wind patterns at middle latitudes in the Southern Hemisphere, this type of seasonal upwelling is less common. Refer to figure 6.18.

> Why is there a year-round upwelling along the west coasts of South America and Africa?
>
> Explain the seasonal upwelling and downwelling along the west coast of the United States.

Eddies 7.10

As the Gulf Stream moves away from the coast, it develops a meandering path. At times, the western edge of the Gulf Stream develops indentations that are filled by cold water from the Labrador Current side. These indentations pinch off and become **eddies,** or packets of water moving with a circular motion. These eddies are displaced to the east and south of the current boundary and transfer cold water into warm water. Likewise, bulges at the western edge of the Gulf Stream are filled with warm water. When these bulges are cut off, they drift to the west and north of the Gulf Stream, into cold water. This process is illustrated in figure 7.20. Figure 7.21 is a satellite view of Gulf Stream eddies as they form along the boundary of the current. Current eddies dissipate the energy of flowing water by friction, preventing the current from increasing its speed as wind energy tries to drive it faster and faster. Daily and weekly current patterns differ greatly from the long-term average current flows shown on most ocean current charts, because these charts do not include the presence of eddies, meanders, and other short-term variations.

Large and small eddies exist in all parts of the oceans and at all depths; they range from tens to several hundred kilometers in diameter. Each eddy contains water with specific chemical and physical properties and maintains its identity as it wanders through the oceans, eventually mixing with the surrounding water.

> What are eddies and where are they found?
>
> How is warm water transferred into cold water and vice versa along the edges of the Gulf Stream?

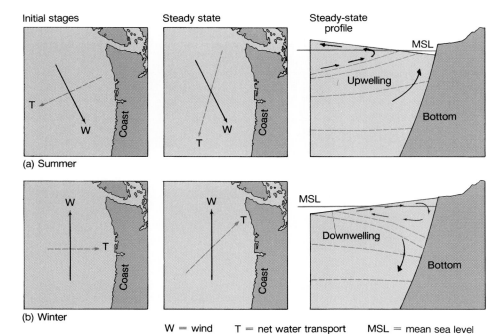

(a) Summer

(b) Winter

W = wind   T = net water transport   MSL = mean sea level

Figure 7.19

Initially, the Ekman transport is 90° to the right of the wind along the northwest coast of North America. As water is transported (a) away from the shore in summer, or (b) toward the shore in winter, a sea surface slope is produced. This slope creates a gravitational force that alters the direction of the Ekman transport to produce a balance of forces and a new steady-state direction of the current.

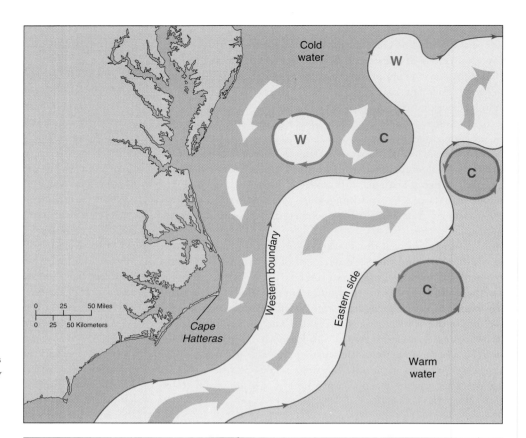

Figure 7.20

The western boundary of the Gulf Stream is defined by sharp changes in current velocity and direction. Meanders form at this boundary after the Gulf Stream leaves the U.S. coast at Cape Hatteras. The amplitude of the meanders increases as they move downstream (a and b). Eventually the current flow pinches off the meander (c). The current boundary re-forms, and isolated rotating eddies of warm water (W) wander into the cold water, while cells of cold water (C) drift through the Gulf Stream into the warm water (d).

More Information:
www.pmel.noaa.gov/bering/pages/env_cur.html

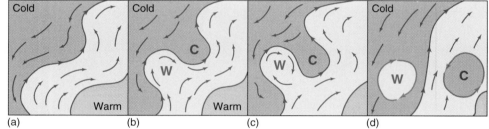

(a) (b) (c) (d)

Figure 7.21

A satellite image of the sea surface reveals the warm (orange and yellow) and cold (green and blue) eddies that spin off the Gulf Stream. (Reddish blue areas at the top are the coldest waters.) These eddies may stir the water column right down to the ocean floor, kicking up blizzards of sediment.

More Information: www.pmel.noaa.gov/bering/pages/env_cur.html

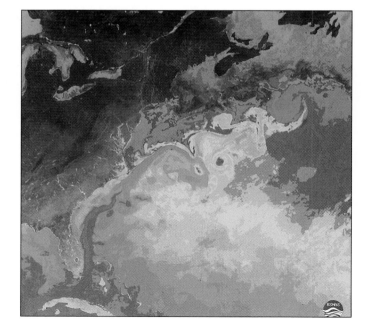

Measuring the Currents   7.11

Current measurements may be made by following a parcel of moving water. The water may be labeled with dye that can be photographed from the air, or the currents may be followed with buoys (fig. 7.22) designed to float at the surface. Buoy positions may be tracked by satellite and a series of pictures or position fixes may be used to calculate the speed and direction of the current. Instruments that drift at predetermined depths use sound to communicate their positions to a surface buoy or ship; or they may also rise to the surface to signal their positions back to the research vessel or a satellite (refer back to fig. 7.11). Another approach to tracking currents is described in the Item of Interest at the end of this chapter.

Currents are also measured by determining the speed and direction of the water as it passes a fixed point. The workhorse **current meters** used by oceanographers during the last 20 years included a rotor to measure the current's speed and a vane to measure its direction (fig. 7.23a). If the current meter was lowered from a stationary vessel in shallow water, the measurements could be returned to the ship by a cable or they could be stored by the meter for reading upon retrieval. In deep water this type of meter can also be attached to an anchored buoy system (fig. 7.24a) that is entirely submerged and so not affected by winds or waves. The floats and meters are retrieved by sending a sound signal to a special acoustical link that detaches the float and meter from the anchor (fig. 7.24b).

Newer electronic current meters have no vane or rotor (fig. 7.23b). These meters emit narrow, fixed-frequency sound beams in four directions. The returning echoes from small particles in the moving water are monitored for their frequencies. If the water moves toward the sound source, the echo frequency is higher, but if the water moves away from the sound source, the sound frequency is lower. The change in echo frequency relates to water speed and is known as the Doppler Effect. To obtain current direction and speed the measurements from all four directions are combined, and the result is checked against the internal compass of the current meter. The data can be stored in the meter or transmitted to a ship or surface buoy by wire or acoustic signal. Some current meters have sensors that measure additional water properties; see fig. 7.23b.

> What measurements are necessary to define a current?
>
> What are the advantages of electronic current meters? Are there any disadvantages?

Figure 7.22

Surface buoys (red) carrying instruments are deployed from the research vessel *Thomas G. Thompson*.

(a)

(b)

Figure 7.23

(a) An internally recording Aanderaa current meter. The vane orients the meter to the current while the rotor determines current speed.
(b) A Doppler current meter sends out sound pulses in four directions. The frequency shift of the returning echoes allows the detection of the current. These meters are also equipped with salinity-temperature-depth sensors as well as instruments for measuring water turbidity and oxygen. The data may be stored internally and collected at another time.

More Information: www.aanderaa.com/

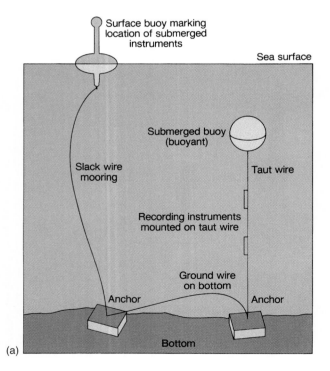
(a)

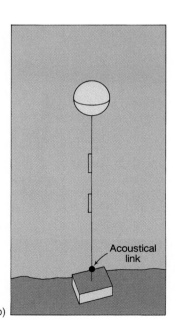
(b)

Figure 7.24

Taut wire moorage. (a) Recovery is accomplished by retrieving the surface buoy and hauling in the wire. If the surface buoy is lost, it is possible to grapple for the ground wire. (b) In this system, a sound signal disconnects the anchor, and the equipment floats to the surface.

Energy Sources 7.12

Because of the transparency and heat capacity of water, large amounts of solar energy can be stored in the ocean. Temperature and salinity variations with depth represent potential sources of energy and may be alternatives to the consumption of fossil fuels. Two possible techniques for extracting this energy are discussed under Thermal Energy and Harnessing the Currents.

Thermal Energy

Ocean thermal energy conversion or **OTEC** depends on the difference in temperature between ocean surface water and water at 600 to 1000 m (2000–3300 ft) depth. There are two different types of OTEC systems: (1) closed cycle, which uses a contained working fluid with a low boiling point, such as am-monia or Freon, and (2) open cycle, which directly converts seawater to steam.

In a closed system (fig. 7.25a) the warm surface water is passed over the evaporator chamber containing the ammonia or Freon. The working fluid is vaporized by the heat derived from the warm seawater.

The vapor builds up pressure that is used to spin a turbine, which generates power. After the pressure has been released, the working fluid is passed through a condenser, where it is cooled by cold water pumped up from depth. Cooling the ammonia or Freon returns it to its liquid state, and it is pumped back to the evaporator to repeat the cycle.

In the open system (fig. 7.25b), warm seawater is converted to steam in a low-pressure vacuum chamber, and the steam is used as the working fluid. Since less than 0.5% of the incoming water is turned into steam, large quantities of warm water must be used. The steam passes through a turbine and is condensed by using cold water from depth to cool the condenser.

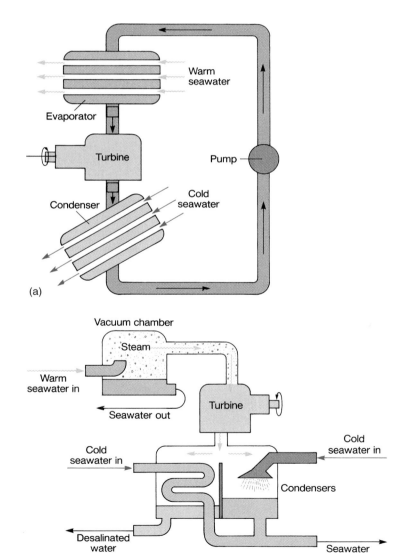

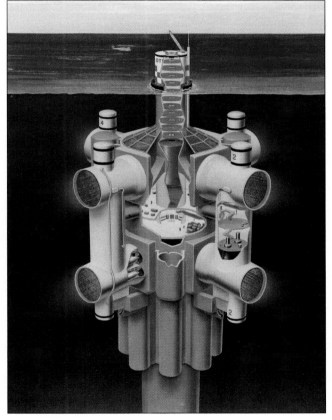

(c)

Figure 7.25

(a) The simplified working system of an OTEC closed-system electrical generator. (b) The simplified working system of an OTEC open-system electrical generator. (c) The proposed Lockheed OTEC Spar, designed for generating electrical energy from the ocean temperature gradient.

More Information: bigisland.com/nelha/

The condensation product is desalinated water, another source of revenue in regions lacking both power and fresh water, for example the Caribbean and Pacific island nations.

OTEC plants using either system can be located onshore, offshore (fig. 7.25), or on a ship that moves from place to place. But, OTEC requires at least a 20°C difference in temperature between surface and depth to generate power in excess of that needed to run its own pumps. This means that the most efficient power plants would be located at latitudes between approximately 25°N and 25°S where seawater temperature differences over depth average 22°C.

Engineers studying the potential and feasibility of OTEC estimate that about 0.2 MW of usable power can be extracted per km^2 of tropical ocean surface, which is about 0.07% of the average absorbed solar energy. The process is considered to be environmentally benign, but the pumping of cold water from depth to the sea surface brings nutrients from depth, which are liberated into the nutrient-poor tropical surface water. This artificial upwelling of cold nutrient water changes the productivity of the surface water, stimulating marine plant growth and interfering with the life process of those surface animals that are dependent on tropical temperatures.

The Natural Energy Laboratory of Hawaii at Keahole Point is a land-based, open-system OTEC plant. It produces a net of 100 kW of power per day using the temperature differences between the water at the ocean's surface and water about 800 m (2500 ft) below the surface. It also produces 26,000 L (7000 gal) of desalinated water per day. It began producing power in 1992 and is the world's only system pumping cold water from depth. The cost of the plant is about ten times the cost of a coal or oil burning plant, but the OTEC fuel is free. Maintenance costs are about the same or less than those for a fossil fuel plant.

The nutrient-rich cold water pumped from depth is used for aquaculture projects including production of oysters, shrimp, fish, and various seaweeds. It is also used as a coolant to provide air-conditioning for the buildings at Keahole Point at one tenth the cost of electricity.

Harnessing the Currents

The massive oceanic surface currents of the world are untapped reservoirs of energy. Their total energy has been estimated at 2.8×10^{14} (280 trillion) watt-hours. Harnessing the energy from open-ocean currents would require the use of turbine-driven generators anchored in place. Large turbine blades would be driven by the moving water just as windmill blades are moved by the wind.

The cost of constructing, mooring, and maintaining current-driven power-generating devices in the open sea makes them noncompetitive with other sources of power at this time. Also, most wind-driven oceanic currents generally move too slowly or are found too far from where the power is needed. However, if ocean currents were to be developed as energy sources, the Florida Current and the Gulf Stream are the most likely to be tapped. These are swift and continuous currents moving close to shore in areas where there is a demand for power.

> What is the difference between a closed-cycle and open-cycle OTEC system?
>
> In what areas of the world is OTEC a suitable energy resource?
>
> Why is energy unlikely to be derived from open-ocean currents in the foreseeable future?
>
> Why can energy produced by either OTEC or ocean currents be considered solar energy?

Summary

The surface water changes its density with changes in salinity and temperature that are keyed to latitude. The densities at depth are more homogeneous.

If the density increases with depth, the water column is stable; unstable water columns overturn and return to a stable distribution. Neutrally stable water columns are mixed vertically by winds and waves. Vertical circulation that is driven by changes in surface density is known as thermohaline circulation. In the open ocean, temperature is more important than salinity in determining the surface density. Salinity is the more important factor in areas that are close to shore and freshwater discharge from the land.

The oceans are layered systems. In the Atlantic Ocean, waters of differing densities are formed at the surface at different latitudes. They sink and flow north or south at differing depths. Arctic Ocean layers are controlled by the seasonal freezing and thawing of the sea ice. The water layers of the Pacific Ocean lose their identity in the large volume of this ocean; their movements are sluggish. The water layers of the Indian Ocean are also less distinct than those of the Atlantic, with no water that corresponds to that formed in the northern latitudes.

For many years water bottles and thermometers have been sampling salinities and sea temperatures from stationary ships. Today's CTD sensors record many more variables and monitor them continuously with depth and time. Some devices are towed by moving ships; others are independent long-term and long-range sensors that move with the current.

Winds push the surface water 45° to the right of their direction in the Northern Hemisphere and 45° to the left in the Southern Hemisphere. Water below the surface receives less energy, moves more slowly, and is deflected still further from the wind. Averaged over the depth of the motion, Ekman transport is 90° to the right (Northern Hemisphere) or left (Southern Hemisphere) of the surface wind.

Large surface current gyres occur in each ocean. Southern Hemisphere gyres rotate counterclockwise; Northern Hemisphere gyres rotate clockwise. The currents of the northern Indian Ocean change with the seasonal monsoon.

The elevated water surface in the center of a gyre is caused by the balance between an outward pressure force and an inward force driving Ekman transport. The surface water in the Sargasso Sea is isolated by this current flow. Ocean currents transport very large volumes of water that flow faster when the current is forced to flow through a narrow space. Currents on the western side of Northern Hemisphere oceans are stronger and narrower than currents on the eastern side.

Water sinks and downwelling occurs at the convergence of surface currents; water rises at zones of surface current divergence. There are major oceanic areas of convergence and divergence located at specific latitudes.

Eddies occur at all depths, wander long distances, and gradually lose their identity.

Seasonal upwellings and downwellings occur in coastal areas that have changing wind patterns and an alternating coastal flow of water onshore and offshore due to the Ekman transport. Thermohaline circulation and wind-driven surface currents are closely related and occur on local and world scales.

A variety of instruments are available to measure currents, either by following the water or by measuring the water's speed and direction as it moves past a fixed point.

Ocean thermal energy conversion (OTEC) is a method of extracting energy from the oceans; a land-based plant operates in Hawaii. Currents represent another possible source of energy, although costs make this source noncompetitive at present.

4. Why would a decrease in the rate of formation of North Atlantic deep water create a decrease in the volume of water moving northward in the Gulf Stream?
5. During what seasons would you expect to find stable conditions in the surface layers of oceans in the tropics, at the poles, and at temperate latitudes? What causes the stability in each case?

Suggested Readings

Currents

Ackerman, J. 2000. New Eyes on the Oceans. *National Geographic* 198(4):86–115.

Bahr, F., and P. Fucile. 1995. Seasoar — A Flying CTD. *Oceanus* 38(1):26–27.

Bowditch, N. 1984. *American Practical Navigator*, Vol. 1. U.S. Defense Mapping Agency Hydrographic Center, Washington, DC. 1414 pp. Ocean currents are covered in Chapters 31 and 32. (Deep-ocean storms and eddies.)

Broecker, W. S. 1995. Chaotic Climate. *Scientific American* 273(5):62–69. (Relates currents and climate.)

Feder, T. 2000. Argo Begins Systematic Global Probing of the Upper Oceans. *Physics Today* 53(7):50–51.

Ingraham, W. Jr., C. Ebbesmeyer, and R. Hinrichsen. 1998. Imminent Climate and Circulation Shift in Northeast Pacific Ocean Could Have Major Impact on Marine Resources. *EOS* 79(16):197 and 201.

Kunzig, R. 1996. In Deep Water. *Discover* 17(12):86–96.

MacLeish, W. H. 1989. The Blue God, Tracing the Mighty Gulf Stream. *Smithsonian* 19(11):44–59.

MacLeish, W. H. 1989. Painting a Portrait of the Stream from Miles Above and Below. *Smithsonian* 19(2):42–55. (More about the Gulf Stream.)

Richardson, P. L. 1993. Tracking Ocean Eddies. *American Scientist* 81(3):261–72.

Whitworth, III, T. 1988. The Antarctic Circumpolar Current. *Oceanus* 31(2):53–58.

Zimmer, C. 1995. The North Atlantic Cycle. *Discover* 16(1): 77.

Energy

Avery, W. H., and C. Wu. 1994. *Renewable Energy from the Ocean — A Guide to OTEC*. Oxford University Press, New York. 474 pp.

International OTEC/DOWA Association Conference Report. 1994. *Sea Technology* 35(5):53–54.

Loupe, D. 1991. The Food Factor. *Sea Frontiers* 37(2):22–27. (Projects associated with OTEC plants.)

Key Terms

Antarctic bottom water, 160

Antarctic intermediate water, 160

conductivity-temperature-depth sensor (CTD), 162

convergence, 169

current meter, 173

divergence, 169

downwelling, 159

eddies, 171

Ekman spiral, 165

Ekman transport, 165

gyre, 166

North Atlantic deep water, 160

ocean thermal energy conversion (OTEC), 175

overturn, 159

South Atlantic surface water, 162

thermohaline circulation, 159

upwelling, 159

western intensification, 166

Critical Thinking

1. Why is the density layering of the Atlantic Ocean more pronounced than the density layering of the Pacific Ocean?
2. Why is there a net northward flow of surface water across the equator in the Atlantic Ocean but not in the Pacific Ocean?
3. Why are surface convergences found at the centers of the large current gyres in both the northern and southern Atlantic and Pacific oceans?

internet references

worldwide websites

Ocean Conditions

Sea Surface Temperature Maps
www.dnr.qld.gov.au/longpdk/lpsst.htm

U.S. JGOFS Home Page
www1.whoi.edu/jgofs.html

Argo Home Page
www.argo.ucsd.edu/

World Ocean Circulation Experiment (WOCE)
www-ocean.tamu.edu/WOCE/uswoce.html

Parallel Ocean Program (POP) Simulation
www.scivis.nps.navy.mil/~braccio/global.html

Equipment and Technical Information

Sea-Bird Electronics, Inc.
www.seabird.com/

Ocean Sensors—Intelligent Information
www.oceansensors.com/

CTD Systems
www.soc.soton.ac.uk/OTD/seasys/otd2.html

Seasoar Development
chowder.ucsd.edu/~lyn/seasoar.html

Seasoar Labs—Home Page
matisse.whoi.edu/seasoarhome.html

Gliders and Slocum
www.webbresearch.com/slocum.htm

Aanderaa Instruments
www.aanderaa.com/

North Atlantic Oscillation
www.cru.uea.ac.uk/tiempo/floor2/data/nao.htm

Pacific Decadal Oscillation (PDO)
tao.atmos.washington.edu/pdo/

Pacific Decadal Oscillation Index
www.iphc.washington.edu/staff/hare/html/
decadal/post1977/pdo1.html

Ocean Currents

Primer on Ocean Currents
www.whoi.edu/coastal-briefs/Coastal-Brief-94-05.
html

Ocean Currents
www.athena.ivv.nasa.gov/curric/oceans/drifters/
ocecur.html

Ocean Planet: Ocean Currents
seawifs.gsfc.nasa.gov/OCEAN_PLANET/HTML/
oceanography_currents_1.html

Currents and Eddies
www.pmel.noaa.gov/bering/pages/env_cur.html

TOPEX/Poseidon—Science Images
topex-www.jpl.nasa.gov/discover/dyn_topo_
arrows.html

Ocean Thermal Energy Conversion (OTEC)

What Is Ocean Thermal Energy Conversion
www.nrel.gov/otec/what.html

History of OTEC
www.nelha.org//otec.html

A Guide to OTEC
www.jhuapl.edu/public/books/avery.htm

The Natural Energy Laboratory of Hawaii
bigisland.com/nelha/

In May 1990 a severe storm in the North Pacific caused the Korean container ship, *Hansa Carrier,* to lose overboard twenty-one deck-cargo containers, each approximately forty feet long (fig. 1). Among the items lost were 39,466 pairs of NIKE brand athletic shoes on their way to the United States. Six months to a year later these began washing up along the beaches of Washington, Oregon, and British

and dates for some 1600 shoes that had been found between northern California and the Queen Charlotte Islands in British Columbia. Additional beachcombers were asked for information, and Ebbesmeyer mapped the times and locations where batches of 100 or more shoes had been found (fig. 3). Next Ebbesmeyer visited Jim Ingraham at NOAA's National Marine Fisheries Service's offices in Seattle to study

The Great Sneaker Spill

Columbia (fig. 2). They were wearable after washing and having the barnacles and the oil removed, but the shoes of a pair had not been tied together for shipping, and pairs did not come ashore together. As beach residents recovered the shoes (some with a retail value of $100 a pair), swap meets were held in coastal communities to match the pairs.

In May of 1991 oceanographer Curtis Ebbesmeyer of Seattle, Washington read a news article on the beached NIKES. He was intrigued and realized that 78,932 shoes was a very large number of drifting objects compared to the 33,869 drift bottles used in a 1956–59 study of North Pacific currents. He contacted Steve McLeod, an Oregon artist and shoe collector who had information on locations

his computer model of Pacific Ocean currents and wind systems north of 30°N latitude. Using the spill date (May 27, 1990), the spill location (161°W; 48°N), and the dates of the first shoe landings on Vancouver Island and Washington State beaches between Thanksgiving and Christmas 1990, Ebbesmeyer found that the shoe drift rates agreed with the computer model's predicted currents.

News of Ebbesmeyer and Ingraham's interest in the shoe spill reached an Oregon news reporter and was then circulated nationally. Readers sent letters describing their own shoe finds. Reports of single shoes were valuable, because each shoe had within it a NIKE purchase order number which could be traced to a specific cargo container. Ebbesmeyer was able to determine from these numbers that only four of the five shoe-containers broke open, so that only 61,820 shoes were set afloat.

The computer model and previous experiments with satellite-tracked drifters showed that there would have been little scattering of the shoes as the ocean currents carried them eastward and approximately 1500 miles from the spill site to shore, but the shoes were found scattered from California to northern British Columbia. The north-south scattering is related to coastal currents that flow northward in winter carrying the shoes to the Queen Charlotte Islands and southward in spring and summer bringing the shoes to Oregon and California. In the spring of 1992 three sneakers from one of the containers were found at Pololu, at the north end of the island of Hawaii, indicating that they were making their way across the Pacific.

Because it takes about four and a half years for an object to drift completely around the North Pacific Current gyre, the shoes were expected to arrive on Japanese beaches during 1994–95, but no shoes have been reported on Japanese beaches.

Other Scenarios

Ebbesmeyer was interested in seeing where the shoes might have gone if they had been lost on the same date but under different conditions in other years. The computer allowed simulations for May 27 of each year from 1946 to 1991. Figure 4 shows the wide variation in model-predicted drift routes. If the shoes had been lost in 1951, they would have traveled in the loop of the Alaska Current. If they had been lost in 1982, they would have been carried far to the north during the very strong El Niño of 1982–83, and if lost in 1973 they would have come ashore at the Columbia River.

Tub Toys and Hockey Gloves Come Ashore

A similar situation occurred in January 1992 when twelve cargo containers were lost from another vessel in the North Pacific at 180°W, 45°N. One of these containers held 29,000 small, floatable, bathtub

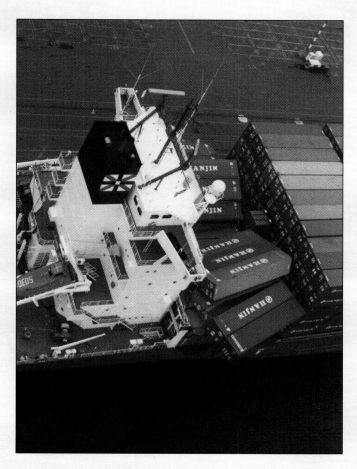

Figure 1

After the storm the *Hansa Carrier* docked in Seattle.

179

Figure 2

Shoes and toys after their rescues from Pacific cargo spills.

Figure 3

Site where 80,000 Nike shoes washed overboard on May 27, 1990, and dates and locations where 1300 shoes were discovered by beachcombers (dots at upper right). Drift of the shoes is simulated with a computer model (colored plume).

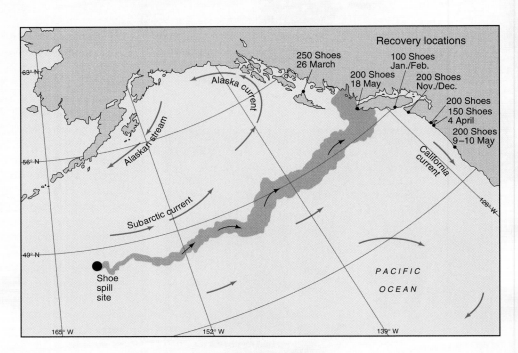

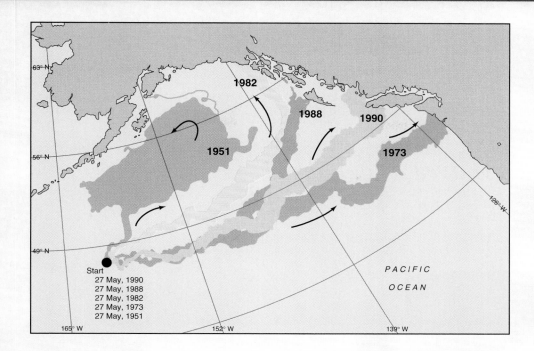

Figure 4

Projected drift tracks for the sneakers for the years 1951, 1973, 1982, 1988, and 1990 based on computer modeling of ocean currents and weather.

From Curtis E. Ebbesmeyer and W. James Ingraham, Jr., Shoe Spill in the North Pacific *in American Geophysical Union EOS, Transactions, Vol. 73, No. 34, August 25, 1992, pp. 361, 365.*

toys. Plastic blue turtles, yellow ducks, red beavers, and green frogs began arriving on beaches near Sitka, Alaska in November of 1992. Advertisements asking for news of toy strandings were placed in local newspapers and the Canadian lighthouse keepers newsletter. Beachcombers reported a total of about 400 of these toys.

None of the toys have been found south of 60°N latitude, suggesting that once the toys reached the vicinity of Sitka, they drifted to the North. This time the computer model showed that if the toys continue to float with the Alaska Current, they will move with the Alaska stream on through the Aleutian Islands into the Bering Sea. Some may continue through the Bering Sea into the Arctic Ocean to the vicinity of Point Barrow, and from there drift north of Siberia with the Arctic pack ice. Eventually some plastic turtles, ducks, beavers, or frogs may come to rest on the coast of western Europe. If the toys turn south they may merge with the Kuroshio Current and be carried past the location where they were spilled.

Two thousand five hundred cases of hockey gloves (34,300 gloves) were lost from a container ship in the North Pacific after a December 1994 fire. In August of 1995 a fishing vessel found seven gloves 800 miles west of the Oregon coast and by January of 1996 the barnacle-covered gloves began to arrive on Washington State beaches. The most northerly glove sighting came from Prince William Sound, Alaska in August of 1996. The gloves are expected to follow the tub toys along the coast of Alaska and into the Arctic.

Internet Resources

www.agu.org/sci_soc/ducks.html

www.marinewatch.com/duckies.html

chapter eight

Waves and Tides

A seat in this ... on a bucking bronco, and, by the same token, a bronco is not much smaller. The craft pranced and reared, and plunged like an animal. As each wave came, and she rose for it, she seemed like a horse making at a fence outrageously high. The manner of her scramble over these walls of water is a mystic thing, and, moreover, at the top of them were ordinarily ... the foam racing down from the ... a new leap, and a leap from ... bumping a crest, she would ... a long incline and arrive ... of the next menace.

A steep wave breaks in the open ocean.

... the ... fact that ... one ... as important and just as nervously anxious to do something ... way of swamping boats. In a ten-foot din ... idea of the resources of the sea in the line of waves that is not probable to the average experience, which is never at sea in a dingey. As each salty wall of water approached, it shut all else from the view of the men in the boat; and it was the final outburst of the ocean, the last effort of the grim water. There was a terrible grace in the move of the waves, and they came in silence, save for the snarling of the crests.

Stephen Crane,
from The Open Boat

Learning Objectives

After reading this chapter, you should be able to

- Recognize the parts and properties of waves.
- Compare deep-water and shallow-water waves.
- Follow the development of waves from creation to breaking.
- Describe interactions between waves.
- List characteristics of special wave types.
- Describe tides as observed along the coast.
- Understand the forces causing tides.
- Explain the modifications that produce observed tides.
- Give examples of energy from waves and tides.

We have all seen water waves. Gentle ripples move across a pond, and breaking waves provide exciting surf at the beach. Storm waves crash against our coasts and force commercial vessels to slow their speed and lengthen their sailing time. Special waves associated with earthquakes have killed thousands of coastal inhabitants and severely damaged their cities and towns. Surfers search for the perfect wave, and ancient peoples navigated by the patterns that waves form.

The tides, best known as the rise and fall of the sea around the edge of the land, are also waves, the longest of the water waves on this planet. Far out at sea, tidal changes go unnoticed; along the shores and beaches, the tides govern many of our commercial and recreational activities.

How a Wave Begins   8.1

If you stand on the beach looking out across a perfectly flat surface of water and throw a stone into the water, the stone strikes the water and displaces, or pushes aside, the water surface. As the stone sinks, the displaced water flows back into the space left behind, and as this water rushes back from all sides, the water at the center is forced upward. The elevated water falls back, causing a depression below the surface, which is refilled, starting another cycle. This process produces a series of waves that move outward and away from the point of disturbance along the boundary between two fluids: air and water.

If the stone thrown into the water is small, the waves are small, and the force that causes the water to return toward its undisturbed surface level is the **surface tension** of the water surface (described further in table 5.3, chapter 5). All very small water waves are affected by surface tension.

The most common force for creating water waves is the wind. As the wind blows across a smooth water surface, the friction between the air and the water stretches the surface, resulting in small wrinkles or ripples. The surface becomes rougher, and it is easier for the wind to grip the roughened water surface and add more energy, increasing the size of the waves. As the waves become larger, the force returning the water to its level state changes from surface tension to gravity (fig. 8.1).

What is required to create a wave at the sea surface?

Why are surface tension and gravity necessary to form a wave?

Why are ripples important in the creation of waves?

Anatomy of a Wave   8.2

The highest part of the wave that is elevated above the undisturbed sea surface is called the **crest;** the lowest part that is depressed below this surface is called the **trough** (fig. 8.2). The distance between two successive crests or two successive troughs is the length of the wave, or its **wavelength.** The **wave height** is the vertical distance from the top of the crest to the bottom of the trough. Sometimes the term

Figure 8.1
Wind-generated storm waves at sea.

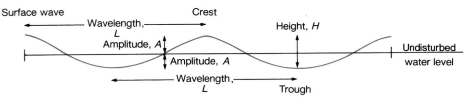

Figure 8.2

A profile of surface waves.

amplitude is used; the amplitude is equal to one-half the wave height, or the distance from either the crest or the trough to the undisturbed water level. A wave is characterized not only by its length and height (or amplitude), but also by its **period.** The period is the time required for two successive crests or two successive troughs to pass a point in space. If you are standing on a piling and start a stopwatch as the crest of a wave passes and then stop the stopwatch as the next crest passes, you have measured the period of the wave. If there is a continuing input of energy, waves increase in length, period, and height; the speed with which they travel across the sea surface also increases.

> Sketch a wave and label its parts.
>
> What is the relationship between the height and amplitude of a wave?
>
> Explain the period of a wave.

Wave Motion and Wave Speed   8.3

As a wave form moves along the water surface, it sets particles of water in motion. Out at sea, where the waves move the water surface quietly up and down, the water is not moving toward the shore. Such an ocean wave does not represent a flow of eventual dissipation, which may occur at sea or against the land.

As the wave crest approaches, the surface water particles rise and move forward. Immediately under the crest the particles have stopped rising and are moving forward at the speed of the crest. When the crest passes, the particles begin to fall and to slow in their forward motion. As the trough advances, the particles slow their falling rate and start to move backward, until at the bottom of the trough they move only backward. As the remainder of the trough passes, the water particles begin to slow their backward speed and start to rise again. This motion (rising, moving forward, falling, reversing direction, and rising again) creates a circular path, or **orbit,** for the water particles. Use figure 8.3 to follow the motion of water particles in their orbits.

It is this orbital motion of the water particles that causes a floating object to bob, or move up and down, forward and backward, as the waves pass under it. This motion affects a fishing boat, swimmer, sea gull, or any other floating object on the seaward side of the surf zone. The surface water particles trace an orbit with a diameter equal to the height of the wave. This same type of motion is present in the water particles below the surface, but as less energy of motion is present, the orbits become smaller and smaller with depth. At a depth equal to one-half the wavelength, the orbital motion has decreased to almost zero (fig. 8.3). Submarines dive during rough weather for a quiet ride, since the wave motion does not extend far below the surface.

A wave's speed along the sea surface may be computed using the wave's length and period. Celerity is the term traditionally used by oceanographers to identify the wave speed of individual waves at the sea surface. The speed or celerity (C) of a surface-water wave is equal to the wavelength (L) divided by its period (T):

$$\text{Speed} = \text{Wavelength/Wave period or } C = L/T$$

Once a wave has been formed, the speed at which the wave moves may change, but *its period remains the same* (period is de-

Figure 8.3

The moving wave form sets the water particles in motion (note the arrows in the diagram). The diameter of a water particle's orbit at the surface is determined by wave height. Below the surface, the diameter decreases and orbital motion ceases at a depth (D) equal to one-half the wavelength.

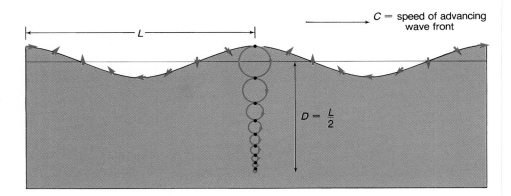

termined by the generating force). The oceanographer at sea determines the wave period (T) by direct measurement and calculates the wave speed (C) and the wavelength (L) using equations from simple wave theory; see section 8.4.

Explain the relationship between flow of water and wave motion in waves seaward of the surf.

What is the maximum diameter of a wave's water particle orbit?

How do the orbits of water particles change with depth?

How are wavelength and wave period related to the speed of a wave?

Deep-Water Waves 8.4

"Deep-water" has a precise meaning for the oceanographer studying waves. To be a **deep-water wave,** the water must be deeper than one-half the wave's length. Under this condition, the orbits of the waves do not reach the sea floor.

The wavelength in meters (L) of a large deep-water wave is related to gravity $(g, 9.81 \text{ m/s}^2)$ and the square of the wave period in seconds (T). Using simple wave theory it can be shown that

$$L = (g/2\pi) \, T^2 = 1.56 \, T^2.$$

This equation demonstrates that long-period waves also have long wavelengths and when used with the equation, $C = L/T$, it shows the relationship between the wave speed in meters per second (C) and the period in seconds (T):

$$C = 1.56 \, T.$$

Most waves observed at sea are caused by the wind, and they travel in a direction determined by the wind. These waves are formed in local storm centers or by the steady winds of the trade and westerly wind belts. In an active storm area covering thousands of square kilometers, the winds are not steady but turbulent, varying in strength and direction. Storm-area winds flow in a circular pattern about the low-pressure **storm center,** creating waves that move outward and away from the storm in all directions. The storm center may also move across the sea surface, following the waves and increasing their heights, supplying energy for a longer time and over a longer distance.

In the storm area, the sea surface appears as a jumble and confusion of waves of all heights, lengths, and periods. Small waves ride the backs of larger waves. This turmoil of mixed waves is called a "sea" by sailors. The turbulent winds of the storm area transfer energy at different intensities and rates to the sea surface, resulting in waves with a variety of periods and heights.

Among the waves escaping from a storm center are some with long periods and long wavelengths. These waves have a greater speed than waves with short periods and short wavelengths. The faster, longer waves gradually move through and ahead of the shorter, slower waves; waves with different periods travel at different speeds. This process is called **sorting** or **dispersion.** Groups of these faster waves move as **wave trains,** or packets of similar waves with approximately the same period and speed. Near the storm center the waves are not yet sorted, but farther away from the storm the faster, long-period wave trains are ahead of the slower, short-period waves. This process is shown in figure 8.4.

When the longer waves are sorted, they appear as a regular pattern of crests and troughs moving across the sea surface. These long, uniform waves, called **swell,** (fig. 8.5) lose their energy very slowly. Groups of large, long-period waves have been traced across the entire length of the Pacific Ocean from Antarctica to Alaska.

Careful observation of a group of waves or a wave train from a stone thrown into the water shows that waves constantly form on the inside of the train as it moves across the water. As each wave joins the inside of the ring, a wave is lost from the outside. The outside wave's energy is lost in advancing the wave form into undisturbed water. Therefore, each individual wave in the group moves faster than the leading edge of the wave train. The wave train moves at a **group speed** that is one-half the speed of the individual waves.

Group speed = 1/2 Wave speed = Speed of energy transport

or

$$V = C/2.$$

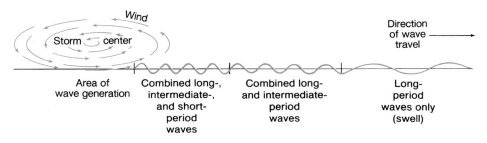

Figure 8.4

Dispersion. The longer waves travel faster than the shorter waves. Waves are shown moving in only one direction in the diagram.

Figure 8.5

Waves of uniform length and period, known as swell, approaching the coast at the entrance to Grays Harbor, Washington.

Define a deep-water wave.

Why do storm centers produce waves of varying sizes?

Why do waves appear more regular when they are far away from a storm center?

Compare the motion of a single wave with that of its group as it escapes a storm center.

Wave Interaction 8.5

When wave trains from one storm meet other similar trains of swell moving away from other storm centers, they pass through each other and continue on. If the crests or the troughs of the two different wave trains are in phase or coincide, the profiles are additive and the amplitude of the crests and troughs increases. This is known as constructive interference. When the crests of one wave train coincide with the troughs of another, the waves are out of phase, and the wave trains cancel each other due to destructive interference. Figure 8.6 shows constructive and destructive interference between waves traveling in either the same or opposite direction. However, wave trains may intersect at any angle, and many possible interference patterns may result. If the two wave trains intersect each other sharply as at a right angle, then a checkerboard pattern can be formed (fig. 8.7). Two or more intersecting wave trains can phase together and suddenly develop large waves unrelated to local storms. If these waves become too high, they may break, lose some of their energy, and create new, smaller waves.

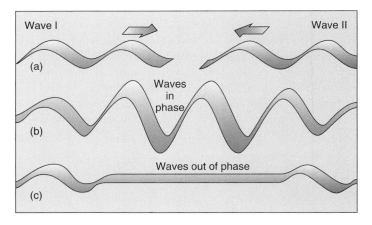

Figure 8.6

(a) Waves approach each other. (b) If the crests of the approaching waves coincide, the height of the combined waves increases, and the waves are in phase; this is constructive interference. (c) If the crests of one wave and the troughs of the other wave coincide, the waves cancel, and the waves are out of phase; this is destructive interference.

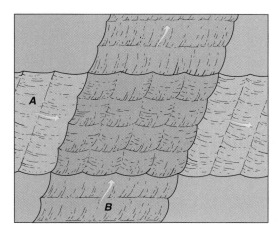

Figure 8.7

The waves from *A* and *B* create a checkerboard of peaks and troughs.

If two identical waves are moving in the same direction, how does the water surface appear if the crests and troughs are (a) in phase, or (b) out of phase?

Which of the situations in the above question is an example of (a) constructive interference, (b) destructive interference?

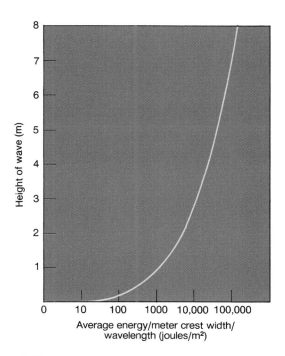

Figure 8.8

Wave energy increases rapidly with increases in wave height. Average wave energy is calculated per unit width of crest and averaged over the wavelength *(L)* and the depth *(L/2)*.

Wave Height   8.6

Three important factors control wind-wave height: (1) wind speed (how fast the wind is blowing), (2) wind duration (how long the wind blows), and (3) **fetch** (the distance over water that the wind blows in a single direction). If the wind speed is very small, large waves are not produced, no matter how long the wind blows over an unlimited fetch. If the wind speed is great, but it blows for only a few minutes, no high waves are produced despite unlimited wind strength and fetch. Also, if very strong winds blow for a long period over a very short

fetch, no high waves form. Only when all three of these factors combine do spectacular wind waves occur at sea.

The average energy of a wave is directly related to the square of the wave height (fig. 8.8). Oceanographers can relate the height and period of storm-formed wind waves to wind speed, duration of wind blow, and fetch. In the open sea, where fetch and duration of blow are not limiting, wind speed alone determines the significant wave height of wind waves (table 8.1). The significant wave height is defined as the average wave height of the one-third highest waves in a series of observations. For example, if a height record of 120 successive wave heights is made and the wave heights are arranged in order, from the lowest to the highest, the significant wave height is the average height of the highest forty waves. Maximum wave

table 8.1	The Relationship between Wind Speed and Wave Height						
Average Wind Speed		**Significant Wave Height**	**Significant Wave Period**	**Significant Wa Speed**	**Maximum Wave Height**	**Minimum Fetch**	**Minimum Wind Duration**
Knots	**m/s**	**Meters**	**Seconds**	**m/s**	**Meters**	**km**	**Hours**
10	5.1	1.22	5.5	8.58	2.19	129	11
20	10.2	2.44	7.3	11.39	4.39	240	17
30	15.3	5.79	12.5	19.50	10.43	1017	37
40	20.4	14.33	18.0	28.00	25.79	2590	65

height (table 8.1) is related statistically to significant wave height.

From early times, those who went to sea recognized that there was a relationship between wind speed and wave height. From this knowledge grew the Universal Sea State Code used by the U.S. Navy today, but originally suggested by British Admiral Sir Francis Beaufort in 1806 (table 8.2). Thus it is also called the Beaufort scale of sea state.

Unrelated to local conditions, large waves or **episodic waves** can suddenly appear at sea. Episodic waves have a height equal to a seven- or eight-story building (20–30 m) and may move at a speed of 25 m per seconds (60 mph or 50 knots), with a wavelength approaching a kilometer (one-half mile). They occur most frequently near the edge of the continental shelf, in water about 200 m (660 ft) deep, and in certain geographic areas with strong wind, wave, and current patterns. These episodic waves may be the result of constructive interference between large wave systems passing through each other. The area where the Agulhas Current sweeps down the east coast of South Africa and meets the storm waves from the Southern Ocean is noted for these waves (fig. 8.9). In the North Atlantic, strong northeasterly gales send large storm waves into the edge of the northward-moving Gulf Stream near the edge of the continental shelf, producing large waves. The shallow North Sea also appears to provide suitable conditions for extremely high episodic waves during severe winter storms.

The best-documented giant wave at sea occurred in 1933 when the USS *Ramapo*, a Navy tanker, encountered a severe storm in the Pacific Ocean. The wave, measured against the ship's superstructure by the officer on watch, was 34.2 m (112 ft) high. The period of the wave was measured at 14.8 seconds, and the wave speed was calculated at 27 m (90 ft)/s.

> Why does an eight-foot-high wave do more damage than a four-foot-wave?
>
> Where in the world's oceans would you expect to find consistently high waves?

Wave Steepness 8.7

There is a maximum possible wave height for any given wavelength. This maximum value, determined by the ratio of the wave's height to the wave's length, is the measure of the **steepness** of the wave:

$$\text{Steepness} = \frac{\text{Height}}{\text{Length}}$$

or

$$S = H/L$$

If the ratio of the height to the length exceeds 1:7, the wave becomes too steep and breaks. Under these conditions, the angle formed at the wave crest approaches 120°; the wave be-

table 8.2	Universal Sea State Code	
Sea State Code	**Description**	**Average Wave Heights**
SS0	Sea like a mirror; wind less than one knot.	0
SS1	A smooth sea; ripples, no foam; very light winds, 1–3 knots, not felt on face.	0–0.3 m 0–1 ft
SS2	A slight sea; small wavelets; winds light to gentle, 4–6 knots, felt on face; light flags wave.	0.3–0.6 m 1–2 ft
SS3	A moderate sea; large wavelets, crests begin to break; winds gentle to moderate, 7–10 knots; light flags fully extend.	0.6–1.2 m 2–4 ft
SS4	A rough sea; moderate waves, many crests break, whitecaps, some wind-blown spray; winds moderate to strong breeze, 11–27 knots; wind whistles in the rigging.	1.2–2.4 m 4–8 ft
SS5	A very rough sea; waves heap up, forming foam streaks and spindrift; winds moderate to fresh gale, 28–40 knots; wind affects walking.	2.4–4.0 m 8–13 ft
SS6	A high sea; sea begins to roll, forming very definite foam streaks and considerable spray; winds a strong gale, 41–47 knots; loose gear and light canvas may be blown about or ripped.	4.0–6.1 m 13–20 ft
SS7	A very high sea; very high, steep waves with wind-driven overhanging crests; sea surface whitens due to dense coverage with foam; visibility reduced due to wind-blown spray; winds at whole gale force, 48–55 knots.	6.1–9.1 m 20–30 ft
SS8	Mountainous seas; very high-rolling breaking waves; sea surface foam-covered; very poor visibility; winds at storm level, 56–63 knots.	9.1–13.7 m 30–45 ft
SS9	Air filled with foam; sea surface white with spray; winds 64 knots and above.	13.7 m and above 45 ft and above

Figure 8.9

A giant wave breaking over the bow of the *ESSO Nederland* southbound in the Agulhas Current. The bow of the supertanker is about 25 m (82 ft) above the water.

comes unstable, cannot maintain its shape, collapses, and breaks (fig. 8.10).

Small unstable, breaking waves are quite common. When wind speeds reach 8 to 9 meters per second (16 to 18 knots), waves known as "whitecaps" can be observed. These waves have short wavelengths (about 1 m), and as each wave reaches its critical steepness it breaks and is replaced by another wave produced by the rising wind. There is rarely sufficient wind at sea to force long waves to their maximum steepness. More frequently, the tops of high waves are torn off by the wind and cascade down the wave faces.

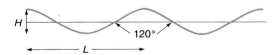

Figure 8.10

Wave steepness. When H/L approaches 1:7, the wave's crest angle approaches 120° and the wave breaks.

Relate the steepness of a wave to the wave's breaking. Why do waves at sea rarely break?

Shallow-Water Waves   8.8

As a deep-water wave moves into shallow water, the water particle orbits become flattened circles or ellipses (fig. 8.11). The waves begin to "feel the bottom," the orbits are compressed, and the forward speed of the wave is reduced due to interaction with the bottom.

Remember that (1) the speed of any wave (deep, intermediate, or shallow) is equal to the wavelength divided by its period, and (2) the period of a wave does not change. Therefore, when the intermediate or shallow wave feels bottom, it slows, and the accompanying reduction in the wavelength and speed results in increased height and steepness as the wave's energy is condensed in a smaller water volume. Figure 8.11 shows the transition of wave properties as the water depth changes.

When the wave enters water with a depth of less than one-twentieth the wavelength, the wave becomes a **shallow-water wave.** The shallow-water wave's length and speed are controlled only by the water depth. Here the speed and wavelength are determined by the square root of the product of the Earth's acceleration due to gravity (g, 9.81 m/s^2) and depth (D):

$$C = \sqrt{gD} \text{ or } 3.13\sqrt{D};$$
$$L = \sqrt{gD}\ T \text{ or } 3.13\sqrt{D}\ T.$$

In shallow-water waves, the elliptical orbits of the water particles become flatter with depth until at the sea floor only a back-and-forth motion remains (fig. 8.12). The horizontal dimension of the orbit remains unchanged in shallow water. The group speed, V, of shallow-water waves is equal to the speed, C, of each wave in the group. Shallow-water waves are formed when deep-water waves enter coastal waters; they are also present in the open ocean when wavelengths are much greater than water depth. See tsunamis (sec. 8.10) and tides (sec. 8.12).

Waves are refracted, or bent, as they begin to feel the bottom and change wavelength and speed. When waves approach the beach at an angle, one end of the wave crest comes into shallow water and begins to feel bottom while the other end is in deeper water. The shallow-water end moves more slowly than the part in deeper water. The result is that the wave crests bend, or **refract,** and tend to become parallel to the shore (fig. 8.13).

Along an irregular coastline, there is often a submerged ridge seaward from a headland and a depression in front of a bay. When shallow-water waves approach this coastline, the part over the ridge slows down more than the part on either side. The central area at the mouth of the bay is usually deeper than the areas to each side, and so the advancing waves slow down more on the side than in the center. This pattern is shown in figures 8.14 and 8.15. More energy is expended on a unit length of shore at the point of the headland because of increased wave height, while the lower-energy environment of the bay provides sheltered water.

A straight, smooth vertical barrier in water deep enough to prevent waves from breaking reflects the waves. The barrier may be a cliff, steep beach, breakwater, bulkhead, or other structure. The reflected waves pass seaward through the incoming waves to produce interference patterns, and steep, choppy seas often result. If the waves reflect directly back on themselves, the resulting waves appear to stand still, rising and falling in place.

Another phenomenon associated with waves as they approach the shore is **diffraction.** Diffraction is caused by the spread of wave energy sideways to the direction of wave travel. If waves move toward a barrier with a small opening (two landmasses separated by a channel or an opening in a breakwater), some wave energy passes through the small opening. Once

Figure 8.11

Deep-water waves become intermediate waves and then shallow-water waves as depth decreases and wave motion interacts with the sea floor.

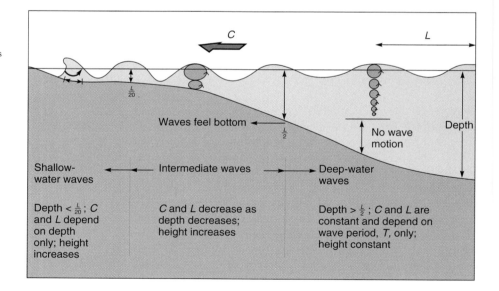

Figure 8.12

Shallow-water wave particles move in elliptical orbits. The orbits flatten with depth due to interference from the sea floor.

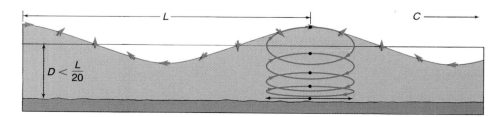

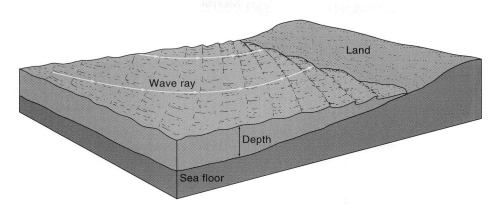

Figure 8.13

Waves moving inshore at an oblique angle to the depth contours are refracted. One end of the wave reaches a depth of $\frac{L}{2}$ or less and slows while the other end of the wave maintains its speed in deeper water. Wave rays drawn perpendicular to the crests show the direction of wave travel and the bending of the wave crests.

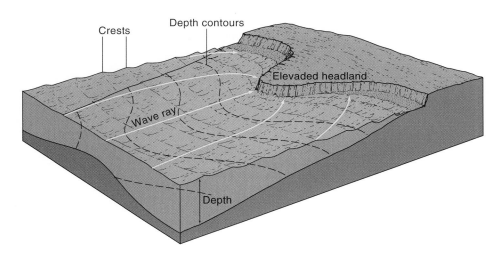

Figure 8.14

Waves refracted over a shallow submerged ridge focus their energy on the headland. The converging wave rays show the wave energy being crowded into a smaller volume of water, increasing the energy per unit length of wave crest as the height of the wave increases.

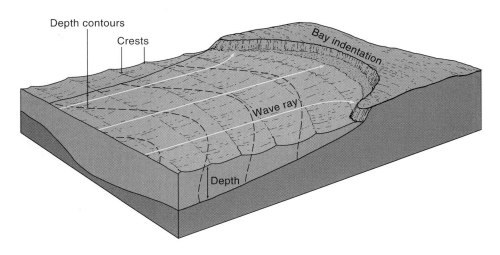

Figure 8.15

Waves refracted by the shallow depths on each side of the bay deliver lower levels of energy inside the bay. The diverging wave rays show the energy being spread over a larger volume of water, decreasing the energy per unit length of wave crest as the wave height decreases.

through the opening, the wave crests decrease in height, radiating outward from the gap. This effect is shown in figure 8.16. If the waves approach a barrier without an opening and move past its end, diffraction can still occur, because energy is transported at right angles to the wave crests as the waves pass the end of the barrier. It is extremely important that harbors be designed to minimize the wave energy entering the harbor to prevent damaging the moored vessels.

In areas of the world where the winds blow steadily and from one direction, as in the trade wind belts, waves at sea are very regular in their direction of motion, allowing a vessel to maintain a constant course relative to the wind and waves. Because waves change speed, shape, and height with water depth, and because waves change direction and pattern due to refraction, diffraction, and reflection, it is possible to deduce the presence of shoals, bars, islands, and coasts from the changes in

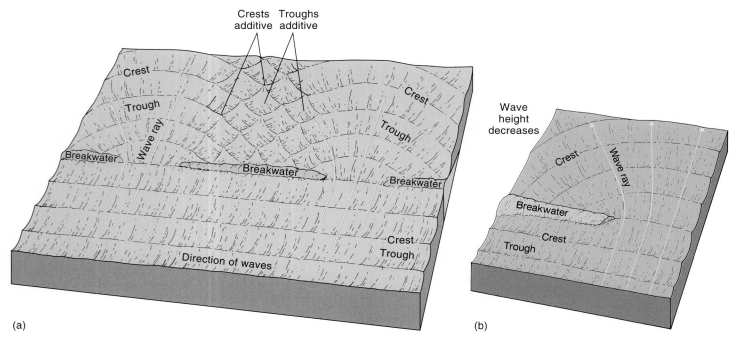

Crests additive   Troughs additive

Crest

Trough

Wave ray

Breakwater

Breakwater

Breakwater

Crest
Trough

Direction of waves

Crest

Trough

Wave height decreases

Crest

Wave ray

Breakwater

Crest

Trough

(a)

(b)

Figure 8.16

(a) Diffraction patterns produced by waves passing through narrow openings; note the interference pattern. (b) Diffraction occurs behind the breakwater.

wave pattern. Although these changing patterns are subtle, they can be detected far from land by those who live with the sea and become sensitive to its variations. The Polynesians of the past made long canoe voyages using their knowledge of wave patterns. This understanding plus their knowledge of star positions and their observations of cloud forms over land and sea allowed them to sail many hundreds, even thousands, of kilometers across the open ocean to reach their destinations.

> How does a deep-water wave change when it enters water with a depth of less than one-twentieth the wavelength?
>
> Which property of the waves does not change?
>
> Why do waves change direction in shallow water?
>
> Compare reflection and diffraction of waves.
>
> Why are waves higher around a point of land extending into the sea?

The Surf Zone   8.9

The surf zone is the shallow area along the coast in which the waves slow rapidly, steepen, break, and disappear in the turbulence and spray of expended energy. The width of this zone is variable and is related to both the length and height of the ar-

riving waves and the changing depth pattern. Longer, higher waves, which feel the bottom before shorter waves, become unstable and break farther offshore in deeper water. If shallow depths extend offshore for some distance, the surf zone is wider than it is over a sharply sloping shore.

Breakers form in the surf zone because the wave motion at depth is affected by the bottom. Orbital motion is slowed and compressed vertically; as the wavelength shortens, the wave steepens and breaks. The two most common types of breakers are plungers and spillers (fig. 8.17).

Plunging breakers form on narrow, steep beach slopes. The curling crest outruns the rest of the wave, curves over the air below it, and breaks with a sudden loss of energy and a splash. The more common spilling breaker is found over wider, flatter beaches, where the energy is dissipated more gradually as the wave moves over the shallow bottom. This action results in the less dramatic wave form, consisting of turbulent water and bubbles flowing down the collapsing wave face. On some plunging breakers there is a slow curling-over of the crest beginning at a point on the steepest part of the crest and moving along the crest as the wave approaches shore at an angle. The result is the "tube" so sought after by surfers (fig. 8.18). Local wind strength and direction may also affect the breaking of waves.

> Why does a gently sloping beach protect the shore from heavy storm waves?
>
> Which wave maintains its wave form longer, a spiller or a plunger?

(a)

(b)

Figure 8.17

Breaking waves. (a) A plunger loses energy more quickly than (b) a spiller.

Tsunamis 8.10

Very long waves produced by sudden movements of the Earth's crust, or earthquakes are known as **tsunamis** (fig. 8.19). Tsunami is from the Japanese, meaning harbor wave. Tsunamis or seismic sea waves are also incorrectly called tidal waves.

If an underwater area of the Earth's crust is suddenly displaced, it may cause a sudden rise or fall in the sea surface above it. This results in waves with extremely long wavelengths 100 to 200 kilometers (54 to 108 nmi) and long periods as well (10–20 min). Since the average depth of the oceans is about 4000 m (13,000 ft), tsunamis are shallow-water waves. They radiate out from the point of the seismic disturbance at a speed determined by the ocean's depth and move across the oceans at about 200 m per second (400 mph). As shallow-water

Figure 8.18

A surfer rides the tube of a large curling wave in Hawaii.

waves, tsunamis may be refracted, diffracted, or reflected in midocean by islands and changes in depth.

When a tsunami leaves its point of origin, it may have a height of 1 to 2 m (3 to 6 ft), but this height is distributed over its extremely long wavelength. A vessel in the open ocean may not see the wave and is in little or no danger if a tsunami passes. When its path is blocked by a coast or island, the wave behaves like any other shallow-water wave, and its height builds rapidly. The loss of energy is also rapid, and a tremendous amount of moving water races up over the land causing large water level changes in harbors and river mouths as well as destructive currents.

The leading edge of the tsunami wave group may be either a crest or a trough. If a trough arrives at the shore before the first crest, sea level drops rapidly, exposing the sea floor with its plants and animals. People have drowned after following the receding water to inspect the marine life, for they find, too late, that there is a wall of advancing water that they cannot outrun. The Pacific Ocean, ringed by crustal faults and volcanic activity, is the birthplace of many tsunamis. The Aleutian Trench produced the 1946 tsunami that heavily damaged Hilo, Hawaii, killing more than 150 people. In 1957 Hawaii was hit again, but due to early warning and evacuation, no lives were lost. The 1964 Alaska earthquake produced tsunamis that

(a)

(b)

(c)

Figure 8.19

(a,b,c) Sequential photos of the major wave of a tsunami at Laie Point, Oahu, Hawaii, from an earthquake in the Aleutian Islands, March 9, 1957. The highest wave in Hawaii was 3.6 m above sea level (11.8 ft).

caused severe damage in Alaska and areas on the west coast of Vancouver Island and the northern coast of California.

More recently, in September 1992 a coastal submarine Earthquake created a tsunami that struck Nicaragua causing extensive damage and killing 170 people, mostly children. In December of this same year an earthquake in the Flores Island region of Indonesia created a 25 m (82 ft) tsunami that killed 1000 people in the town of Maumere and swept away two-thirds of the population of the island of Babi. In July of 1993 a near-shore earthquake of 7.8 magnitude on the Richter scale occurred near Okushiri Island, off the southwest coast of Hokkaido, Japan causing a tsunami that ran 15 to 30 m (50 to 100 ft) up the beach over a 20 km (12 mi) length of shoreline. This event caused about $600 million in damages and claimed more than 185 lives (see figure 8.20).

A severe tsunami struck the northwest coast of Papua, New Guinea in July of 1998. A close-to-shore earthquake caused a 2 meter (6 ft) vertical drop along an underwater fault causing a series of large waves. These 7 to 15 m (22–50 ft) waves swept across the beach destroying four fishing villages (fig. 8.21). Debris and bodies were deposited in the lagoon and the mangrove trees beyond. More than 2,000 persons were killed or reported missing.

Tsunamis also appear in the Caribbean Sea, which is bounded by an active island arc system, and in the Mediterranean Sea and along the west coast of South America.

> Why are tsunamis not seen in the open ocean?
>
> Why is the tsunami a shallow-water wave in the deep sea?

Standing Waves 8.11

Standing waves are shallow-water waves that reflect back on themselves. They occur in ocean basins, partly enclosed bays and seas, and estuaries. A standing wave can be demonstrated by slowly lifting one end of a container partially filled with water and then rapidly but gently returning it to a level position. If this is done, the surface alternately rises at one end and falls at the other end of the container. The surface oscillates about a point at the center of the container, the **node,** and the alternations of low and high water (troughs and crests) at each end are the **antinodes** (fig. 8.22a). The wavelength is twice the length of the basin.

By tilting the basin back and forth at a faster rate, it is possible to produce a wave with more than one node, as shown in figure 8.22b. In the case of two nodes, there is a crest at either end of the container and a trough in the center. This arrangement alternates with a trough at each end and the crest in the center; the wavelength is equal to the basin length.

Figure 8.20

Aonae at the southern tip of Okushiri Island in the Sea of Japan after the tsunami and fire of 1993. The area on the left extending to the tip of the island was leveled despite the protecting sea wall. Debris left in the streets by the surging water prevented fire control equipment from reaching the burning buildings.

Figure 8.21

The tsunami that struck the northwest coast of Papua, New Guinea, 17 July, 1998, destroyed a chain of villages along 30 km (18 mi) of beach. The eroded beach and the shattered palms were all that was left of the communities.

Standing waves in bays or inlets with an open end behave somewhat differently than standing waves in closed basins. A node is usually located near the entrance to the open-ended bay, so that one-quarter of a wavelength is present in the bay. There is little or no rise and fall of the water surface at the entrance node, but the tidal currents pulsing in and out of the entrance may produce a large rise and fall at the antinode at the closed end of the bay, if the natural period of oscillation of the basin is close to the tidal period. Multiple nodes may also be present in open-ended basins. The dimensions of each individual basin determine the wavelength and period of its standing wave.

Standing waves can be triggered by seismic events that suddenly tilt or oscillate the water in natural and artificial basins, even swimming pools. If storm winds create a change in water level at one side of a basin, the surface may oscillate when the wind ceases until friction causes the oscillation to die out. The water in the basin oscillates at a period governed by the shape and size of the basin. A standing wave occurring in a natural basin is called a **seiche,** and the process of oscillation is called seiching. Severe storms moving rapidly across the Great Lakes often produce a tilted water surface that oscillates as the storm moves on; water level fluctuations of about 0.5 m (1.5 ft) occur at the shore due to seiching.

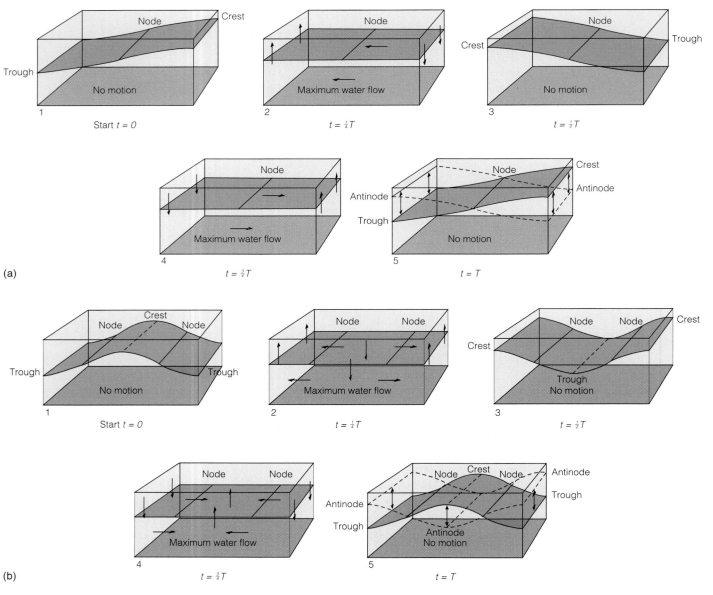

Figure 8.22

(a) A standing wave in a basin oscillating about a single node. (b) A standing wave oscillating about two nodes. The time for one oscillation is the period of the wave, T.

Tides can also produce standing waves or seiches in oceanic and coastal basins if the oscillation period of the basin is a multiple of the tidal period, for example the Bay of Fundy, a coastal basin in southeastern Canada; see section 8.13. Tide waves oscillating in the open ocean and the Coriolis effect acting on the currents cause the tide wave to rotate as it oscillates. This rotation reduces the tide wave node to a point around which the tide wave crest rotates, producing what is known as a rotary standing tide wave.

> How is a standing wave different than a wind wave?
>
> Compare standing waves in open and closed basins.
>
> What is a seiche?

The Tides 8.12

Tidal Patterns

The **tides** are best known as the rise and fall of the sea around the edges of the land. In some coastal areas there is a regular pattern of one high tide and one low tide each day; this is a **diurnal tide.** In other areas there is a cyclic high water–low water sequence that is repeated twice in one day; this is a **semidiurnal tide.** In a semidiurnal tidal pattern, both high tides reach about the same height, and both low tides drop to about the same level with each cycle. A tide in which the high tides regularly reach different heights and the low tides drop regularly to different levels is called a **mixed tide.**

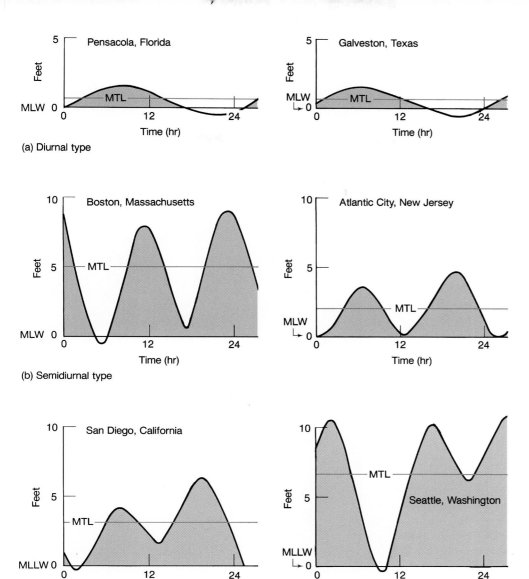

(a) Diurnal type

(b) Semidiurnal type

(c) Mixed type

Figure 8.23abc

Tide types and tidal ranges vary from one coastal area to another. The zero tide level equals mean low water (MLW) or mean lower low water (MLLW), as appropriate. MTL equals mean tide level. All tide curves are for the same date.

Curves for typical tides at some U.S. coastal cities are shown in figure 8.23.

Tidal Levels

In a uniform diurnal or semidiurnal tidal system, the greatest height to which the tide rises on any day is known as **high water,** and the lowest point to which it drops is called **low water.** In a mixed tide it is necessary to refer to **higher high water, higher low water, lower high water,** and **lower low water.** Tidal measurements taken over many years are used to calculate the average or mean tide levels, also mean high waters and low waters.

In areas of uniform diurnal or semidiurnal tide patterns, the zero depth reference on navigation charts is usually equal to mean low water; in regions with mixed tides, mean lower low water is used. Occasionally the low tide falls below the mean value, producing a **minus tide.** A minus tide exposes shoreline usually covered by seawater; it can be a hazard to boaters but is cherished by clam diggers and students of marine biology.

As the water level along the shore increases in height, the tide is said to be rising or flooding; a rising tide is a **flood tide.** When the water level drops, the tide is falling or ebbing; a falling tide is an **ebb tide.**

Tidal Forces

When tides are studied mathematically in response to the laws of physics, they are known as **equilibrium tides.** To simplify the study of relationships between the oceans and the Moon and the Sun, this method uses tides on an Earth model covered with a uniform layer of water. The tides are also studied as they occur naturally; these tides are called **dynamic tides.** Oceanographers study dynamic tides modified by the landmasses, the geometry of the ocean basins, and the Earth's rotation.

The effect of the Sun's and Moon's gravity and the rotation of the Earth on tides is most easily explained by studying equilibrium tides on a smooth water-covered sphere. In this discussion consider the Earth and the Moon as a single unit, the Earth-Moon system that orbits the Sun; refer to figure 8.24a. The Moon orbits the Earth, held by the Earth's gravitational force acting on the Moon (B). There is also a force acting to pull the Moon away from the Earth and send it spinning out into space (B'); this is considered a **centrifugal force** in the following discussion. Forces (B) and (B') must be equal and opposite to keep the Moon in its orbit. Likewise, the Moon's gravitational force acting on the Earth (C) must be balanced by the centrifugal force (C'). B' and C' are caused by the Earth-Moon system rotating about an axis at the center of the system's mass (fig. 8.24b), which is a point 4640 k (2500 mi) from the Earth's center along a line between the Earth and Moon. The Earth-Moon system is held in orbit about the Sun by the Sun's gravitational attraction (A). A centrifugal force again acts to pull the Earth-Moon system away from the Sun (A'). To remain in this orbit, the Earth-Moon system requires that the gravitational forces be equal and opposed to the centrifugal forces (fig. 8.24c).

The Moon's gravitational force is stronger on particles on the side of the Earth closest to the Moon, while the centrifugal force is stronger on particles on the side of the Earth farthest from the center of mass of the Earth-Moon system. This distribution of forces tends to pull surface particles away from the center of the Earth and creates the tide-raising force field on the Earth. The stronger gravitational force of the Moon acting on a unit mass of water at the Earth's surface closest to the Moon is proportional to Mr/R^3; M is the mass of the Moon, r is the radius of the Earth, and R is the distance between the centers of the Earth and the Moon. Because the water covering of the Earth is liquid and deformable, the Moon's gravity moves water toward a point under the Moon, producing a bulge in the water covering. At the same time, the centrifugal force acting on the surface opposite the Moon creates a similar bulge of water. The Earth model develops two bulges with two depressions in between the bulges, or two crests and two troughs, or two high tide levels and two low tide levels (fig. 8.25). The wavelength is equal to half the circumference of the Earth.

Another way to understand the tide-raising forces is to think of the Sun and the Moon exerting a gravitational force that continuously displaces the Earth from a straight-line path causing the Earth to maintain its orbit. In this case the gravitational force, or **centripetal force,** of the Sun and the Moon holds the Earth in orbit. The orbit of the Earth about the Sun is shown in fig. 2.12. In the case of the Moon, the Earth is held in an orbit about the common axis of rotation of the Earth-Moon system (fig. 8.24b).

In each case the gravitational or centripetal force is constant and is equal to the average gravitational force of the Sun or the Moon acting at the Earth's center. The difference between the average gravitational attraction at the Earth's center and the gravitational attraction at individual points on the Earth's surface produces the tide-raising forces. The two methods of considering the tide-raising forces yield the same results.

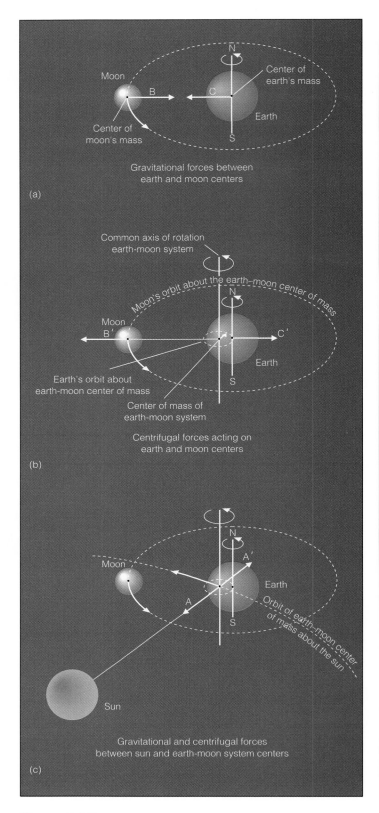

Figure 8.24

Gravitational and centrifugal forces act to keep the Earth-Moon system in balance.

Tidal Day

The Earth makes one rotation in 24 hours, and the bulges or crests in the water covering tend to stay under the Moon as the Earth turns. A point on Earth that is initially centered under a bulge or high tide passes to a trough or low tide, to another high tide, to another low tide, and back to the original high tide position during one revolution of the Earth.

While the Earth turns, the Moon is also moving eastward along its orbit about the Earth. After 24 hours the Earth point that began directly under the Moon is no longer directly under the Moon. The Earth must turn another 12°, requiring an additional 50 minutes, to bring the starting point on Earth back in line with the Moon. Therefore, a lunar **tidal day** is not 24 hours long, but 24 hours and 50 minutes. This also explains why corresponding tides arrive at any location about one hour later each day (fig. 8.26).

The Sun also acts to produce a tide wave. Because the Sun is far away, the gravitational and centrifugal forces that cause tides are only 46% of those associated with the Moon. The length of the tidal day of the Sun tide is 24 hours, but it does not greatly affect the time of the average tidal day, 24 hours 50 minutes.

Spring Tides and Neap Tides

Relative to a point on Earth it takes the Moon 29½ days to complete one orbit. Using figure 8.27, follow the motions of the Earth, Moon, and Sun during this period. During the period of the new Moon, the Moon and Sun are on the same side of the Earth so that the high tides or bulges produced independently by each reinforce each other. Tides of maximum height and depression produced during this period are known as **spring tides.** In a week's time, the Moon is in its first quarter and is located approximately at right angles to the Earth and Sun. At this time, the crests of the Moon tide coincide with the troughs of the Sun tide to produce low-amplitude tides known as **neap tides.** The tides follow a four-week cycle with spring tides every two weeks and a period of neap tides occurring in between the spring tides. This spring-neap pattern is shown in figure 8.28.

Diurnal Tides

If the Moon or Sun stands north or south of the Earth's equator, one bulge or crest is found in the Northern Hemisphere and the other in the Southern Hemisphere. Under these conditions,

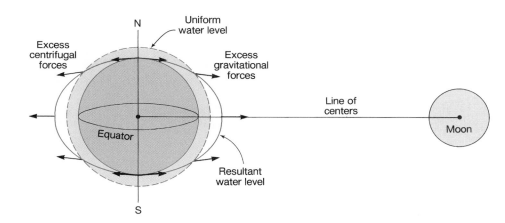

Figure 8.25

Distribution of tide-raising forces on the earth. Excess lunar gravitational and centrifugal forces distort the Earth model's water envelope to produce bulges and depressions.

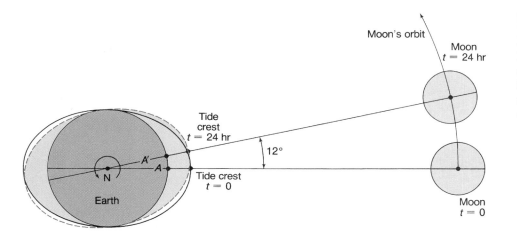

Figure 8.26

Point *A* requires 24 hours to complete one Earth rotation. During this time the moon moves 12° east along its orbit, carrying with it the tide crest. To move from *A* to *A'* requires an additional 50 minutes to complete a tidal day.

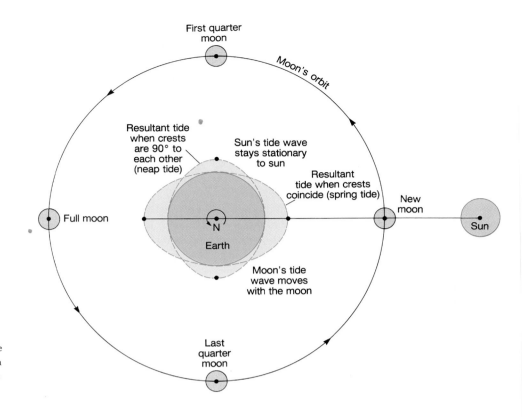

Figure 8.27

Spring tides result from the alignment of the Earth, Sun, and Moon during the full Moon and the new Moon. During the Moon's first and last quarters, neap tides are produced.

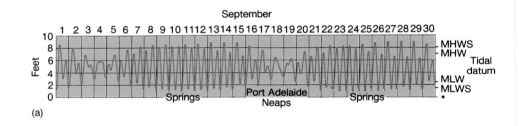

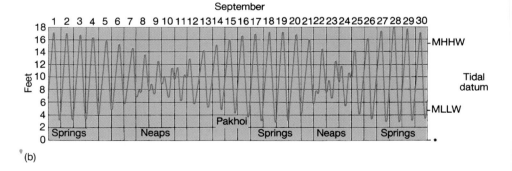

Figure 8.28

Spring and neap tides alternate during the tides' monthly cycle. MHWS is mean high water spring tides; MLWS is mean low water spring tides. (a) A semidiurnal tide from Port Adelaide, Australia. (b) A diurnal tide from Pakhoi, China.

point A in figure 8.29 passes through only one crest or high tide and one trough or low tide each tidal day; a diurnal tide is formed. Compare the diurnal tide at point A with the semidiurnal tide at point B in this figure.

The Sun moves from $23\frac{1}{2}°$N (the Tropic of Cancer) to $23\frac{1}{2}°$S (the Tropic of Capricorn) and returns each year. Therefore, tides are more diurnal during the summer and win-

ter solstices. The Moon moves from $28\frac{1}{2}°$N to $28\frac{1}{2}°$S and returns in a period of 18.6 years. Occasionally the Sun and the Moon both reach their most northern or southern points in these cycles at the same time; when this happens, tides become more strongly diurnal at midlatitudes. Because of the Earth's elliptical orbit about the Sun, the Earth is closest to the Sun during the Northern Hemisphere's winter, and tides

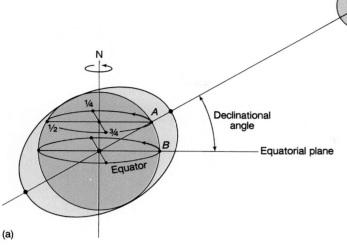

(a)

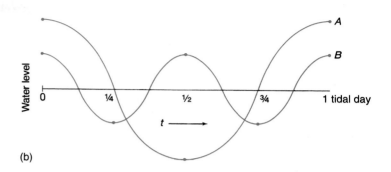

(b)

Figure 8.29

The declination of the Moon produces a diurnal tide at latitude *A* and a semidiurnal tide at latitude *B*. Fractions indicate portions of the tidal day.

tend to be more extreme (higher highs, lower lows) at that time due to the Sun's greater gravitational effect. When the Moon's orbit brings it closer to the Earth, the tides are also more extreme.

Draw and label the tide stages of a diurnal, semidiurnal, and mixed semidiurnal tide.

What is the difference between an ebb tide and a flood tide?

Explain how gravity and centrifugal force cause the ocean tides.

Why is a tidal day longer than a solar day?

Explain spring tides and neap tides.

Why is a tide sometimes more diurnal or semidiurnal than at other times?

Real Tides in Real Oceans   8.13

Although equilibrium tides allow us to understand the basic concepts associated with the oceans' tides, investigating the actual tides that occur in the discontinuous ocean basins requires the dynamic approach. Because continents separate the oceans from each other, the tide wave is discontinuous except in the Southern Ocean around Antarctica. The tide wave has a long wavelength as compared to the oceans' depths; therefore tides behave as shallow-water waves.

The shallow-water tide wave moves in theory at about 200 m/sec (400 mph), but at the equator the Earth moves eastward under the tide wave at more than twice the speed at which the tide can travel freely. Therefore, friction displaces the tide crest to the east of its expected position under the Moon. This eastward displacement continues until the friction force is balanced by the portion of the Moon's attractive force that pulls the tide wave westward. These two forces hold the tide crest in a position to the east of the Moon rather than directly under it. This process is illustrated in figure 8.30.

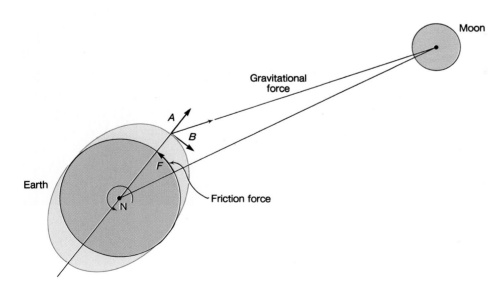

Figure 8.30

The crest of the tide wave is displaced eastward until the *B* component of the gravitational force balances the friction (*F*) between the Earth and the tide wave. Component *A* of the gravitational force is the tide-raising force. Component *B* causes the tide wave to move as a forced wave.

(a)

(b)

Figure 8.31

(a) Low tide and (b) high tide at The Rocks Provincial Park, Hopewell Cape, New Brunswick, Canada. The tidal range at the head of the bay exceeds 10 meters.

The tide wave is also reflected from the edges of land-masses and mid-ocean ridges, refracted by the change in water depth, and diffracted as it passes through gaps between land masses. In some basins, the tide wave is reflected from the edge of the continents and becomes a standing wave. In this case the scale and duration of tidal motion cause the Coriolis effect to deflect the moving water as the water flows from the crest to the trough. This causes the crests of standing-wave type tides to rotate in their ocean basins. Together these factors produce the dynamic tides. In addition, different kinds of tides interact with each other along their boundaries, and the results are exceedingly complex.

Along the coasts in narrow bays, a standing wave with the node at the entrance to the bay and the antinode at the head of the bay may produce an extremely high **tidal range** at the head of the bay, if the water in the bay naturally oscillates at a period close to the tidal period. For example, the Bay of Fundy in southeast Canada has a tidal range near the entrance node of about 2 m (6 ft) and a range of 10.7 m (35 ft) at the head of the bay (fig. 8.31).

When water moves into a coastal region on the rising tide and out of it on the falling tide, **tidal currents** form. These currents may be extremely swift and dangerous as they move through narrow channels into large bays and harbors. When the tidal current changes from an ebb to a flood or vice versa, there is a period of **slack water,** during which the tidal currents slow, stop, and then reverse. Slack water may be the only time that a vessel can safely navigate a narrow channel with swiftly moving tidal currents, sometimes in excess of 20 km per hour (10 knots).

List reasons for the differences in behavior between equilibrium tides and natural tides.

How does a tidal current differ from an open-ocean surface current?

Predicting Tides and Tidal Currents 8.14

Because of all the natural combinations of tides and the factors that affect them, it is not possible to predict the Earth's tides from knowledge of the Sun and Moon forces alone. Accurate, dependable, daily tidal predictions are made by combining local measurements with astronomical data. Water-level recorders are installed at coastal sites, and the rise and fall of the tides are measured over a period of years. Primary tide stations make these water-level measurements for at least 19 years to allow for the 18.6-year period of the Moon. From these data, mean tide levels are calculated. Oceanographers separate the tide record into components with magnitudes and periods that match the tide-raising forces of the Sun and Moon. They are then able to isolate the local effect from the astronomical data. This local effect is used with astronomical data to predict future tides in the area.

The National Oceanic and Atmospheric Administration (NOAA) of the U.S. Department of Commerce has long been the source of published tide tables for North and South America, Alaska, Hawaii, and the coast of Asia. Tidal data is now available on the Internet and private companies have taken over the publication of NOAA tide predictions. These tables give the dates, times, and water levels for high and low water at primary tide stations and give the correction values required to convert these data for use at auxiliary stations. Private companies also publish tidal current data previously published by NOAA; current data are derived from predicted tide data and field observations.

Why are astronomical data not sufficient to predict tides for a specific area?

Tidal Bores 8.15

In some areas of the world, large-amplitude tides cause large and rapid changes in water volume along shallow coasts, bays, or river mouths. Under these conditions, the tide level increases more rapidly than the tide wave can normally move. The tide wave forms a spilling, breaking wave front that moves rapidly into the shallow water or up the river (fig. 8.32). This broken wave front appears as a wall of turbulent water and is called a **tidal bore,** which produces an abrupt change in water level as it passes. The bores are usually less than a meter in height but can be as much as 8 m (26 ft) high, as in the case of spring tides on the Qiantang River of China. The Amazon, Trent, and Seine Rivers have bores. Fast-rising tides also send bores across the sand flats surrounding Mont-Saint-Michel in France. These

Figure 8.32
The fast-rising tide in the Bay of Fundy produces a tidal bore that sweeps across the shallows.

bores can be hazardous, because they suddenly flood areas that have been open stretches of beach only minutes before.

Explain the hazards of tidal bores.

Energy from Waves and Tides   8.16

A tremendous amount of energy exists in ocean waves. The power of all waves is estimated at 2.7×10^{12} watts, which is about equal to 3000 times the power-generating capacity of Hoover Dam. Unfortunately for human needs, this energy is widely dispersed and is not constant at any given location or time.

Wave energy can be harnessed in three basic ways: (1) using the changing level of the water to lift an object; (2) using the orbital motion of the water particles or the changing tilt of the sea surface to rock an object to and fro; and (3) using the rising water to compress air in a chamber. In each case, the wave motion may be used either directly or indirectly to turn a generator and generate electricity.

The 1991 estimate for total world energy produced by all operating wave-energy systems is less than 5.0×10^{5} watts. One energy system allows waves to rush into a tapered channel pushing the water to an elevation of 2 to 3 m (6–10 ft). The water spills over the channel and down through a turbine, which produces electricity. This system is used to generate power by a 7.5×10^{4}-watt plant on Scotland's Isle of Islay, a 3.50×10^{5}-watt plant at Toftestalen, Norway, and two 1.5×10^{5}-watt plants under construction, one in Java and the other in Australia.

Along a wave-exposed coast, air traps can be installed so that the crest of a wave moving into the trap compresses a large volume of air, forcing it through a one-way valve. This compressed air powers a turbine to produce energy. The trough of the wave allows more air to enter the trap, readying it for compression by the following wave crest (fig. 8.33). This system is in use in northern Norway and western Ireland.

Average wave power along the coast of Great Britain is calculated at about 5.5×10^{4} watts per meter of coastline. If the wave energy could be completely harnessed along 1000 km (620 mi) of coast, it would generate enough power to supply 50% of Great Britain's present power needs. Along the northern California coast, waves are estimated to expend 2.3×10^{10} watts annually. It is thought that 20% of this power could be harvested.

However, for wave-energy systems to help us in our quest for new energy sources requires the thoughtful consideration of certain questions. If all the wave energy were extracted from the waves in a coastal area, what effect would this action have on the shore area? If the nearshore areas are covered with wave energy absorbers, what will the effect be on other ocean uses? Since the individual units collect energy at a slow rate, can they collect enough energy over their projected life span to exceed the energy used to fabricate and maintain them? Harvesting wave energy is not without an effect on the environment; it may be neither cost nor energy effective, and its location may present enormous problems for installation, maintenance, and transport of energy from sites of energy generation to sites of energy use.

The possibility of obtaining energy from the tides exists in coastal areas with large tidal ranges. For thirty years 5.4×10^{10} watt-hours per year of commercial power have been generated by a tidal power station on the Rance River estuary in France. The province of Nova Scotia in Canada began its power station project in the tidal estuary of the Annapolis River in 1984 (fig. 8.34). Tidal ranges at the Annapolis site vary from 8.7 m (29 ft) during spring tides to 4.4 meters (14 ft) during neap tides. Annual production from the unit is between 3–4 $\times 10^{10}$ watt-hours per year. Plants of this kind require building a dam across a bay or estuary so that seawater can be held in the bay at high tide and released through turbines as the out-

Figure 8.33

Each rise and fall of the waves pumps pulses of compressed air into a storage tank. A smooth flow of compressed air from the storage tank turns a turbine that generates electricity.

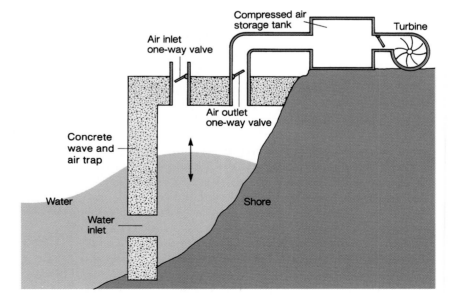

Figure 8.34

The Annapolis River tidal power project is the first tidal power plant in North America. It was completed in 1984.

side tide level drops. These systems generate power either on the ebb tide or both the ebb and flood tides (fig. 8.35).

However, there are few places in the world where the tidal range is sufficient, 7 m (23 ft) or more, and where natural bays or estuaries can be dammed at their entrances at reasonable cost and effort. Neither are these places necessarily located near the population centers that need the power. The costs of the installations and the power distribution, in addition to periodic low power production because of the changing tidal amplitude over the tide's monthly cycle, make this type of power expensive in comparison to other sources.

Explain the sources of power in waves and tides.

Where is commercial electric power being generated by waves and tides at the present time?

What are the reasons for and against the production of electrical power from waves and tides?

Table 8.3 is a summary of the characteristics of the various waves discussed in this chapter.

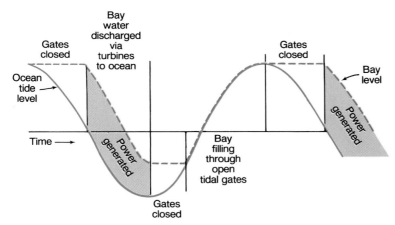

(a) Single-action power cycle; ebb only

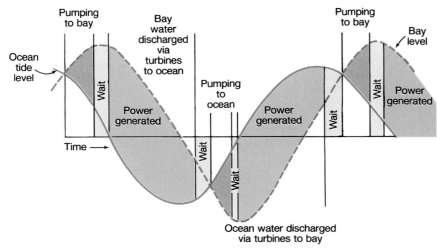

(b) Double-action power cycle; ebb and flood

Figure 8.35

(a) A single-action tidal power system. Power is generated on the ebb tide. (b) A double-action tidal power system. Power is generated on the ebb and flood tides.

table 8.3 Wave Summary

Type	Cause	Period	Class of Wave	Open-Ocean Height	Coastal Height	Wave Speed	Group Speed	Orbit Type
Wind wave	Wind	0.1–20 s	Deep-water $D > L/2$	Up to 30 m (100 ft)	Up to 6 m (20 ft)	$C = 1.56T$ $C^2 = 1.56L$ 0–111 km/hr (0–60 kts)	$V = C/2$	Circle
			Shallow-water $D < L/20$	—	Up to 6 m (20 ft)	$C = \sqrt{gD}$	$V = C$	Ellipse
Tsunami	Seismic upheaval	10–30 min	Shallow-water $D < L/20$	Small	10–18 m (30–60 ft)	$C = \sqrt{gD}$ 740 km/hr (400 kts)	$V = C$	Ellipse
Standing	Wind, tsunamis, tides, pressure changes	10 min– 30 hrs	Normally shallow-water $D < L/20$	0–2.5 m (0–8 ft)	0–15 m (0–50 ft)	None	None	None
Tide	Gravity of Sun and Moon	12 hrs 25 min and 24 hrs 50 min	Shallow-water $D < L/20$	0–1.5 m (0–4 ft)	0–15 m (0–50 ft)	$C = \sqrt{gD}$ >740 km/hr (>400 kts)	None	Ellipse

Summary

When the water's surface is disturbed, a wave is formed. The highest point of a wave is the crest; the lowest point is the trough. The wavelength is the distance between two successive crests or troughs. The wave height is the vertical distance between the crest and the trough. Wave period measures the time required for two successive crests or troughs to pass a location. The moving wave form causes water particles to move in orbits. A wave's speed is related to wavelength and period.

Deep-water waves occur in water deeper than one-half the wavelength. Wind waves are formed in storm centers as deep-water waves with a period that never changes. As waves move out from the storm center, the longer, faster waves move through the shorter, slower waves and form groups of waves. This process is known as sorting or dispersion. Waves of uniform characteristics are known as swell. The speed of a group of waves is half the speed of the individual waves in deep water. Swells from different storms cross, cancel, and combine with each other as they move across the ocean.

Wave height depends on wind speed, wind duration, and fetch. Single large waves unrelated to local conditions are called episodic waves. The energy of a wave is related to its height. When the ratio of the height to the length of a wave, or its steepness, exceeds 1:7, the wave breaks.

Shallow-water waves occur when the depth is less than one-twentieth the wavelength. The speed of a shallow-water wave depends on the depth of the water. As the wave moves toward shore and decreasing depth, it slows, shortens, and increases in height. Waves coming into shore are refracted, reflected, and diffracted. The patterns produced by these processes helped peoples in ancient times to navigate from island to island.

In the surf zone, breaking waves produce a water movement toward the shore. Breaking waves are classified as plungers or spillers.

Tsunamis are seismic sea waves. They behave as shallow-water waves and can produce severe coastal flooding and destruction. Standing waves or seiches form when displaced water oscillates in a basin.

The three tidal patterns are diurnal, semidiurnal, and mixed. Rising tides along coasts create flood currents; falling tides create ebb currents.

Tide-raising forces are explained as a balance between gravitational and centrifugal forces or by centripetal forces only. A semidiurnal tide is a tide wave with two crests and two troughs per tidal day. A diurnal tide has one crest and one trough per tidal day. The tidal day is 24 hours 50 minutes long. Spring tides have the greatest range between high and low water; they occur at the new and full Moons. Neap tides have the least tidal range; they occur at the Moon's first and last quarters. When the Moon or Sun stands far to the north or south of the equator, the tides become more strongly diurnal.

Dynamic tides are the actual tides as they occur in the ocean basins. The tide wave is a discontinuous shallow-water wave that oscillates in some ocean basins as a standing wave with motions that persist long enough to be acted on by the Coriolis effect. The tide wave is reflected, refracted, and diffracted along its route. In narrow basins, tides may oscillate about a node at the entrance to the bay and the antinode at the head of the bay.

A rapidly moving tidal bore is caused by a large-amplitude tide wave moving into a shallow bay or river.

Tidal heights and currents are predicted from astronomical data and actual local measurements. Annual tide and tidal current tables are published from NOAA data.

It is possible to harness the energy of the waves by using water-level changes or the changing surface angle. Difficulties include cost, location, environmental effects, and lack of wave regularity. Tidal power plants are in use in France and Canada. Few places have large enough tidal ranges and suitable locations for tidal dams.

Key Terms

amplitude, 184
antinode, 194
centrifugal force, 198
centripetal force, 198
crest, 183
deep-water wave, 185
diffraction, 190
dispersion (sorting), 185
diurnal tide, 196
dynamic tide, 197
ebb tide, 197
episodic wave, 188
equilibrium tide, 197
fetch, 187
flood tide, 197
group speed, 185
higher high water, 197
higher low water, 197
high water, 197
lower high water, 197
lower low water, 197
low water, 197
minus tide, 197
mixed tide, 196
neap tide, 199

node, 194
orbit, 184
period, 184
refract, 190
seiche, 195
semidiurnal tide, 196
shallow-water wave, 190
slack water, 203
spring tide, 199
standing wave, 194
storm center, 185
surface tension, 183
swell, 185
tidal bore, 203
tidal current, 203
tidal day, 199
tidal range, 203
trough, 183
tsunami, 193
wave height, 183
wavelength, 183
wave period, 184
wave steepness, 188
wave train, 185

Critical Thinking

1. If you were ocean-sailing at night between islands in the trade wind belt, how could you use the waves to help you maintain your course?
2. Compare a tsunami and a storm surge (chapter 6). How are they the same and how are they different?
3. If a group of mixed waves is generated in a sudden storm, why does it take more time for the wave group to pass an island far from the storm center than to pass a nearby island?
4. Explain why a tide is a wave.
5. At what time of year in the Northern Hemisphere would you expect the highest high tides? Why?

Suggested Readings

Waves
Alper, J. 1993. A Wave Is Born. *Sea Frontiers* 39(1):22–27.
Bascom, W. 1980. *Waves and Beaches: The Dynamics of the Ocean Survey*, rev. ed. Doubleday, Garden City, N.J. 366 pp.
Changery, M. J., and R. G. Quayle. 1987. Coastal Wave Energy. *Sea Frontiers* 33(4):259–62.
Folger, T. 1994. Waves of Destruction. *Discover* 15(5):66–73.
Hokkaido Tsunami Survey Group. 1993. Tsunami Devastates Japanese Coastal Region. *EOS* 74(37):145, 156–67.
Inamura, F. et al. 1997. Irian Java Earth Quake and Tsunami Cause Serious Damage. *EOS* 78(19):197, 201.
Junger, S. 1997. *The Perfect Storm*. Harper Paperbacks.
Kawata, J. et al. 1999. Tsunami in Papua New Guinea Was as Intense as First Thought. *EOS* 80(9):101 and 104–05.
Lockridge, P. A. 1989. Tsunami: Trouble for Mariners. *Sea Technology* 30(4):53–57.
McCreedie, S. 1994. Tsunamis: The Wrath of Poseidon. *Smithsonian* 24(12):28–39.
Yeh, H., et al. 1993. The Flores Island Tsunamis. *EOS* 74(33):369, 371–73.

Tides
Garrett, C., and L. R. M. Maas. 1993. Tides and Their Effects. *Oceanus* 36(1):27–37.
Lynch, D. K. 1982. Tidal Bores. *Scientific American* 247(4):146–56.

Energy
Dutton, G. 1992. Catch a Wave for Clean Electricity. *The World & I* July: 290–97.
Greenberg, D. A. 1987. Modeling Tidal Power. *Scientific American* 257(5):128–31.
McCormick, M. E. 1986. Ocean Wave Energy Conversion. *Sea Technology* June: 32–34.

internet references

worldwide websites

Waves

Neptune's Web—An Oceanographic Voyage in Learning
pao.cnmoc.navy.mil/educate/neptune/neptune.htm

Global Analysis Wave Map
www.oceanweather.com/data/global.html

Welcome to the WWA Web Site
www.oceanor.no/wwa/

Marine Prediction Center—Home Page
www.ncep.noaa.gov

H.M.R.C. European Wave Energy
www.ucc.ie/ucc/research/hmrc/ewern.htm

Tsunami

Tsunami!: The WWW Tsunami Information Resource
www.geophys.washington.edu/tsunami/welcome.
html

Pacific Marine Environmental Laboratory, Tsunami
Program
www.ngdc.noaa.gov/seg/hazard/resource/
geohaz/tsu_pmel.html

NOAA Websites—Sciences—Ocean—Tsunamis
www.websites.noaa.gov/guide/sciences/ocean/
tsunamis.html

Tides from Space

Ocean Tides from TOPEX/Poseidon Altimetry
outside.gsfc.nasa.gov/SCIDOC/SH94/E_sanchez.
html

TOPEX/Poseidon Main Screen
topex-www.jpl.nasa.gov/

Tide Prediction

Scripps Tide Predictor Services
scilib.ucsd.edu/sio/tide/

Tide/Current Predictor Site—Scripps Institution of
Oceanography
tbone.biol.sc.edu/tide/sitesel.html

Tide Tables
library.uncg.edu/depts/docs/us/tide.html

NOAA, Our Restless Tides
www.opsd.nos.noaa.gov/restless1.html

Wave, Tide, and Ocean Energy

Wave Energy Activities at Oceanor
oblea.oceanor.no/wave_energy/

Wave Energy Articles on the Net
www.waveenergy.dk/blgartwebuk.htm

Alternative Energy Sources—Wave Energy
www.schwaben.de/home/kepi/waves1.htm

Water Power, Dams, Waterwheels, OTEC, Wave and
Ocean Power
hydroelectricity.hypermart.net/tidal.html

Tidal Energy
www.iclei.org/efacts/tidal.htm

Tsunami Prediction and Warning Centers are located in Hawaii and Alaska; they were built after a 1946 tsunami struck Hawaii. Today's computer models can predict tsunamis from seismic data, but they are not yet able to determine the actual presence of a tsunami that is still close to its source. A model may predict a tsunami that does not exist or it may fail to predict a true tsunami. Twenty tsunami warnings

(Japan and Chile) that warn in about five minutes are useful less than 100 km (62 miles) from the source. All three systems use location and earthquake data to trigger warnings and then rely on coastal tide stations for verification of the waves.

Because no two harbors or estuaries are exactly the same, the wave data transmitted from one tide station does not forecast tsunami

Tsunami Warning Systems

have been issued since 1996, based primarily on seismic data. Fifteen of these warnings were false alarms.

At the present time, the Pacific-wide system can issue a warning within about one hour of a seismic event. To receive the warning in time to notify and evacuate shore populations, an area must be located more than 750 km (465 mi) from the source. U.S. systems in Alaska, Hawaii, and one each in Japan, Russia, and French Polynesia can give a ten minute warning to those living along shores that are 100 to 750 km (62–465 mi) from the source. Local systems

characteristics in locations only a short distance away. Until accurate deep-ocean tsunami measurements are available, it will not be possible to forecast site-specific information in hazardous areas.

In 1996 the states of Alaska, Washington, Oregon, California, and Hawaii joined with three federal agencies—the National Oceanic and Atmospheric Administration (NOAA), the U.S. Geological Service (USGS), and the Federal Emergency Management Administration (FEMA)—to fund the National Tsunami Hazard and Mitigation Program (see figure 1). At present two ocean stations are

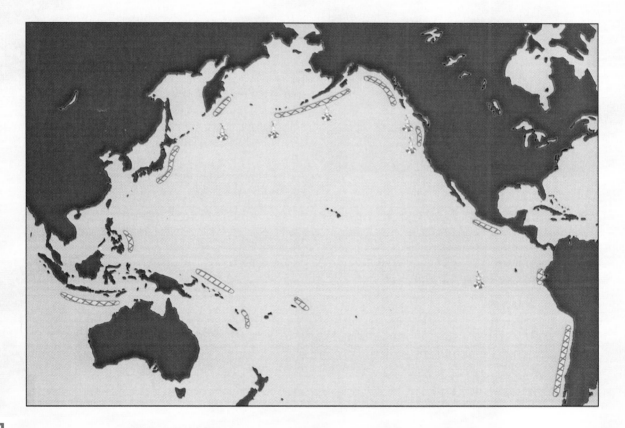

Figure 1

NOAA Deep Ocean Tsunami Measurement Program. Tsunami reporting stations are shown as small buoys. Two ocean stations (one south of the Aleutian Islands and one west of the Oregon coast) are presently in place; ten more stations are planned for the future. Undersea earthquake zones are associated with the subduction zones (shown in yellow and red) surrounding the Pacific Ocean.

Courtesy of National Tsunami Hazard and Mitigation Program, PMEL, HOAA.

More Information: www.pmel.noaa.gov/tsunami/pro_array.html

in place; bottom-mounted pressure sensors (long-wave detectors) linked to surface buoys (fig. 2) transmit data from tsunamis to passing satellites. The satellites relay the information to the hazard and mitigation program centers. One buoy is located south of Kodiak Island; it monitors waves generated south of Kodiak Island, in the Aleutian Trench, or in the Gulf of Alaska. The second buoy is off central Oregon, and monitors the Cascadia subduction zone, which lies between the Gorda–Juan de Fuca Ridge and the west coast of the U.S. It is anticipated that six more ocean sensors and thirty-eight additional seismometers will be installed in future years.

These warning systems will provide information and allow time for evacuation while the tsunami is still at sea. However, if a tsunami is formed close to a coastal area, residents may have only minutes to seek safety. The Cascadia subduction zone is very close to the west coast of the United States and Canada; more than half a million people live in potential tsunami flooding areas. Residents of Oregon and northern California would have less than five minutes warning after a large earthquake in this zone; coastal areas to the north might have twenty to forty minutes. Those who reside in tsunami-prone areas need to remember that any local earthquake is the signal to rapidly move away from the immediate coastline.

Reading

Bernard, E. 1998. Program Aims to Reduce Impact of Tsunamis on Pacific States. *EOS* 79(22):258, 262–3.

Internet Resources

Tsunami Program and Reporting Systems
www.pmel.noaa.gov:80/tsunami/home.html
www.pmel.noaa.gov/tsunami/sat_buoy.html

Tsunami Warning System
www.geophys.washington.edu/tsunami/general/
warning/warning.

Figure 2

A tsunami detection buoy in place off the western coast of the United States. The buoy is linked to a sensor resting on the sea floor. Data are transmitted from the buoy to a satellite and relayed from the satellite to tsunami warning centers.

More Information: www.pmel.noaa.gov/tsunami/s9709_11.html

chapter nine

Coasts, Estuaries, and Environmental Issues

A high tide and storm waves combine to erode the beach and damage homes.

Learning Objectives

After reading this chapter, you should be able to

- Understand the differences between primary and secondary coasts and explain their features.
- Describe the principal features of a beach.
- Describe different types of beaches.
- Understand the processes that form a sand beach.
- Relate sediment transport in a coastal circulation cell to beach formation.
- Relate shoreline changes to changes in sea level.
- Define an estuary and describe the different types.
- Compare a bay in a high-evaporation environment with an estuary.
- Understand estuary circulation and how water is exchanged with the ocean.
- Describe the effects of people and their structures on beaches and beach processes.
- Appreciate the importance of coastal wetlands.
- Discuss degradation of marine coastal environments and give examples.

The history of life has been a history of interaction between living things and their surroundings. To a large extent, the physical form and the habits of the earth's vegetation and its animal life have been molded by the environment. Considering the whole span of earthly time, the opposite effect, in which life actually modifies its surroundings, has been relatively slight. Only within the moment of time represented by the present century has one species—man—acquired significant power to alter the nature of his world.

Rachel Carson

from Silent Spring, 1962

e visit the coast to see the ocean, to play along the beach, to enjoy the constant interaction between moving water and what appears to be the stable land. On any visit the tide may be high or low, the logs and drift may have changed position, and the dunes and sandbars may have shifted—coasts and beaches are dynamic, not static. No two regions are exactly the same, nor do they change in the same way. Along these coasts are indentations where tidal water from the oceans and freshwater drainage from the land meet and mix. The processes that regulate this mixing are not constant: river flow and tides vary; severe storms and river flooding occur from time to time. Coastal areas share certain characteristics, but each is also unique.

People play an important role in coastal areas, because people and their structures change these areas more than they realize. As the world's population grows, and as more and more people gather in these areas to live, work, and play, they affect them to a greater and greater degree. Solving the problems of these areas is difficult because it is dominated by natural events and complicated by many voices. However, whether you live by the ocean or not, the ocean is a part of your world, and you need to consider the situations that are discussed in this chapter.

The Coast 9.1

The **coasts** of the world are the areas where the land meets the sea. Coastal areas are not static features of the earth's environment. The coasts that we see today have undergone dramatic changes over long periods of time. Tides, winds, waves, earthquakes, changes in sea level, changes in the position of the continents, changes in climate all shape and reshape the world's coasts and will continue to do so.

The coast includes cliffs, dunes, beaches, and sometimes the hills and plains that form the edge of the land. The distance to which the coast extends inland varies and is determined by local geography, climate, and vegetation as well as the perception of the limits of marine influence in social customs and culture. The coast is most generally described as the land area that is or has been affected by marine processes such as tides, winds, currents, and waves. The seaward limit of the coast usually coincides with the beginning of the beach or shore, but sometimes includes nearby offshore islands.

The term **coastal zone** includes the open coast as well as the bays and estuaries found in coastal indentations. This phrase incorporates both land and water areas, and has become the standard term used in legal and legislative documents affecting U.S. coastal areas. Under these circumstances the coastal zone's landward boundary is defined by a distance (often 200 ft) from some chosen reference (usually high water); the seaward boundary's limit is defined by state and federal laws.

> What are the boundaries of a coast?

Types of Coasts 9.2

Not all coasts show the same characteristics. Coastal features are related to the tectonic history of an area and also to the area's location on a lithospheric plate. When along a coast, a divergent plate boundary is displaced from its spreading center and becomes tectonically inactive, the coast changes slowly with geologic time. The continental shelf is wide and the coastal region may sink gradually as the crust ages and cools. The east coast of the continental United States is associated with a divergent plate boundary. At a convergent plate boundary, collision between plates along subduction zones or motion along a transverse fault modifies the coast by raising or lowering the plate edge, often leaving a narrow continental shelf. Volcanism is associated with this type of plate activity. The northwest coasts of the United States and Alaska are located along such a plate boundary.

Coasts are considered **emergent** if they are rising relative to sea level. This occurs if sea level falls, if the coast is uplifted by tectonic processes, or if there is a decrease in the weight on the coastal crust (for example, receding glaciers). Coasts are considered **submergent** if they sink relative to sea level. This happens if sea level rises, if the land is lowered by tectonic processes, or if the weight on the crust increases (for example, glacier deposits, river sediment deposits). Compaction of coastal sediments also leads to submergence. An emergent coast shows old beaches and shore features above present sea level (fig. 9.2d), while along a low-lying, submergent coast the water moves inland and reclaims the land (fig. 9.1a). Changes in sea level in excess of

100 m (300 ft) have occurred as the volume of the ocean has changed in response to changes in the amounts of water stored in the continental ice sheets.

To distinguish between the many types of coasts, table 9.1 and figures 9.1 and 9.2 use the system of the late Francis P. Shepard of the Scripps Institution of Oceanography. Classification depends on large-scale features created by tectonic, depositional, erosional, volcanic, or biological processes. All coastal areas belong in one of two major categories: (1) **primary coasts** that owe their character and appearance to processes that occur

table 9.1 Coastal Types

Primary Coasts	Features	Examples
Erosional Examples		
Drowned river valleys (ria coasts) Figure 9.1a	Multibranched shallow areas; sea level has risen or the land has sunk.	Chesapeake or Delaware Bay systems, Rio Plata, Argentina.
Fjord coasts Figure 9.1b	Glacially carved drowned channels. Deep and elongate.	Mountainous coasts of previously glaciated temperate and subpolar regions; Alaska, Greenland, New Zealand, Scandinavia.
Depositional Examples		
Deltaic coasts Figure 9.1c	Heavy river deposition in shallow coastal regions.	Mouths of rivers such as the Mississippi, Nile, Ganges, and Amazon.
Glacial moraine coasts	Deposits left by glaciers.	Long Island and Cape Cod, Baltic Sea, Sea of Okhotsk.
Dune coasts Figure 9.1d	Sand deposits on coasts by wind from the beach or wind from the land.	Central and northern Oregon coast. West coast of Sahara Desert and west coast of Kalahari Desert, Africa.
Volcanic Examples		
Lava coasts	Areas of active volcanoes with lava that reaches the sea.	East coast island of Hawaii.
Cratered coasts Figure 9.1e	Craters of coastal volcanoes at sea level filled with seawater.	Hawaiian islands, and islands and coastal areas of the Aegean Sea.
Tectonic Examples		
Fault coasts Figure 9.1f	Coastal faults filled with seawater.	Tomales Bay, California, coast of Scotland, Gulf of California, Red Sea.
Secondary Coasts	**Features**	**Examples**
Erosional Examples		
Regular cliffed coasts Figure 9.2a	Irregular coast made more uniform by wave erosion.	Hawaii, southern California, southern Australia, Cliffs of Dover.
Irregular cliffed coasts Figure 9.2b	Coast made more irregular by wave erosion.	Sea stack coasts of northern Oregon and Washington, New Zealand, Australia.
Depositional Examples		
Barrier coasts Figure 9.2c	Shallow coast with abundant sediments bordered by barrier islands and sand spits.	East coast of the United States, Gulf of Mexico, Netherlands.
Beach plains Figure 9.2d	Flat areas of elevated coastal seashore. May be terraced by wave erosion during uplift.	Northern California, Sweden, New Guinea.
Mudflats and salt marshes Figure 9.2e	Sediments deposited in wave-protected environments.	Common in many areas.
Built by Marine Organisms		
Coral reef coasts	Shallow tropical waters; corals produce massive coastal structures.	Tropical regions of the Pacific, Indian, and Atlantic oceans.
Mangrove coasts Figure 9.2f	Tropical and subtropical areas with freshwater drainage.	Florida, northern Australia, Bay of Bengal, other tropical areas.
Marsh grass coasts Figure 9.2g	Shallow, protected areas of fine sediments where plants root in intertidal land.	Eastern United States, Canada, coast of Europe, temperate climates.

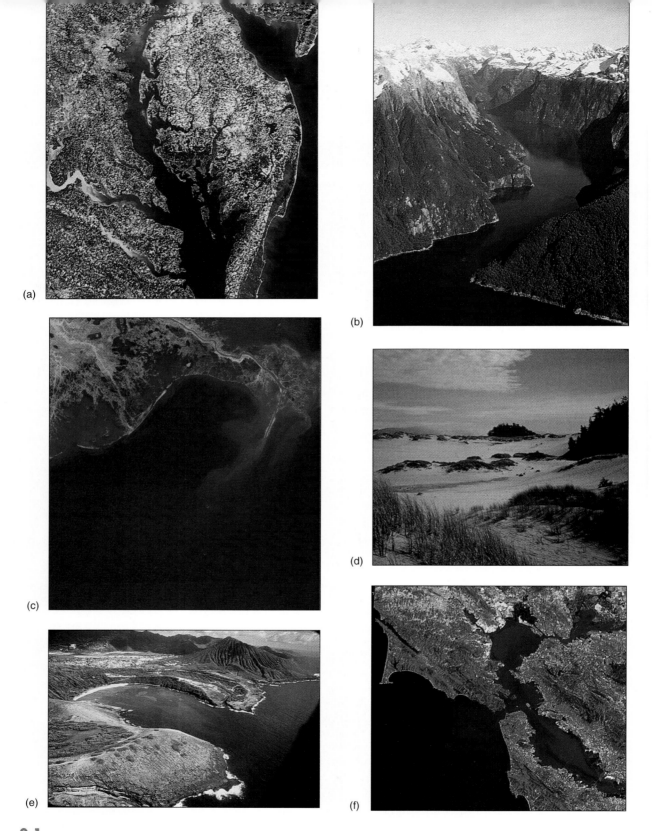

Figure 9.1

Primary coasts: (a) Delaware Bay (upper right) and Chesapeake Bay (center) are examples of drowned river valleys. (b) The narrow channel of a fjord, Milford Sound, New Zealand. (c) The Mississippi River delta. Because the river is being kept in its channel by levees, river-borne sediments are continuing to the river's mouth and being deposited in deep water instead of spreading out and replenishing the delta. (d) Coastal sand dunes near Florence, Oregon. (e) A volcanic crater coast. Hanauma Bay, Hawaii, is a crater that has lost the seaward portion of its rim. (f) The San Andreas Fault runs along the California coast to the west of San Francisco, separating Point Reyes from the mainland. Bolinas Bay and Tomales Bay (upper left) are fault bays produced by displacement along the fault line.

More Information: marine.usgs.gov/

Figure 9.2

Secondary coasts: (a) A cliffed coast made more regular by erosion, New Zealand. (b) Sea stacks are common features along the southern coast of Australia; these stacks are known as the Twelve Apostles. (c) Sea Island, Georgia, is a barrier island that has been extensively developed. The shallow bay between the island and the coast is visible on the left. (d) A series of pocket beaches cut an elevated marine terrace along the California coast. The flat elevated land is an old beach plain. (e) A protected mudflat shore along Puget Sound, Washington. (f) Mangrove trees along the shore of Florida Bay. (g) Coastal saltwater marsh, Gaspé, Québec, Canada.

More Information: marine.usgs.gov/

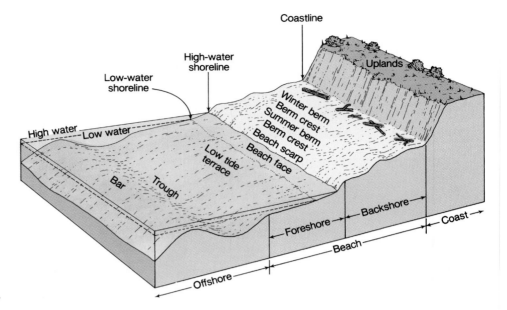

Figure 9.3

A typical beach profile with associated features.

at the land-air boundary, and (2) **secondary coasts** that owe their character and appearance to processes that are primarily of marine origin. Primary coasts have not yet been substantially altered or modified by marine processes. Features of secondary coasts may have been originally land-derived, but are now distinctly the result of ocean processes.

> Under what circumstances do emergent and submergent coasts form?
>
> How do primary and secondary coasts differ?
>
> Give examples of primary coasts; explain their features.
>
> Give examples of secondary coasts; explain their features.

Anatomy of a Beach 9.3

The **beach** is an accumulation of sediment (sand or gravel) that occupies a portion of the coast. The beach is not static, but is ever changing and dynamic, because the beach sediments are constantly being moved seaward, landward, and along the shore by nearshore wave and current action. Above the high tide mark, there may be dunes or grass flats spotted with occasional drift logs left behind by an exceptionally high tide or severe storm.

A beach profile (fig. 9.3) represents a cut through a beach perpendicular to the coast, and in this case shows the major features that might appear on any beach. Use figure 9.3 while reading this section, remembering that a given beach may not have all these features, for each beach is unique. The **backshore** is the dry region of a beach that is submerged only during the highest tides and severest storms; the **foreshore** extends

out past the low tide level, and the **offshore** extends from the shallow-water areas seaward of the low tide level to the outer limit of wave action on the bottom.

Berms appear in the area where foreshore and backshore meet. A berm is formed by wave-deposited material and has a relatively flat top like a terrace, or it may have a **berm crest** that runs parallel to the beach. When two berms are found on a beach (fig. 9.4), the berm higher up the beach is the winter or storm berm, formed during severe storms when the waves reach high up the beach and pile up material along the back-

Figure 9.4

Berms on a gravel beach. The winter storm berm with drift logs is located high on the beach. The summer berm is between the winter berm and the beach face in the foreground.

Figure 9.5

A wave-cut scarp on a gravel beach.

shore. The berm closer to the water is the summer berm, produced by the gentle waves of spring, summer, or fall that do not reach as far up the beach. Once the winter berm is formed, it is not disturbed by the less intense storms during the year, but the waves that produce a winter berm erase the summer berm.

Between the berm and the water level there may be a wave-cut **scarp** at the high-water level. The scarp is an abrupt change in the beach slope that is caused by the cutting action of waves at normal high tide (fig. 9.5). Berms and scarps do not form between normal high and low tide levels because of the wave action and continual rise and fall of the water over the foreshore. The evenly spaced, crescent-shaped depressions known as **cusps** (fig. 9.6) that are found along sand and cobble beaches of both quiet coves and exposed shores are a puzzle. Coastal scientists have been attempting to understand the formation of cusps by installing instruments along the beaches of North Carolina's Outer Banks. They have also removed the cusps and monitored their return. The results have been inconclusive and the answer to the formation of cusps remains elusive.

Figure 9.6

The action of waves forms a series of cusps along the beach at Point Reyes, California. How or why the cusps form is not clearly understood. Notice the cliffed coast made regular by the eroding waves.

More Information: www.frf.usace.army.mil/SandyDuck/Gallery/SandyDuck-ctl/slide1.html

The portion of the foreshore below the low tide level is often flat, forming a **low tide terrace;** the upper portion between high and low tide levels has a steeper slope and is known as the **beach face.** The slope of a beach face is related to the particle size of the loose beach material and the wave energy.

Seaward of the low tide level, in the offshore region, there may be **troughs** and **bars** running parallel to the beach. Troughs and bars change seasonally as beach sediment moves seaward, enlarging the bars during winter storms, and then shoreward in summer, diminishing the same bars.

Beaches do not occur along all shores. Along rocky cliffs, no beach area may be exposed between low and high tide levels. However, there may be small pocket beaches between headlands, each separated from the next by rock headlands or cliffs (see fig. 9.2d).

Distinguish between backshore, foreshore, and offshore.

What is a berm? Distinguish between a summer berm and a winter berm.

Why is a beach face without features?

What is the function of bars along a beach?

Beach Types   9.4

Beaches are also described in terms of (1) shape and structure, (2) composition of beach material, (3) size of beach materials, and (4) color. In the first category, beaches are described as wide or narrow, steep or flat, and long or discontinuous (pocket beaches). A beach area that extends outward from the main beach and then turns and parallels the shore is called a **spit** (fig. 9.7). If a spit extends to an offshore island, it is called a **tombolo.** A spit extending offshore in a wide, sweeping arc is often called a hook. See figure 9.8 for an example of a tombolo and figure 9.9 for an example of a hook.

Along some coasts bars form parallel to the shore in shallow water where there are abundant sediments. When these bars accumulate sufficient material to break the sea surface they become **barrier islands.** Once a barrier island is formed, vegetation helps to stabilize the sand and increase the island's elevation by trapping sediments and accumulating organic matter. Barrier islands may also form when rising sea level inundates a dune coast and the elevated dunes are separated from the low-lying mainland. Elongate barrier islands along the East and Gulf Coasts of the United States protect low-lying coasts from direct wave damage. Here towns and recreation sites have been built on the larger islands; see figure 9.2c.

Figure 9.7

A spit has formed across the entrance to Sequim Bay, Washington.

Figure 9.8

A spit connected to an offshore island forms a tombolo.

Figure 9.9

Ediz Hook is a long spit forming a natural breakwater at Port Angeles, Washington (infrared false color photo).

Materials that form beaches include shell, coral, rock, and lava. Shingles are flat, circular, smooth stones formed from layered rock when wave action slides the stones back and forth with the water movement. When stones roll rather than slide, round stones or cobbles are formed. The size of the beach particles is described by the terms sand, mud, pebble, cobble, and boulder; refer to chapter 4.

Beaches are formed from whatever loose material is present in the environment. Land materials are brought to the coast by the rivers or come from local cliffs by wave erosion. The strong wave action along some beaches erodes the softer materials and leaves rock structures such as the "pancake rocks" in figure 9.10 and the large boulders in figure 9.11.

Coral and shell particles broken by the pounding action of the waves are carried onshore by the moving water. In Hawaii there are white sand beaches derived from coral, and black sand beaches derived from lava. Green sands can be found in areas where a specific mineral (olivine or glauconite) is available in large enough quantities, and pink sands occur in regions of sufficient shell material.

Distinguish between a spit, a hook, and a tombolo.

What forms a beach?

Figure 9.10

Wave action exposes these layered limestone rock formations known as "pancake rocks," found along the beaches of the Tasman Sea on the South Island of New Zealand.

Figure 9.11

Mud balls are washed out of the cliffs of many beaches. These "Moeraki" boulders, hardened by the deposition of calcite minerals, are among the world's largest, approximately 2 m (6.5 ft) in diameter. The shape of the boulders is not due to wave action. They are found on the east coast of New Zealand's South Island.

Beach Processes   9.5

A sand beach exists because there is a balance between the supply and the removal of the material that forms it. A sand beach that does not appear to change with time is not necessarily static, but is more likely to represent a dynamic equilibrium, with equal supply and removal of beach material.

The gentle waves of summer move sand up the beach and deposit it, where it remains until the winter season. The large storm waves of winter remove the sand from the beach and transport it back offshore to the sandbars. This process lowers the beach each winter and covers it with sand during the summer (fig. 9.12).

Waves moving toward a beach produce a current in the surf zone that moves water onshore and along the beach. The landward motion of water is called the **onshore current,** and this current transports suspended sediment toward the shore. Waves usually do not approach a shore with their crests completely parallel to the beach but instead strike the shore at a slight angle. This pattern sets up a current in the surf zone that moves down the beach; this is the **longshore current.** Both onshore and longshore flow in the surf zone are illustrated in figure 9.13. The turbulence that occurs as the waves break in the surf zone tumbles the beach material into suspension in the water. The wave-produced longshore current displaces this wave-suspended sediment down the shore in the surf zone, producing a **longshore transport.** In addition the uprush and backwash, or swash, of water from each breaking wave moves sediment in a zigzag pattern down the beach face as a part of the longshore transport (fig. 9.13). Along both coasts of North America, the predominant longshore transport steadily moves the sediments in a southerly direction.

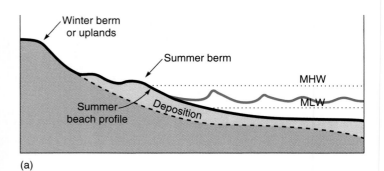

(a)

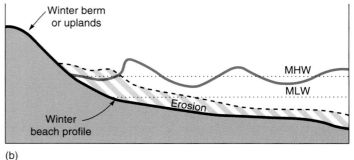

(b)

Figure 9.12

Seasonal beach changes. (a) The small waves and average tides of summer move sand from offshore to the beach and form a summer berm. (b) The high waves and storm tides of winter erode sand from the beach and store it in offshore bars. Winter conditions remove the summer berm, leaving only the winter berm on the beach.

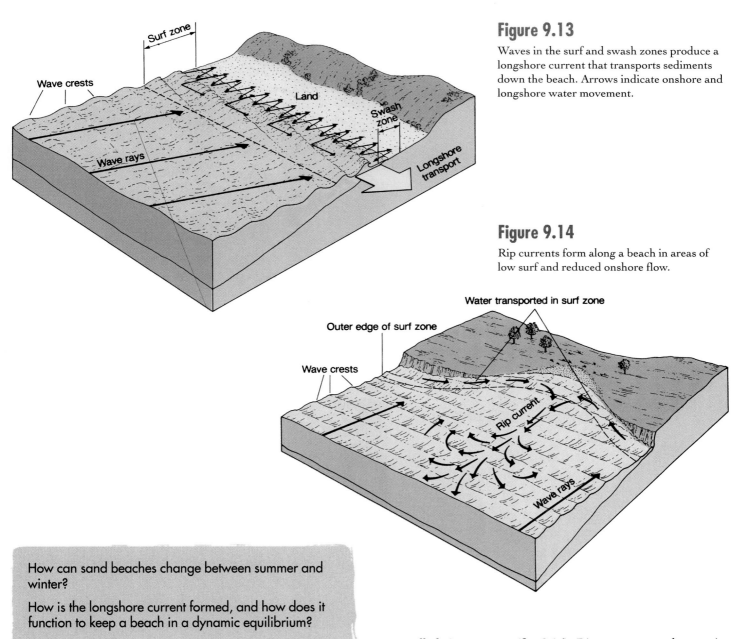

Figure 9.13

Waves in the surf and swash zones produce a longshore current that transports sediments down the beach. Arrows indicate onshore and longshore water movement.

Figure 9.14

Rip currents form along a beach in areas of low surf and reduced onshore flow.

How can sand beaches change between summer and winter?

How is the longshore current formed, and how does it function to keep a beach in a dynamic equilibrium?

Coastal Circulation Cells   9.6

The onshore current accumulates water as well as sand against the beach. This water must return seaward one way or another. A headland may deflect the longshore current seaward, or the water flowing along the beach can return seaward in areas of quieter water with smaller waves and less onshore flow.

Regions of seaward return are frequently narrow and some distance apart, located above troughs or depressions in the sea floor. The returning water flow must be swift in order to carry enough water beyond the surf zone to balance the flow toward the beach. These narrow areas of rapid seaward flow

are called **rip currents** (fig. 9.14). Rip currents can be a major hazard to surf swimmers; in the spring of 1994, five swimmers were drowned and six others hurt in a single rip current system at American Beach, Florida. To escape a rip current one must swim parallel to the beach, or across the rip current, and then return to shore.

Rip currents are visible as streaks of discolored turbid water extending seaward through the clearer water of the outer surf zone. The discoloration of the water is from the particles stirred up by waves and carried along the beach until they reach the rip current.

Just seaward of the surf, the rip current dissipates into an eddy. Some of the sediment is deposited in deeper water and is lost from the down-beach flow. However, some is returned to shore with the onshore transport on either side of the rip current. The path followed by sediment from its source to its place of deposit is called a **coastal circulation cell** (fig. 9.15). In a major coastal circulation cell, the seaward transport of sediment results in deposits far enough offshore to prevent any recycling. At the end point of such a circulation cell, the longshore current is deflected away from the beach, and the sand is carried and deposited offshore, often moving through submarine canyons into the offshore basins. This action removes the sand from its journey along the coast, and the sand beaches disappear below this point until a new source contributes sand to form the next major coastal circulation cell.

South of Point Conception in southern California, oceanographers recognize four distinct major coastal circulation cells (fig. 9.16). Each cell begins and ends in a region of rocky headlands where submarine canyons are found offshore. (Refer to chapter 4 for a review of submarine canyons.)

Figure 9.15

A coastal circulation cell extends down the coast from the sediment source to the area of deposit.

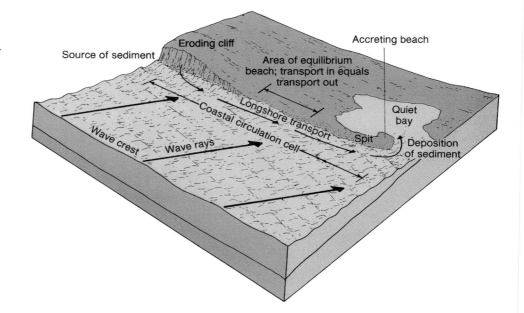

Figure 9.16

Major coastal circulation cells along the southern California coast. Each cell starts with a sediment source and ends where beach material is transported into a submarine canyon.

From D. L. Inman and J. D. Frautschy, "Littoral Processes and the Development of Shorelines, 1965," pp. 511–536. Coastal Engineering, ASCE, New York. Reprinted by permission of the authors.

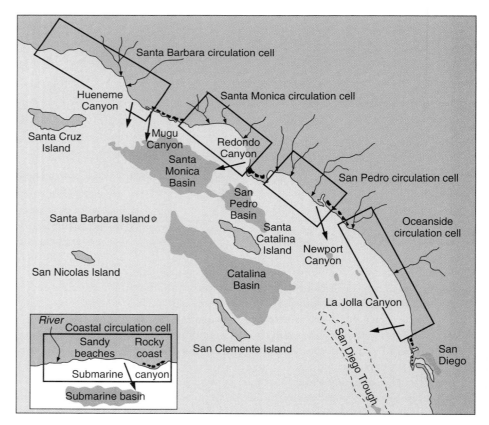

Rising Sea Level 9.7

Sea level has risen about 15 cm (6 in) during the past century, but the rate of rise has recently increased, and it is thought that sea level may gain another 30 cm (1 ft) in the next fifty years because of melting land ice and warming ocean water. If such an increase happens, the edge of the coast will move inland from hundreds to thousands of meters along low-lying coasts, and coastal erosion will be accelerated.

The increasing rate of sea-level rise has been primarily blamed on the greenhouse effect discussed in chapter 6. If the amount of CO_2 in the atmosphere doubles during the next 100 years, there are estimates that the ocean surface water will warm an additional 2° to 4°C. Under these circumstances it is thought that the greatest warming will occur at the high latitudes, where continental ice masses will melt and increase the volume of the oceans. The rise in sea level caused by land ice melting is estimated at between 31 and 110 cm (1–3.5 ft) by the year 2100. The expansion of the warming water will also affect sea level by raising it an additional 60 cm (24 in) for each degree of increase in the average ocean temperature.

Barrier islands protect the main part of the coast by absorbing the storm wave energy. As the barrier beaches are overwashed by storm waves there is a slow migration of the islands shoreward. Under rising sea level conditions the main shore may also be slowly inundated. Despite this recession of the coast, the processes maintain the barrier islands, the intervening sheltered bay, and the main coastline.

During intense storms, barrier islands and any dwellings, towns, or recreational facilities built on these islands sustain the damage that the continental coast is spared. The islands may be flooded and breached in some places, and old channels may be filled in other places. Because of their susceptibility to frequent and severe storm damage as well as rising sea level, the 1982 Coastal Barrier Resources Act prohibits federal subsidies for repairing roads, bridges, piers, and flood insurance coverage for homes built after 1983 in certain designated undeveloped barrier-island areas. States from Maine to North Carolina to Washington have regulated construction of seawalls and other permanent structures in the coastal zones.

Estuaries 9.8

An **estuary** is a portion of the ocean that is semi-isolated by land and diluted by freshwater drainage. An estuary's circulation, water exchange with the ocean, and current patterns are determined by the tides, the river flow, and the geometry of the inlet. Structure and geologic history of an inlet may be used to classify estuaries; however, this method does not give a clear indication of the dynamic behavior of an estuary as a body of water. Therefore, circulation and distribution of salinity are used to distinguish between the four basic types of estuaries, listed in table 9.2 and shown in figure 9.17a–d.

Salt wedge estuaries at the mouths of rivers are strongly stratified, as measured by salinity-density changes over depth. These estuaries are never well mixed vertically, except above the boundary between the river's surface and the upper surface of the salt wedge. Other estuary types have various degrees of vertical stratification: (1) the weakly stratified or well-mixed, (2) the partially stratified and mixed, and (3) the **fjord,** highly stratified or poorly mixed. The degree of stratification is governed by the speed of tidal currents and associated turbulence, the rate of freshwater addition, the roughness of the bottom topography, and the geometry of the estuary. Some parts of a complex estuary system may function as one type of estuary, and other parts may function as other estuary types.

The net circulation of an estuary, out (or seaward) at the top and in (or landward) at depth, exchanges water between the estuary and the ocean. Tidal flow that oscillates with the rise and fall of the tide produces the energy for mixing and should not be confused with the net flow in estuaries. The net circulation carries wastes and accumulated debris seaward and disperses them in the larger oceanic system. It is also through this circulation that organic materials and juvenile organisms produced in the estuaries and their marsh borderlands are moved seaward, while nutrient-rich salt water is brought inward at depth to replenish the estuary's water and salt content.

Areas of High Evaporation 9.9

Bays and seas located near latitudes 30°N and S have low precipitation and high evaporation rates. Although they may have rivers, these areas are not estuaries if they do not receive sufficient fresh water to become diluted. Instead, evaporation increases the surface water salinity and density, causing the surface

table 9.2 Types of Estuaries Summary

Type	Characteristics	Examples
Salt Wedge Estuary In mouths of rivers flowing strongly and directly into seawater. Figure 9.17a	Has tilted, sharply defined boundary between salt and fresh water. Boundary moves up and down stream with changes in tide and river flow. Salt water mixes upward into river flowing seaward above the boundary. Salt distribution and salt transfer governed by river flow. Strong density stratification over saltwater wedge.	Columbia, Mississippi, and Hudson Rivers. In the mouth of any reasonable size river entering an ocean or a saltwater bay.
Well-Mixed Estuary In shallow bays with large intertidal volumes, strong tidal mixing, and weaker river flows. Figure 9.17b	Salt content varies in the horizontal. Little or no vertical change in salt, and no strong density stratification. Weak net flow of water seaward at all depths. Salt moves toward river by mixing, not by flow. Also in constricted channels with strong turbulent tidal flow.	Chesapeake Bay, Delaware Bay, Humboldt Bay, California.
Partially-Mixed Estuary Deeper than shallow bays with both strong tidal flows and relatively high rates of freshwater discharge. Figure 9.17c	Salt content varies both vertically and horizontally. Moderate density stratification. Salt moves toward the river by strong net seawater flow at depth and some horizontal mixing. Vertical mixing and flow combine and form well-developed net seaward surface flow. Strong density stratification near surface when freshwater input high. Net flows assure good exchange of water with the sea.	San Francisco Bay, Puget Sound, Strait of Georgia, British Columbia.
Fjord-Type Estuary Long, narrow, deep, glacially carved channel, often with shallow entrance sill. Weak tidal currents except at the entrance. Figure 9.17d	Water at depth uniform in salt content, isolated from sea by sill, and exchanged very slowly. Surface layer very dilute with strong changes in salt content and density at sill depth. Strong stratification at sill depth acts as false bottom to seaward-moving surface layer. Net seaward flow strong in surface layer. Water at depth may stagnate.	Southeast Alaska, British Columbia, Norway, Greenland, New Zealand, Chile, and Argentina.

water to sink and accumulate at depth, where it flows seaward to exit into the ocean. The ocean water is less dense than the outflowing high-salinity water, and ocean water flows into the sea or bay at its surface (fig. 9.18). Inflow and outflow from these seas and bays are the reverse of an estuary's circulation, and they are sometimes called "inverse estuaries." The Red Sea and the Mediterranean Sea are examples of such areas. These evaporative seas with their continuous density-driven overturn often have a better exchange with the ocean than do true estuaries.

Compare the circulation of a bay in a high-evaporation environment to a partially mixed estuary.

Water Exchange 9.10

The time required for an estuary to exchange its water with the ocean is its **flushing time.** If the net circulation is rapid and the total volume of an estuary is small, flushing is rapid. A rapidly flushing estuary has a high carrying capacity for wastes, because the dissolved or suspended wastes are moved quickly out to sea and diluted. A slowly flushing estuary risks accumulating wastes and building up high concentrations of land-derived pollutants. The flushing time of estuaries is determined by dividing their volume by the rate of net seaward flow. Understanding the circulation and flushing time of our estuary systems is essential to maintaining them as healthy, productive, and useful bodies of water.

Some bays and harbors that do not qualify as estuaries because they have no freshwater inflow may be flushed by tidal action. On each change of the tide, a volume of water equal to the area of the bay times the water level change between high and low waters exits and enters the bay. When this volume leaves the bay on the ebb tide, it may be carried down the coast by a prevailing coastal current. With the next tidal rise, an equivalent volume of different ocean water enters with the rising tide. Here flushing time is measured in the number of tidal cycles required to remove and replace the average volume of the bay. However, the flushing is often not complete because if coastal currents are weak, part of the ebbing water may recycle back into the bay on the next flood tide. In addition, the seawa-

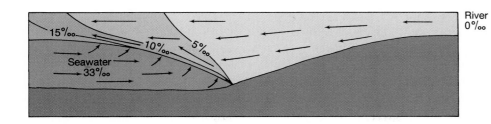

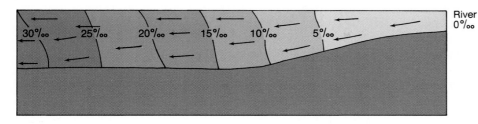

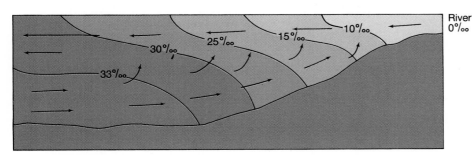

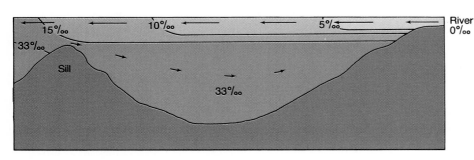

Figure 9.17

(a) The salt wedge estuary. The high flow rate of the river holds back the lesser flow of salt water. The salt water is drawn upward into the fast-moving river flow. Salinity is given in parts per thousand (‰).

(b) The well-mixed estuary. Strong tidal currents distribute and mix the seawater throughout the shallow estuary. The net flow is weak and seaward at all depths. Salinity is given in parts per thousand (‰).

(c) The partially-mixed estuary. Seawater enters below the mixed water that is flowing seaward at the surface. Seaward surface net flow is larger than river flow alone. Salinity is given in parts per thousand (‰).

(d) The fjord-type estuary. River water flows seaward over the surface of the deeper seawater and gains salt slowly. The deeper layers may become stagnant due to the slow inflow rate of salt water. Salinity is given in parts per thousand (‰).

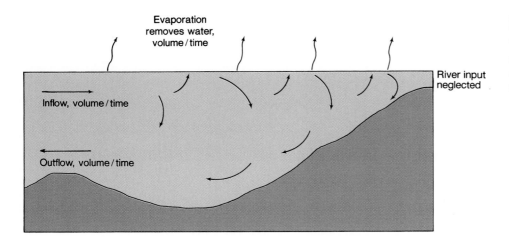

Figure 9.18

The evaporative sea. Evaporation at the surface removes water, and river flow is negligible. Seawater flows inward at the surface, and the seaward flow is at depth.

ter entering the bay may not mix completely with the bay water on the rising tide and may return to the sea on the next ebb tide without displacing bay water.

Recycling may also occur in partially-mixed estuaries with two-layer flow. If tidal current turbulence is strong at the estuary entrance, some of the seaward moving surface layer may be mixed down into the deeper inflowing seawater. This action increases the flushing time by decreasing the net rate of seaward-moving water, and flushing time is determined by the portion of the seaward flow that does not recycle. This recycling of surface water also has the advantage of aerating the deeper parts of the estuary by mixing oxygen-rich surface water downward.

> Define flushing time.
>
> Explain the role of recycling in estuary circulation.

Environmental Issues   9.11

Historically, people have depended on access to salt water for trade and transport and access to fresh water for their living requirements; this has resulted in the establishment of major population centers along the shores of estuaries and coastal seas. In the United States today, over one-half of the population lives within 80 km (50 mi) of the coasts (including the Great Lakes). This population, with its necessary industries, energy-generating facilities, and waste-treatment plants has created a tremendous burden on coastal zones.

In the United States, the 1960s and 1970s produced the environmental laws and agencies that presently monitor and control many practices in these oceanographically complex, biologically sensitive, and economically important areas. In the 1980s a policy based on environmental principles coincided with a series of environmental disasters, and the legacy of the 1990s is now being written. There appears to be a greater concern for acting with care and judgment in coastal areas, but as the economic burden of protecting the environment increases, these rules may be relaxed or changed. There is also a legacy from past practices, for in many cases the industries responsible for environmental problems no longer exist. In these cases the often difficult and always expensive cleanup becomes a burden to the taxpayers rather than to the perpetrators.

Beaches

People like beaches; their desire for property along this sensitive fringe of land is great. People change beaches, and one way in which they do so is by damming rivers to control floods and generate power. The sand that once moved down these rivers to supply coastal beaches is now deposited in the lakes behind the dams. An important source of material to the continued balance of the beaches has been removed. Although the sediment supply to the beaches is reduced, the longshore current continues to carry away suspended beach material, and the beaches are eroded.

To protect harbors and coastal areas from the force of waves, people build breakwaters and jetties. Breakwaters are usually built parallel to the shore, while jetties extend seaward to protect or partially enclose a water area. The protected areas behind jetties and breakwaters are quiet; materials carried by the longshore current settle here, for they are no longer held in suspension by wave action. The result is less material available for a beach further down the coast. Along some coasts, people trying to maintain their beaches place groins of rock or timber perpendicular to the beach, figure 9.19. These groins trap sand moving in the longshore transport, and the result is a buildup of sand on the up-current side of the groins and loss of sand on the down-current side.

Harbors are often constructed without taking into consideration all the factors governing longshore transport. They fill with the sand that falls out in their quiet waters and require constant dredging to keep them open. California's Santa Barbara harbor, shown in figure 9.20 is one example. In this case the jetty extending seaward on the harbor's north side acts as a groin trapping sediment. When this area behind the groin is filled, the sediment sweeps southward along the jetty and settles out in the quiet water behind the jetty's south end until it is dredged and pumped back into the longshore current further down the coast.

In their efforts to maintain their homes and beaches, many people have constructed bulkheads or seawalls to prevent the loss of their property. These expensive structures are made of timbers, poured concrete, or large boulders. They armor the land in hopes of preventing erosion by storm waves, high tides and boat wakes. Many times these structures fail, and often they make the erosion problem worse. If only a portion of the coastline is protected, wave energy may be concen-

Figure 9.19

These groins have been placed along the beach at Hastings in southern England to trap sand moving along the coast. Beach height difference at this groin is 4 m (12 ft).

Figure 9.20

Santa Barbara harbor as seen from the south. In the foreground, a dredge removes the sediment that has built up in the protected, quiet water behind the jetty. At the upper left, sand accumulates against the jetty and widens the beach.

trated at the ends of the seawall and the breaking waves move behind the wall and erode the shore. Because the seawall causes the waves to expend their energy over a very narrow portion of the beach, severe erosion of the beach materials at the foot of the seawall may occur, and the seawall may collapse. Seawalls may reflect waves that combine with incoming waves to cause higher more damaging waves at the seawall or at some other place along the shore.

Eroding beaches are often artificially maintained by supplying the beach with sand and gravel mined from upland areas or dredged from offshore bars. Such programs are expensive but may be necessary to compensate for the erosion of beach sediment. If these programs are to be successful, the particle sizes of the material must be carefully chosen to maintain the beach under its own particular wave-energy conditions. If too fine a material is used, it is eroded swiftly and transported to some other area where it piles up, producing unwanted shoaling or even smothering bottom-dwelling organisms.

The natural shore allows waves to expend their energy over a wide area, eroding the land but maintaining the beaches, all a part of nature's processes of give and take. Time and time again, coastal facilities have been built in response to the demands of people, only to trigger a chain of events that results in newly created problems of either erosion or deposition. Forty-three percent of the shoreline of our national and

territorial areas, excluding Alaska, is now losing more sediment than it receives. Our beaches are eroding and disappearing into the sea.

Wetlands

The value of shore and estuarine areas as centers of biological productivity and nursery areas for the coastal marine environment is well known to biologists and oceanographers. Freshwater and saltwater marshes and swamps, known as **wetlands,** border coasts and estuaries providing nutrients, food, shelter, and spawning areas for marine species including such commercially important organisms as crabs, shrimp, oysters, clams, and many kinds of fish.

Along the muddy shores of tropical and subtropical lagoons and estuaries, salt-tolerant mangrove trees (fig. 9.2f) protect the shore from wave erosion and storm damage. On and among their tangled roots the mangroves provide specialized habitats for fish, crustaceans, and shellfish. In recent years mangroves have been logged for timber, wood chips, and fuel; mangrove swamps have been cleared and filled to provide land for crops, shrimp ponds, and resorts. The result is that over 50% of the world's mangroves are gone.

In the United States it has been routinely customary to fill coastal wetlands to produce usable ground for industry and

Figure 9.21

The industrialized estuary and wetlands of the Duwamish River at Seattle, Washington.

port facilities, and channels were dredged to permit the entry of deep-draft vessels into ports and harbors (fig. 9.21). The San Francisco Bay estuary is considered the most modified major estuary in the United States. It has a surface area of 1240 square kilometers (480 mi^2), and its river systems drain 40% of the surface area of California. The marshlands of the bay and its river deltas were diked, first to increase agricultural land and then for homes and industries. The conversion of wet land to dry land has reduced the tidal marshes from an area of about 2200 square kilometers (860 mi^2) to 125 square kilometers (49 mi^2). The increased cropland required increased irrigation, and the state built a series of dams, reservoirs, and canals. Today, 40% of the Sacramento and San Joaquin Rivers' flow is removed for irrigation, and another 24% is sent by aqueduct to central and southern California. By the early 2000s, it is expected that San Francisco Bay will receive only 30% of its 1850 freshwater input. Proposals to prevent further degradation of the bay include restrictions on freshwater diversion to agriculture and limits on urban discharge and development.

Each year 100 to 150 square kilometers (40 to 60 mi^2) of Mississippi River wetland marsh disappear under the Gulf of Mexico. The wetlands of the Louisiana coast are a great natural resource, 4 million acres, providing habitat for fish, shellfish, birds, and other animals that are important to the economy, biology, and history of the region. Over time, the delta of a river, such as the Mississippi, loses a portion of its area to erosion and sinking of the land due to compaction. However, if the supply of sediments exceeds the rate of land loss the delta grows and wetlands increase. If the delta does not receive its normal sup-

ply of sediment because of levees, dams, and other flood control efforts, the delta land is lost rapidly.

Industrial growth and diversification as well as increasing populations in coastal areas create enormous pressure for coastal wetlands. Between the mid-1950s and the mid-1980s approximately 20,000 acres of coastal wetlands were lost per year in the contiguous United States. Although U.S. federal and state environmental legislation regulates the development and modification of many of these areas, over half of the coastal wetlands in the conterminous states have been destroyed and the remainder continues under pressure for residential, resort, and marina use (fig. 9.22).

Water Quality

Dumping solid waste and pouring liquid pollutants into estuaries and nearshore waters are now recognized as the wrong solutions to industrial and domestic waste problems. The costs of cleaning up and of building plants to process these wastes are high, requiring increased taxes and increased product prices. While indiscriminate industrial dumping of materials in the estuaries and marine waters surrounding the United States has substantially halted in response to federal laws, using the sea as a dump for trash and garbage continues as a common practice around the world. Probably more than 25% of the mass of all material dumped at sea is dredged material from ports and waterways, and one of the major methods of industrial waste disposal for many nations is still dumping at sea. Pesticides such as DDT (dichloro-diphenyl-trichloro-ethane), long-lived toxic

Figure 9.22

The wetlands of Barnegat Bay, New Jersey, were replaced by a housing and recreational complex.

More Information: www.nwrc.gov/fringe/barriers.html

organic compounds such as polychlorinated biphenyls (PCBs), polycyclic aromatic hydrocarbons (PAHs), and heavy-metal ions such as lead, mercury, zinc, and chromium are still making their way through the environment, even though many of these materials are closely controlled or no longer manufactured. For example, U.S. manufacture, sale, and use of PCBs stopped in 1978 and 1979, but electrical transformers filled with PCBs are still being found, often leaking into the marine environment.

Surface runoff from agriculture carries pesticides and nutrients, which can poison or overfertilize the waters. By producing an excess of plant material, nutrients can be as destructive as pesticides, because the plants eventually die and decay, removing large quantities of oxygen. Lack of oxygen then kills other organisms, which in turn decay and continue to remove oxygen from the water.

Urban surface runoff via storm sewers adds a wide mix of materials, including silt, hydrocarbons from oil, lead from gasoline, residues from industry, pesticides and fertilizers from residential areas, and coliform bacteria from animal wastes. Even the chlorine added to drinking water and used to treat sewage effluent as a bactericide may form a complex with organic compounds in the water to produce chlorinated hydrocarbons that can be toxic to marine organisms.

Considering all the pathways, sources, and types of materials that can be classed as pollutants, their management and eventual exclusion from the coastal zone is an extremely difficult and complex problem. However, efforts to reduce the discharge of toxic materials into the marine environment are showing some signs of success. Improved sewage treatment de-

creases the discharge of nutrients and toxicants, and the decrease in use of leaded gasoline appears to be reducing the amount of lead being deposited in the North American marine environment. Lead carried by the Mississippi River to the Gulf of Mexico has been reduced about 40% in the last two decades.

Other countries may not be able to match the U.S. efforts to reduce the discharge of toxic materials because of economic and political difficulties. Eighty-five percent of the pollutants flowing into the Mediterranean come from the land, and 80% of the sewage flowing into the Mediterranean is untreated. A plan to improve this situation was adopted in 1976, and by 1992 the number of contaminated swimming beaches dropped by 30%, but few nation participants have established adequate guidelines to meet the plan's requirements.

Dead Zones

Each year a large low-oxygen area is found in the Gulf of Mexico adjacent to the Louisiana coast. It appears in February, peaks in summer, and dissipates in the fall. This area has been named a **dead zone** because its oxygen level at depth is too low to support most marine life. Each year nitrogen based fertilizers wash off America's crop lands, and the Mississippi River carries the fertilizers to the Gulf. The amount of nitrogen as nitrate that flowed into the gulf tripled between 1960 and the late 1990s. This overabundance of nitrate stimulates small plant plankton growth; when the plankton die their remains drop to the bottom where bacteria consume oxygen as they break down the organic material. This process lowers the dissolved oxygen in the water and creates the dead zone. Similar problems have been reported in Chesapeake Bay, northern Europe (in the Baltic and Black Seas), Japan, Australia, and New Zealand.

Toxic Sediments

Many toxicants reaching the estuaries do not remain in solution in the water but become adsorbed onto the small particles of clay and organic matter suspended in the water. These particles clump together, then settle out due to their increased size; this increases the concentration of toxins in the sediments. Once the sediments have been contaminated they become a source of continual leaching of toxicants to the water and to organisms living in contact with the bottom. Correcting this kind of contamination requires either the removal of the contaminated sediments, which must be stored in specialized landfills or, when water depth is sufficient, the covering of the contaminated sediments with a thick layer of clean material, isolating the contaminated sediments from both water and organisms. Both of these methods are very expensive and should not be considered unless the original source of the contamination has been removed.

Some of the particulate matter that adsorbs toxicants has a high organic content and forms a food source for marine organisms. In this way, heavy metals and organic toxicants find their way into the body tissues of organisms, where they may accumulate and be passed on to predators and human harvesters. Polycyclic aromatic hydrocarbons (PAHs) from combustion of fossil fuels and other organic materials are known to cause lesions in bottom fish living in contact with contaminated sediments. Shellfish concentrate heavy metals at levels many thousands of times higher than the levels found in the surrounding waters. Between 1953 and 1960 in Minamata, Japan, mercury from a local industrial source was released into a bay from which shellfish were harvested by a local village. The contaminated shellfish caused severe mercury poisoning and death among the villagers who ate the shellfish. The physical and mental degenerative effects of the poisoning were especially severe on children whose mothers had eaten the shellfish during pregnancy. The condition produced by this tragic episode has been named Minamata disease.

Since 1986 the Mussel-Watch program has used tissue analysis from mussel and oyster populations to monitor certain chemicals as indicators of human activity and environmental quality. More than two hundred U.S. coast and estuary sites are used in this ongoing effort. The study has shown more decreases than increases in chemical concentrations between 1986 and 1990 and appears to support the limits being set on discharges of chemical contaminants and the laws supporting these limits.

Plastics

It is estimated that every year more than 77 tons of plastic trash are routinely dumped or lost at sea by naval and merchant ships (fig. 9.23a). The National Academy of Sciences estimates that the commercial fishing industry yearly loses or discards about 135 million kg (298 million lbs) of fishing gear (nets,

ropes, traps, and buoys) made mainly of plastic and dumps another 24 million kg (52 million lbs) of plastic packaging materials. Recreational vessels, passenger vessels, and oil and gas drilling platforms all add their share.

Nobody knows how long plastic remains in the marine environment, but an ordinary plastic six-pack ring could last 450 years. So-called biodegradable plastic available in the United States shows no evidence of solving plastic problems at sea. The light-absorbing molecules that break down the plastic after a few months' exposure to sunlight are unlikely to be very helpful in an environment that keeps the plastic cool and coated with a thin film of organisms that shade it from the light.

Thousands of marine animals are crippled and killed each year by plastics (fig. 9.23b,c,d). As many as 30,000 fur seals a year are estimated to become entangled in lost or discarded plastic fishing nets and choked to death in plastic cargo straps. Seabirds and marine mammals swallow plastics. Porpoises and whales have been suffocated by plastic bags and sheeting. Fish are trapped in discarded netting. Sea turtles eat plastic bags and die. Discarded plastics have become as great a source of mortality to marine organisms worldwide as oil spills, toxic wastes, and heavy metals.

The Marine Plastic Pollution Research and Control Act of 1987 is a U.S. law that prohibits dumping plastic debris everywhere in the oceans; other types of trash may be dumped at specific distances from shore. Ports and terminals are required to provide waste facilities for debris, and enforcement is delegated to the U.S. Coast Guard. There is no international enforcement. It may be that only people educated to act in a responsible manner will reduce the tide of plastics rising around the world.

Oil Spills

The transport of oil and oil products on the high seas and the production of oil from offshore wells create the potential for accidents that release large volumes of oil. The average discharge of oil into the world's oceans and navigable waters during each of the last ten years has been about 47×10^3 metric tons (34×10^6 gal), not including the occasional megaspill of 3.8×10^4 metric tons (111×10^6 gal) or more. See figure 9.24. Spills far at sea are difficult to assess, because direct visual and economic impacts on coastal areas do not occur, and damage to marine life cannot be accurately evaluated. Spills due to the grounding of vessels or due to accidents during transfer or storage occur in coastal areas, where the environmental degradation and loss of marine life can be readily observed. Three oil spills have occurred in the last 25 years that have become ecological landmarks; these are the sinking of the *Amoco Cadiz*, the grounding of the *Exxon Valdez*, and the wartime discharge of crude oil into the Persian Gulf.

In March 1978 the supertanker *Amoco Cadiz* lost its steering in the English Channel and broke up on the rocks of the Brittany coast of France (fig. 9.25a and b). Gale winds and high tides spread the oil over more than 300 km (180 mi) of the French coast. Of the approximately 210,000 metric tons of oil spilled, it is considered that 30% evaporated and was carried in the air over the French countryside, 20% was cleaned from the beaches

Figure 9.23

The impact of plastics on the marine environment: (a) Persistent litter includes plastic nets, floats, and containers. (b) A common murre entangled in a six-pack yoke. (c) A young gray seal caught in a trawl net. (d) A sea turtle with a partially ingested plastic bag. Photos (a), (b), and (c) are from Sable Island in the North Atlantic Ocean, approximately 240 km (150 mi) east of Nova Scotia, Canada.

More Information: www.law.cornell.edu/uscode/33/

by the army and volunteers, another 20% soaked into the sand of the beaches to stay until winter storms washed it away, 10% (the lighter, more toxic fraction) dissolved in the seawater, and still another 20% sank to the sea floor in deeper water, where it will continue to contaminate the area for years to come.

Eleven years later and 40 km (25 mi) out of Valdez, Alaska, the *Exxon Valdez*, loaded with 170,000 metric tons of crude oil, spilled 35,000 tons, or 10.8 million gallons, of oil into Prince William Sound (fig. 9.25c and d). Local contingency plans for cleanup were not prepared to handle a spill of this size; delays as well as lack of equipment and personnel, the rugged coastline, the weather, and the tidal currents of the enclosed area combined to intensify the problem. The oil spread quickly, distributing itself unevenly over more than 2300 square kilometers (900 mi²) of water, and took a large toll of seabirds, marine mammals, fish, and other marine organisms. The oil moving out of Prince William Sound did not wash out to the open sea but was captured by the nearshore currents and repeatedly oiled the rocky wilderness beaches in the weeks that followed. In the cold subarctic conditions of Prince William Sound, the degradation of the oil by sunlight and microbes proceeds more slowly than in warmer regions. Plant and animal populations also recover more slowly in the cold water, where organisms tend to live longer and reproduce more slowly. Researchers inspecting the area three years after the disaster, in 1992, reported that the area was recovering and repopulating with organisms as the oil aged and degraded. They also noted that the areas most intensively cleaned to remove oil are recovering more slowly and with less-balanced populations than the areas in which the oil has been allowed to degrade naturally.

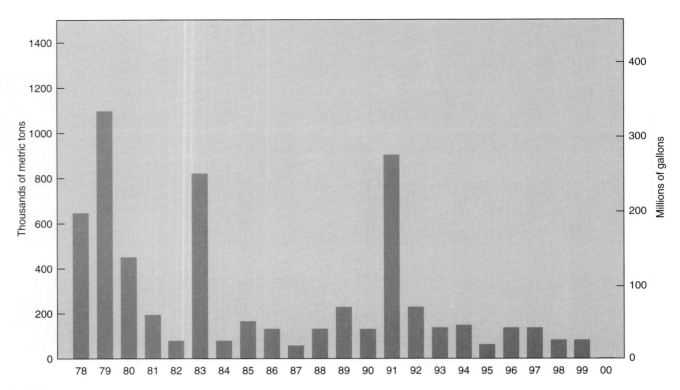

Figure 9.24

World annual oil spillage.

Source: Data from Oil Spill Intelligence Report, *Arlington, MA.*

More Information: www.cutter.com/osir/

The average depth of the Persian Gulf is 40 m (130 ft) and the circulation of its high-salinity water is sluggish. During the eight-year Iran-Iraq war the bombing of Persian Gulf oil facilities produced major spills, including one from an oil rig that poured out 172 metric tons (60×10^3 gal) per day for nearly three months. The Gulf endured an even greater catastrophe during the 1991 Gulf War as an estimated 800,000 metric tons (240×10^6 gal) of crude oil gushed into Gulf waters, the world's greatest oil spill to date. Some of this oil was deliberately released, some came from a refinery at a Gulf battle site, and bombing contributed additional quantities. Changes in the wind in the weeks after this spill stopped the oil from drifting the entire length of the Gulf, but some 570 km (350 mi) of Saudi Arabia's shoreline was oiled (fig. 9.25e). International oil-spill experts rushed to help and were able to protect the water-intake pipes of desalination plants and refineries. About half of the oil evaporated and about 3×10^5 metric tons have been recovered. Some shore areas were bulldozed clean; others became test plots to assess natural recovery. Much has sunk to the bottom of the Gulf, and it is expected to take decades for the damaged areas to recover.

The tragedy of these spills continues to demonstrate that there is no adequate technology to cope with large oil spills, particularly under difficult weather and sea conditions along rugged, remote coastlines. On the average, no more than 15% of the oil spilled under these conditions has been recovered

from any oil spill, and this figure may be an inflated number, because the recovered oil mixture has a high water content.

The technology for oil cleanup at sea includes oil booms and oil skimmers. These devices are useful in confining and recovering small spills in quiet protected waters, but are not very effective in open and exposed areas. The onshore cleanup is often as destructive as the spill, especially in wilderness situations where the numbers of people, their equipment, and their wastes further burden the environment. The toxicity of weathered crude oil is low, and many oil-spill experts believe that cleanup efforts should be concerned not with removing oil from the beaches but with moving oil seaward to prevent it from reaching the beaches. They believe that many of the cleanup methods used along shorelines cause more immediate and long-term ecological damage than leaving the oil to degrade naturally. Habitats treated with hot water take longer to recover than those left untreated; backhoeing and high-pressure washing (100 lbs/in^2) destabilize gravel and sand beaches, killing live animals and working the oil into the sediments; see figure 9.25c. To prevent oil from reaching the beaches, some cleanup experts support newly developed, low-toxicity dispersants; the dispersed oil moves into deeper water where it is diluted and its toxicity lessened.

More data are needed to assess the effectiveness of attempts to speed up natural degradation by adding nutrients to the beaches in order to increase populations of biodegrading

(a)

(b)

(c)

(d)

(e)

(f)

Figure 9.25

(a) The tanker *Amoco Cadiz* aground and broken in two off the coast of France, March 1978. (b) A bay along the French coast was fouled with oil. (c) Hot water under high pressure is used to clean Prince William Sound beaches after the 1989 *Exxon Valdez* oil spill. (d) Thick oil collects among the rocks on a beach in Prince William Sound. (e) Oil along the Persian Gulf coast after the 1991 Gulf War. (f) Cleaning up a 1993 oil spill along the beaches of Tampa Bay, one more time in 1994.

More Information: response.restoration.noaa.gov/seaserver.nos.noaa.gov/projects/oilspills.html

microorganisms. Releasing other strains of bacteria has been tested with limited success; the new organisms may not compete effectively with natural populations.

Once oil is removed from a beach, the first cleaning operation may not be the last. On the east coast of Florida the beaches of Tampa Bay are lined with resorts and condominiums, and the economy depends heavily on these beaches for tourism. A freighter and two oil barges collided here in August of 1993. One barge carried diesel fuel and gasoline that caught fire and burned. The second barge carried heavy fuel oil; this barge sank and released the fuel oil to mix with the sediments on the bottom of the bay. Each time strong winds stir the bay, contaminated sediments are moved onshore to create a new oil contamination problem. Figure 9.25f shows the end of a costly, January 1994, cleanup episode along St. Petersburg beach.

The immediate damage from a large spill is obvious and dramatic; by contrast, the effects of the small but continuous additions of oil that occur in every port and harbor are much more difficult to assess, because they produce a chronic condition from which the environment has no chance to recover. Refined products such as gasoline and diesel fuel are more toxic to marine life than crude oil, but they evaporate rapidly and disperse quickly. Crude oil is slowly broken down by the action of water, sunlight, and bacteria, but the portion that settles on the sea floor moves down into the sediments, where it continues to contaminate for years.

What activities contribute to the degradation of the marine coastal environment?

What is the effect of each structure when built along a beach: (a) breakwater, (b) jetty, (c) seawall?

Why are wetlands important?

What has happened to the world's mangroves?

Why have the majority of U.S. wetlands been altered and eliminated?

What are the sources and pathways for toxicants entering the coastal zone?

Why do coastal sediments have higher toxicant concentrations than the overlying water?

Why are plastics so harmful to marine animals?

Discuss the international effort to reduce plastics in the marine environment. How can it help the situation? How may it not help it?

What happens to spilled oil in the marine environment?

Discuss the technology used for oil-spill cleanup. Why is it not fully effective?

How does constant addition of small amounts of oil compare to the effects of a single large oil spill?

Summary

The coast is the land area that is affected by the ocean; the coastal zone includes coastal bays, estuaries, and inlets. Primary coasts are formed by nonmarine processes and include drowned river valleys, fjords, deltaic coasts, glacial moraines, dune coasts, lava and cratered coasts, and fault coasts. Secondary coasts are modified by ocean processes and include regular and irregular cliffed coasts, barrier coasts, beach plains, mudflats, coral reefs, mangrove, and marsh grass coasts.

Typical beach features include an offshore trough and bar and a beach area with a low tide terrace, beach face, beach scarp, and berms. Winter and summer berms are produced by seasonal changes in wave action. Beaches are described by their shape and the size, color, and composition of beach material.

Beaches exist in a dynamic equilibrium, where supply balances removal of beach material. Gentle summer waves move the sand toward the shore; high-energy winter waves scoop the sand off the beach and deposit it in an offshore sandbar. The breaking waves of the surf zone produce a longshore current that moves the sediment down the shore in longshore transport. Coastal circulation zones are defined by the path followed by beach sediments from source to region of deposit. Rip currents are narrow, fast seaward movements of water and sediment through the surf zone. Rising sea level along low-lying coasts may move the coast a significant distance inland in the next fifty years. Barrier islands are particularly susceptible to coastal flooding and storm damage.

Estuaries are semi-isolated portions of the ocean that are diluted by fresh water. There are four basic estuary types based on circulation patterns and salt distribution. In bays and seas with high evaporation rates, there is continuous overturn, and water at depth flows seaward while ocean water enters at the surface.

Estuaries that flush rapidly have a greater capacity for dissipating wastes than those that flush slowly. In bays and harbors that are flushed only by tidal action, recycling of water and wastes can occur. Partial recycling is also possible in estuaries with two-layered flow.

Artificial structures change beaches; they are often responsible for unwanted beach erosion. Human interference with the natural processes of the coastal zone also results in harbors filled with sand, like that of Santa Barbara, California. Wetlands border coasts and estuaries; they are important areas that often have been destroyed by filling, dredging, and development.

Coastal and estuary water quality is affected by dumping solid waste and liquid pollutants. Low oxygen areas or "dead zones" appear annually in various areas of the world as a result of over-fertilization. Pesticides and long-lived toxicants move through the environment; runoff from agricultural and urban areas does much damage. Toxicants become concentrated in the coastal sediment, and organisms concentrate toxicants and pass them along the marine food chains. Plastic trash is an increasing problem to marine animal life. Oil spills are a special problem in inshore waters, for which there is no adequate cleanup technology.

Key Terms

Critical Thinking

1. Why are multiple berms more likely to be seen on a North American beach between March and August than between September and February?

2. Fjord coasts and drowned river valleys are primary coasts. Explain why their appearances are distinctly different?

3. Estuaries are classified by their net circulation and salt distribution. What other features could be used to describe and classify them?

4. In order to prevent the erosion of a beach, your neighbor wants to build a groin. What effect might the groin have on your neighbor's beach? On your beach if you live upstream of your neighbor? On your beach if you live downstream of your neighbor?

5. If contaminated sediments are dredged from the floor of a harbor to use in a landfill, what hazards to the environment should be considered during the dredging, the transport, and the placement of the sediments?

Suggested Readings

Coasts

Ackerman, J. 1997. Islands at the Edge. *National Geographic* 192(2):2–31. Discusses barrier islands.

Davis, R. A., Jr. 1996. *Coasts, Prentice Hall Earth Science Series.* Prentice Hall, Upper Saddle River, N.J.

Dean, C. 1999. *Against the Tide, The Battle for America's Beaches.* Columbia University Press, N.Y.

Kusler, J. A., W. J. Mitsch, and J. S. Larson. 1994. Wetlands. *Scientific American* 270(1):64–70.

Pilkey, O. H. 1990. Barrier Islands. *Sea Frontiers* 36(6):30–36.

Reid, P., and M. Trexler. 1991. *Drowning the National Heritage.* World Resources Institute, Washington, D.C. 49 pp.

Rützler, K., and I. Feller. 1996. Caribbean Mangrove Swamps. *Scientific American* 274(3):94–99.

Stuller, J. 1994. On the Beach. *Sea Frontiers* 40(6):28–34.

Weber, P. 1984. It Comes Down to the Coasts. *World Watch* 7(2):20–29.

Zedler, J. B., and A. N. Powell. 1993. Managing Coastal Wetlands. *Oceanus* 36(2):19–28.

Estuaries

Boicourt, W. 1993. Estuaries, Where the River Meets the Sea. *Oceanus* 36(2):29–37.

Horton, T. 1993. Chesapeake Bay—Hanging in the Balance. *National Geographic* 183(6):2–35.

Lacombe, H. 1990. Water, Salt, Heat and Wind in the Med. *Oceanus* 33(1):26–36.

Malakoff, D. 1998. Restored Wetlands Flunk Real-World Tests. *Science* 280(5362):371–72.

Ross, J. F. 1996. Seeing the Chesapeake as a Whole. *Smithsonian* 27(4):100–109.

Environment

Atlas, R. 1993. Bacteria and Bioremediation of Marine Oil Spills. *Oceanus* 36(2):71.

Aubrey, D., and M. Connor. 1993. Boston Harbor. *Oceanus* 36(1):61–70.

Beardsley, T. 1997. Death in the Deep. *Scientific American* 277(5):17.

Canby, T. V. 1991. After the Storm. *National Geographic* 180(2):2–32. (Oil spill cleanup in the Persian Gulf.)

Capuzzo, J. 1990. Effects of Wastes on the Ocean: The Coastal Example. *Oceanus* 33(2):39–44.

Hass, P. M., and J. Zuckmann. 1990. The Med is Cleaner. *Oceanus* 33(1):38–42.

Hawley, T. M. 1990. Herculean Labors to Clean Wastewater. *Oceanus* 33(2):72–75. (Technology of sewage treatment.)

Hodgson, B. 1990. Alaska's Big Spill. *National Geographic* 177(1):5–43.

Holloway, M. 1996. Sounding Out Science. *Scientific American* 275(4):82–88. (Prince William Sound seven years after the oil spill.)

Malakoff, D. 1998. Death by Suffocation in the Gulf of Mexico. *Science* 281(5374):190–92.

Nixon, S. 1998. Enriching the Sea to Death. *The Oceans—Scientific American Quarterly* 9(3):48–53.

Parker P. A. 1990. Clearing the Oceans of Plastics. *Sea Frontiers* 36(2):18–27.

Wolfe, D. A., et al. 1994. The Fate of the Oil Spilled from the Exxon Valdez. *Environment, Science and Technology* 28(13):561–67.

internet references

worldwide websites

Coasts

Coasts in Crisis, Table of Contents
pubs.usgs.gov/circular/c1075/contents.html

USGS Western Region Coastal and Marine Geology
walrus.wr.usgs.gov/

Geological Survey Coastal and Marine Geology
Program
marine.usgs.gov/

SandyDuck
www.frf.usace.army.mil/SandyDuck/Gallery/
SandyDuck-CH/slide1.html

SandyDuck
bigfoot.wes.army.mil/c316.html

Fragile Fringe Barrier Islands Page
131.113.58.10:10081/=@=:www.nwrc.gov/
fringe/barriers.html

Louisiana Barrier Islands
marine.usgs.gov/fact-sheets/Barrier/barrier.html

Ocean and Coastal Resources Management Plan
wave.nos.noaa.gov/ocrm/

Office of Ocean Resource and Assessment—ORCA
www-orca.nos.noaa.gov/

Coastal Services Center—NOAA WWW Home Book
www.csc.noaa.gov/

National Geographic Data Center Support To Coastal
Programs
www.ngdc.noaa.gov/coast.html

Rising Sea Level

Sea Level Rise
www.nap.usace.army.mil/cenap-en/slr_links.htm

CTB:Rising Seas
www.greenpeace.org/~climate/ctb/risingseas.html

Global Sea Level Rise
ibis.grdl.noaa.gov/SAT/hist/gsl_rise/gsl.html

Sea Level Data
math.cd-rom-directory.com/cdrom-2.cdprod1/
010/211.Sea.Level.Data.shtml

University of Hawaii Sea Level Center
www.soeast.hawaii.edu/kilonsky/uhslc.html

Sea Level, Pacific Ocean
uhslc.soeast.hawaii.edu/uhslc/volume.html

Sea Level Factors and the Estuary
bellnet.tamu.edu/sealevel.htm

Estuaries

San Francisco Bay and Delta, Water Quality
sfbay.wr.usgs.gov/access/quality.html

The Chesapeake Bay, Maryland Sea Grant
www.mdsg.umd.edu/CB/index.html

Coastal Pollution

Environment Quality and Preservation—USGS
WRCMG
walrus.wr.usgs.gov/environment/

NOAA NOS Fact Sheets—Index
www.nos.noaa.gov/facts/facts_index.html

Natural Hazards and Public Safety
walrus.wr.usgs.gov/hazards/

Navigation and Navigable Waters U.S. Codes
www.law.cornell.edu/uscode/33/

IMO's Home Page
www.imo.org/imo/welcome.htm

National Status and Trends Home Page
seaserver.nos.noaa.gov/projects/nsandt/nsandt.
html

Massachusetts Water Resources Authority
www.mwra.com/

Boston Globe Online/Reprint
www.boston.com/globe/search/stories/reprints/
mwra032000.htm

Modernizing the Sewer System
www.mwra.com/sewer/html/sewditp.htm

Oil Spills

Links2Go
www.links2go.com/more/www.alaska.net/~ospic

NOAA Office of Response and Restoration—Home Page
response.restoration.noaa.gov/

Oil Spill Intelligence Report
www.cutter.com/osir/

Spill Basics: A Primer for Students
www.cutter.com/osir/osirbasc.htm

Oil Spill Emergency Response and Planning
seaserver.nos.noaa.gov/projects/oilspills/oilspill.
html

Marine Sanctuaries

NOAA—National Marine Sanctuaries
whale.wheelock.edu/whalenet-stuff/NOAA_
Sanctuaries.html

National Marine Sanctuaries Headquarters
search.nos.noaa.gov:81/compass

Monterey Bay Aquarium
www.mbayaq.org/

One hundred years after the establishment of Yellowstone as our first national park, the U.S. Congress passed the Marine Protection, Research and Sanctuaries Act of 1972. Today there are twelve marine sanctuaries: five off the U.S. west coast, four off the U.S. east coast including the Florida Keys, and one each in the Gulf of Mexico, Hawaii, and American Samoa (see fig. 1). Together these twelve sites

National Marine Sanctuaries

cover 46,000 km² (18,000 mi²), less than one-half of one percent of the ocean area within U.S. national boundaries. Two more sites are proposed: one in Puget Sound, Washington and one in Lake Huron and still others are under consideration.

Monterey Bay NMS is the largest marine sanctuary stretching along the Pacific coast for 580 km (350 miles) and reaching seaward as much as 90 km (53 miles). This area supports a great diversity of marine life in various habitats including an ocean canyon more than

3,300 m (2 mi) deep. This sanctuary is closely associated with the research and education programs of the Monterey Bay Aquarium and Monterey Bay Research Institute (MBARI). To the south, the Channel Islands NMS is home to a myriad of sea birds, more than twenty kinds of sharks, thousands of sea lions, as well as northern fur seals, elephant seals, and the rare Guadalupe fur seal, which was hunted

nearly to extinction in the 19th century. The edges of the islands are ringed with forests of giant kelp, seaweeds that provide food and shelter for a rich diversity of marine species.

North of Monterey Bay the Gulf of the Farallones NMS lies 50 km (30 mi) west of the Golden Gate. These islands provide a dwelling place for twenty-six species of marine mammals and the largest concentrations of breeding seabirds in the continental U.S. Still farther north the beaches of the Olympic Coast NMS remain almost the

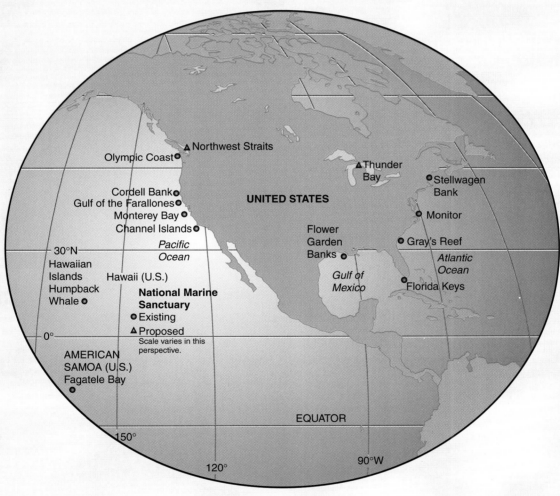

Figure 1

U.S. National marine sanctuaries.

Figure 2

Breaching humpback whale in Hawaiian Islands.

same as they have ever been, an undeveloped wilderness rich in marine life along rocky shores and tide pools.

Hawaiian Islands Humpback Whale NMS is the breeding, calving, and nursing home for much of the North Pacific humpback whale population (fig. 2). The very rare Hawaiian Monk seal is also found here. On the Atlantic coast of the U.S. mainland Stellwagen Bank NMS, a part of Massachusetts Bay east of Boston and north of Cape Cod, lies over an area of gravel and coarse sand left by receding glaciers. This is a very productive area and a feeding ground for humpback, fin, and northern right whales.

Sixty kilometers (100 mi) off the Texas-Louisiana coast in the Gulf of Mexico a pair of salt domes support the northernmost coral reefs in North America. This is Flower Garden NMS, home to manta rays, hammerhead sharks and loggerhead turtles. Off the tip of Florida a 370 km (220 mi) arc of reefs, seagrass beds, and mangroves make up the Florida Keys NMS (fig. 3).

These sanctuaries are multiple-use areas managed by the National Oceanic and Atmospheric Administration (NOAA), a branch of the Commerce Department. Commercial fishing, sportfishing, and spearfishing are permitted in the sanctuaries; so are boating, snorkeling, diving, and other kinds of marine recreation. Shipping traffic is not prohibited, but it is restricted to certain lanes and areas. In general drilling, mining, dredging, dumping waste, and removal of artifacts are forbidden.

The annual budget for the entire sanctuary system is less than 12 million dollars. Staffs are small and sanctuaries depend heavily on volunteers. Educating the public has become a top staff priority with some monitoring of water and populations. The sanctuaries are works in progress that require continued support if they are to protect and restore these coastal environments.

Figure 3

A diver watches a school of grunts in a coral garden in the Florida Keys.

Suggested Reading

Chadwick, D. 1998. Blue Refuges. *National Geographic* 193 (3):2–13.

Internet References

Monterey Bay Aquarium
www.mbayaq.org/

National Marine Sanctuaries—NOAA
whale.wheelock.edu/whalenet-stuff/NOAA_Sanctuaries.html

chapter ten

Oceanic Environment and Production

Large kelps reach toward the sea surface and sunlight along the California coast.

Learning Objectives

After reading this chapter, you should be able to

- Name and describe the major environmental zones of the oceans.
- Describe adaptations made by organisms for life in the ocean.
- Relate photosynthesis and respiration to primary production.
- Discuss the factors that control primary production and explain its significance to all forms of marine life.
- Explain how primary production varies over the world's ocean regions.
- Trace the cycling of nitrogen and phosphorus in the marine environment.
- Relate flows of energy and nutrients through the levels of a trophic pyramid to food chains and food webs.
- Discuss the efficiency of harvesting at various trophic levels.
- Understand chemosynthesis and give examples of marine chemosynthetic communities.
- Discuss the conditions and locations where extremophiles are found.

*t*he oceans and the coastal seas provide a complex variety of environments for living organisms. At the surface, conditions range from polar to tropical, over depth from light to total and constant darkness, and the sea floor may be rock, sand, or mud. Organisms of the marine environment have much in common with the organisms of the land, but they also have very different survival problems and have developed unique solutions to cope with them. This chapter explains some of the characteristics that allow organisms to live in the sea.

In all the environments of all the oceans, organisms prey upon each other in the series of prey-predator relationships called food chains. Understanding these chains, from the minute, microscopic organisms with which they start to the more familiar large carnivores such as tuna, salmon, and seals with which they end, requires a knowledge of the ocean's production, or synthesis of organic material by photosynthesis or chemosynthesis. Most global ocean production is based on the microscopic plantlike organisms that inhabit the surface waters of all the world's oceans. To understand the production of the oceans one must understand the factors that cause variations in the abundance of these tiny plantlike cells.

Environmental Zones   10.1

Because the marine environment is so large and complex, it is divided into subunits called zones. The zone classifications used here are based on the system developed by Joel Hedgpeth in 1957; they are shown in figure 10.1. In marine **ecology** the water environment is the **pelagic zone,** and the seafloor environment is the **benthic zone.** The pelagic zone is divided into the coastal or **neritic zone** above the continental shelf and the **oceanic zone** or deep water away from the influence of land. The oceanic zone is then subdivided into the subzones shown in figure 10.1. The surface, where there is enough light intensity for plant growth, is known as the **photic zone,** and it extends through both the neritic and oceanic zones. The photic zone has a depth of 50 to 100 m (150–300 ft). All the other subdivisions of the pelagic zone are without sunlight; they form the **aphotic zone.**

The benthic environment of coastal regions is subdivided on the basis of depth (fig. 10.1). Tidal fluctuations and waves at the shoreline define the **supralittoral** or splash zone (above high water), the **littoral** or **intertidal zone** (between high and low water), and the **sublittoral** or **subtidal zone** (below low water). These three coastal

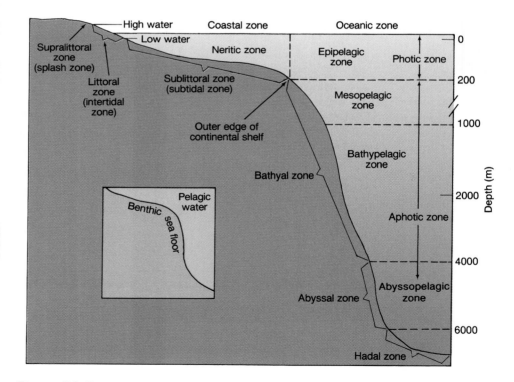

Figure 10.1
Zones of the marine environment.

zones occupy the same area as the benthic photic zone, where there is sufficient light to support plantlike single-celled organisms and benthic plantlike organisms. The oceanic zones are divided by depth and only the epipelagic zone receives sufficient light for photosynthesis. The remaining zones (below 200 m) are cold and dark.

Life in all the zones is influenced by variations in temperature, dissolved gases, substrate, nutrients, and all the factors discussed in previous chapters. The coastal littoral and the pelagic photic zones change greatly with latitude and season, in contrast to regions of greater depth where conditions remain nearly constant. There are great differences between the shallow zones of polar, middle, and tropic latitudes. The photic zone is deeper in the clear waters of the tropics than in the coastal waters of the midlatitudes. Substrate is an important factor in benthic zones; at shallow depths the bottom may be rock, sand, or mud while in the deeper zones there is less variety. The importance of substrate to the distribution of benthic organisms is further discussed in chapter 12.

> Name and describe the major environmental zones of the oceans.
>
> Name and describe the environmental zones of the intertidal region.
>
> Make a list of factors affecting marine environmental zones.

cuttlefish (fig. 10.2b) has a soft, porous, internal shell or cuttlebone. These animals regulate their buoyancy by controlling the relative amounts of gas and liquid within their shells.

Many fish have gas-filled swim bladders that keep them neutrally buoyant. When a fish changes depth, it adjusts the gas pressure in its swim bladder to compensate for the pressure change in the water; this limits its vertical swimming speed. Active, continuously swimming predatory species such as the mackerel, some tuna, and the sharks do not have swim bladders, and bottom fish also lack them.

Floating organisms store their food reserves as oil droplets that decrease their density and retard sinking. Many have developed spines, ruffles, and feathery appendages that increase their surface area and decrease their sinking rate, helping them to float near the sea surface. Large members of the nekton such as whales and seals decrease their density and increase their buoyancy by storing large quantities of blubber, which is mainly low-density fat. Sharks and other varieties of fish store oil in their livers and muscles.

> How are so many marine organisms able to survive without skeletons and rigid structural support?
>
> Give examples of adaptations used by marine organisms to increase their buoyancy.
>
> How does a swim bladder limit the vertical motion of a fish?

Life and the Marine Environment   10.2

Buoyancy and Flotation

Unlike land organisms that require structural strength to support their bodies in air and against gravity, marine organisms are often exceedingly delicate and fragile. The salt water surrounding marine organisms has a density similar to the bodies of many of the organisms, and since the organisms move with the moving water, they do not require structural strength to withstand its flow. The **buoyancy** of objects in salt water is due to the water's density and helps them to stay afloat. It supports the bodies of the bottom-living creatures and lessens the energy expended by the swimmers.

Many organisms have ingenious adaptations to resist sinking. Some jellyfish-type animals secrete gases into a float that enables them to stay near the sea surface. Some plantlike organisms secrete gas bubbles and form gas-filled floats, which help keep their fronds in the sunlit waters while they are anchored to the sea floor. One floating snail produces and stores intestinal gases; another forms a bubble raft to which it clings. The chambered nautilus (fig. 10.2a), a relative of the squid, lives in its last shell chamber and fills the remaining chambers with gas (mainly nitrogen). Another relative of the squid, the

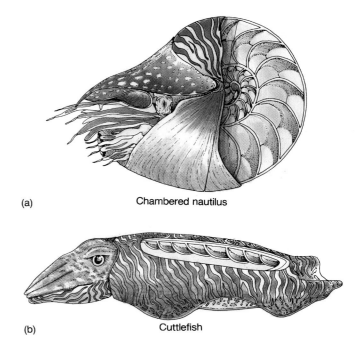

(a)   Chambered nautilus

(b)   Cuttlefish

Figure 10.2

(a) Chambered nautilus. (b) Cuttlefish. The chambered shell that provides buoyancy is shown in each organism.

Fluid Balance

Special problems are posed for living creatures if the salt content of their body fluids differs from the salinity of the water that surrounds them. Water molecules cross the membranes that separate body fluids from seawater, moving along a gradient from a high concentration of water (low salinity) to a low concentration of water (high salinity); this process is known as **osmosis.** Most fish have body fluids less salty than seawater, that is, their body fluids have a water concentration that is greater than seawater. Their tissues tend to lose moisture, and fish must constantly expend energy to prevent dehydration. Fish stay in fluid balance by drinking seawater nearly continually and excreting its salt across their gills. Sharks and rays do not have this problem, because their body fluids have the same approximate salt content as seawater. The body fluids of many bottom-dwelling organisms such as sea cucumbers and anemones are also similar to seawater; there is no concentration gradient across their tissues, and the organisms neither gain nor lose water. The fish and the sea cucumber are compared in figure 10.3.

Some organisms can maintain their salt-fluid balance over only limited salinity ranges; others have remarkable abilities to move between high- and low-salinity water. Salmon spawn in fresh water but move down the rivers as juveniles and live their adult lives in the sea. After several years (the time depends on the species), the salmon return to their home streams. The Atlantic common eel reverses this process by migrating downstream to spawn in the Sargasso Sea. The new generation of eels spends one to three years at sea, then returns to fresh water to live for up to ten years before migrating seaward. Many fish and crustaceans use the low-salinity coastal bays and estuaries as breeding grounds and nursery areas for their young, then as adults they migrate farther offshore into higher-salinity waters.

> Why do marine fish lose moisture from their body tissue?
>
> How does the life cycle of a salmon differ from that of an Atlantic eel?

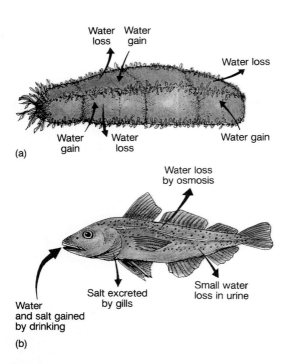

Figure 10.3

(a) The salt concentration of the seawater is the same as the salt concentration of the sea cucumber's body fluids (35‰). The water moving out of the sea cucumber is balanced by the water moving into it. (b) The salt concentration in the tissues of the fish is much lower (18‰) than that of the seawater (35‰). To balance the water lost, the fish drinks salt water, from which the salt is removed and excreted.

Bioluminescence

Sunlight illuminates only the surface waters, but there is another source of light in the oceans. On a dark night, when the wake of a boat is seen as a glowing ribbon, or the water flashes with light as oars dip, or a person's hands glow briefly when a net is hauled in, living organisms are producing the light. This phenomenon is **bioluminescence,** produced by the interaction of the compound lucifern and the enzyme luciferase, a biochemical reaction releasing light with 99% efficiency.

Bioluminescence is triggered by the agitation of the water that disturbs microscopic organisms, causing them to flash and produce glowing wakes and wave crests. Jellyfish also called seajellies glow if they feed on these organisms, and so do one's hands if they come in contact with crushed tissue. Other bioluminescent organisms in the sea include squid, shrimp, and some fish. Many mid-depth and deep-water fish have light-producing organs, some in patterns on their sides, possibly for identification. Others have light-producing organs on their ventral surfaces, making them difficult to see from below, and still others have glowing bulbs dangling below their jaws or attached to flexible dorsal spines acting as lures for their prey. Flashlight fish living in the reefs of the Pacific and Indian Oceans have specialized organs below each eye which are filled with light-emitting bacteria. These fish use the light to see, communicate, lure prey, and confuse predators.

> What is bioluminescence and what kinds of organisms produce it?
>
> How do deep-water fish use bioluminescence?

Color

Some sea animals are transparent and blend with their background, for example jellyfish and most of the small floating animals in the surface layers. In the clear waters of the tropics, where light penetrates to greater depths, bright colors play their greatest role. Some fish conceal themselves with bright color bands and blotches. These colors disrupt the outline of the fish and may draw the predator's attention away from a vital area to a less important region; for example, a black stripe may hide the eye while a false eye spot appears on a tail or fin. Color is also used to send a warning. Organisms that sting, taste foul, bear sharp spines, or have poisonous flesh are often striped and splashed with color, for example sea slugs and some poisonous shellfish. Among fish that swim near the surface in well-lighted water, for example salmon, rockfish, herring, and tuna, dark backs and light undersides are common. This color pattern allows the fish to blend with the bottom when seen from above and with the surface when seen from below (fig. 10.4).

In the turbid coastal waters of temperate latitudes, there is less light penetration, and drab browns and grays conceal animals seen against the plants and the bottom. Cold-water bottom fish are usually uniform in color to match the bottom, or speckled and mottled with neutral colors. Some flatfish have skin cells that expand and contract to produce color changes, allowing them to conceal themselves by matching the bottom type on which they live (fig. 10.5).

Color is thought to be important in species recognition, courtship, and possibly in keeping schools of fish together. However, we do not know how the animals see the colors, and color may play roles that we do not know or understand.

What is countershading and what is its benefit?

Discuss several ways in which marine organisms use color.

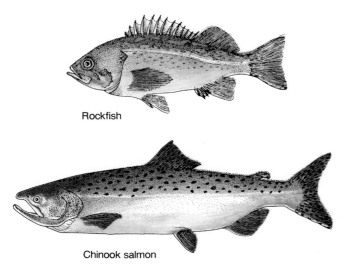

Rockfish

Chinook salmon

Figure 10.4

Viewed from above, the dark dorsal surface of the fish blends with the sea floor; viewed from below, the light ventral surface blends with the sea surface. This type of coloration is known as countershading.

Barriers and Boundaries

Rapid changes in temperature, density, salinity, and light with depth act as barriers, isolating species and populations from one another. As the water deepens, the effects of such barriers decrease as water properties become more homogeneous. In the horizontal direction similar boundaries exist at zones of converging and diverging waters and between fresh water and seawater in coastal areas. When one type of water is moving through or adjacent to another type of water, the boundaries may be sharp; for example, populations that do well in the warm, salty water of the Gulf Stream may die if they are displaced into the cold, less salty Labrador Current flowing between the North American coast and the Gulf Stream.

Ridges may isolate one deep-ocean basin from another, separating water of differing characteristics with different populations. Isolated seamounts with their peaks in shallow water may support specific isolated communities of sea life. Near-surface and surface species of the tropic regions are prevented from moving between the Atlantic and Pacific Oceans by the land barrier of Central America, and Africa acts as a barrier keeping the tropical species of the Indian Ocean from exchanging freely with the tropical species of the Atlantic. The cold water south of both South America and Africa is a barrier to the survival of the tropical species. The barriers created by changes in seafloor type (rock, sand, mud) and the complex interactions that produce various coastal environments are discussed in chapter 12.

Primary Production 10.3

Plantlike organisms and animals that float or drift with the movements of the water are the **plankton.** Plantlike plankton are called **phytoplankton;** they are mainly unicellular (single-celled) organisms. Like land plants, the phytoplankton require

Figure 10.5

The winter flounder resting on a checkerboard pattern shows its use of camouflage.

sunlight, nutrients or fertilizers, carbon dioxide gas, and water for growth. Phytoplankton cells contain the pigment **chlorophyll** that traps the sun's energy for use in **photosynthesis.** The photosynthetic process uses the Sun's energy to convert carbon dioxide and water into sugars or high-energy organic compounds from which the cell forms new materials. The synthesis of organic material from inorganic molecules by photosynthesis is termed **primary production.** The total amount or mass of living matter produced by photosynthesizing organisms is the gross primary production.

Photosynthesis is represented by the equation:

$$6\ CO_2 + 6\ H_2O - \text{sunlight/chlorophyll} \rightarrow C_6H_{12}O_6 + 6\ O_2$$

| 6 | + | 6 | —sunlight/chlorophyll → | 1 | + | 6 |

molecules carbon dioxide | molecules water | | | molecule sugar | | molecules oxygen

The sugars produced by photosynthesis are broken down by the plant cells with the addition of oxygen to yield energy, carbon dioxide, and water in the process known as **respiration.** Respiration provides both plants and animals with energy for their life processes.

$$C_6H_{12}O_6 + 6\ O_2 \rightarrow 6\ CO_2 + 6\ H_2O + \text{life-support energy}$$

1 + 6 → 6 + 6 + life-support energy

molecule sugar | molecules oxygen | molecules carbon | molecules water dioxide

If the primary production that is broken down by respiration to yield energy for life processes is deducted from the total or gross primary production, any remaining gain in phytoplankton mass is termed the net primary production. This net primary production is available to be consumed by animals and decomposed by bacteria.

The amount of living material produced by net and gross primary production can be expressed as the number of organisms produced, their weight, or grams of carbon; this is known as **biomass.** Biomass is usually reported as the dry weight of organic carbon in grams present under a square meter of sea surface or gC/m^2. The rate at which biomass changes, also known as **primary productivity,** is reported as $gC/m^2/\text{time}$. Biomass can also be reported per volume.

The total amount of living material in an area at any one instant in time is the biomass known as **standing crop.** It is the result of growth, reproduction, death, and grazing by other organisms (fig. 10.6). When a population of herbivores is grazing the phytoplankton population at the same rate as the net primary production increases it, a nearly constant standing crop results. See figure 10.12 for satellite views of the global phytoplankton standing crop.

Because of their size and abundance in coastal areas, seaweeds are the most conspicuous photosynthesizers of the oceans. However, they represent only 5 to 10 percent of the total photosynthetic material produced by the oceans. Seaweeds are discussed in more detail in chapter 12.

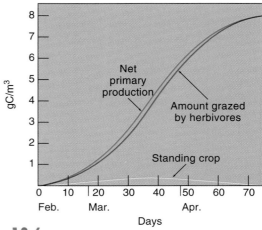

Figure 10.6

When net cumulative primary production is balanced by the grazing of herbivores, both populations increase during the spring, and the standing crop remains nearly constant.

Distinguish between photosynthesis and respiration.

Relate primary production and photosynthesis.

Distinguish between gross primary production and net primary production; define biomass.

Define standing crop and relate it to primary production.

Light and Primary Production   10.4

Phytoplankton growth is controlled in part by the light at the ocean surface. At polar latitudes, the summer days of nearly continuous but low-intensity daylight and the little sunlight available during the rest of the year result in a short period of rapid growth during the midsummer (fig. 10.7a).

At the middle latitudes, the sunlight's intensity and duration vary with the seasons. The spring increase in solar radiation warms the surface and increases the density stability of the water column, helping the phytoplankton to remain at the surface. Figures 10.7b and 10.8 show an initial increase in phytoplankton biomass with the lengthening of daylight hours in the spring. This is followed by a decrease in phytoplankton biomass due to animal grazing. A second increase in the population may occur in the late summer or early fall; this is caused by other phytoplankton species that reproduce later in the year.

Compare figures 10.7a and 10.7b with figure 10.7c, showing the lower level but extended growth of phytoplankton in the tropics. The population increase associated with the seasonal change in tropic-latitude sunlight is slight; the year-round sunlight provides abundant high-intensity solar energy at all times.

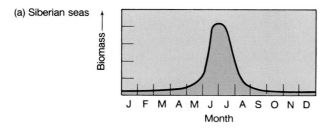

(a) Siberian seas

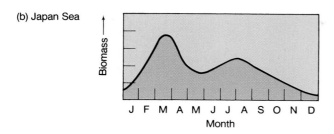

(b) Japan Sea

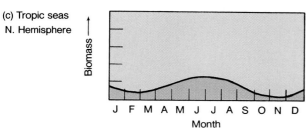

(c) Tropic seas
N. Hemisphere

Figure 10.7

Phytoplankton growth cycles vary with the light available at different latitudes. (a) At high latitudes, growth is limited to a brief period in midsummer. (b) Seasonal light changes at the middle latitudes increase growth in early spring and continue it through the summer. (c) Sunlight levels vary little at tropic latitudes, where phytoplankton growth is nearly uniform throughout the year.

The phytoplankton must remain in the ocean's surface layers to take advantage of the available sunlight. The warm surface layer of low-density water in the tropics exists all year long, and the phytoplankton are not displaced easily into the underlying more dense water. At midlatitudes a low-density, warm surface layer is found only in summer; if the warm weather is late or broken by cool periods, both the formation of the low-density layer and the active growth of the phytoplankton are delayed. During the winter, surface cooling and storms mix the surface layers into deeper water. Refer back to figure 7.4. At polar latitudes the growing season is short; the low level of sunlight warms the sea surface to form only a weakly stable density layer. The addition of low-salinity water from melting sea ice helps to maintain a stable water column at high latitudes.

If light were the sole factor influencing phytoplankton growth, the level of this growth in the tropics would be expected to maintain year-round levels similar to the summer values shown for the polar and middle latitudes. But nutrients as well as light are required if the phytoplankton population is to increase.

How does light control primary production in polar seas and in equatorial seas?

Relate the phytoplankton biomass to sunlight in each of the four seasons at midlatitudes.

Nutrients and Primary Production   10.5

At polar latitudes, winter overturn resupplies the surface waters with nutrients. The growing season is short, and the nutrients are more than sufficient. At these latitudes the availability of light controls phytoplankton growth.

In middle latitudes, the surface nutrients are replaced each year by winter storms and the cooling of the surface water to produce a shallow overturn. Nutrients are available to the phytoplankton cells when the increase in spring sunlight warms the surface and decreases the surface water density. Together the sunlight and nutrient supply trigger the first period of phytoplankton growth. Grazing organisms decrease the phytoplankton and release nutrients for the second phytoplankton increase during the late summer. The phytoplankton bloom continues, although controlled by the rate of nutrient resupply, and the population declines with the reduction in light levels as winter begins. Note the interaction between the sunlight, the stability of the surface water, the nutrients in the surface water, and the phytoplankton biomass in figure 10.8. In the tropics, lack of surface mixing and overturn mean low levels of nutrients and a low plant biomass despite the constant high level of solar energy.

When a plant or an animal dies, organisms known as **decomposers** (bacteria and fungi) release the energy within the organism's tissues to the environment and break down the organism's complex organic molecules to basic molecules, such as carbon dioxide, nutrients, and water. Since no significant amount of new matter comes to the earth from space, living systems must recycle inorganic molecules to form the organic compounds required for their systems and life processes.

Nitrogen is essential in the formation of proteins, and phosphorus is required in energy reactions, cell membranes, and nucleic acids. Nitrogen in the form of nitrates, and phosphorus in the form of phosphates are two water soluble nutrients that are removed from the water by the phytoplankton as their populations grow and reproduce. Nitrogen and phosphorus are cycled into the animal populations as the animals feed on the plants and are returned to the water as organisms die and decay. Excretory products from the animals are also added to the seawater, broken down, and used again by a new generation of plants and animals. The cyclic pathways for nitrogen and phosphorus are illustrated in figures 10.9 and 10.10.

Nitrogen gas is useless to plants; it must be converted to nitrate to be incorporated into plant tissue. This is done by a few species of microorganisms that convert nitrogen gas to

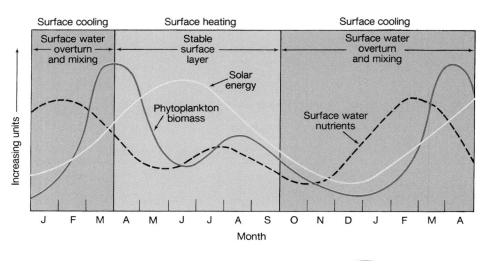

Figure 10.8

Phytoplankton biomass, nutrient supply, and surface water stability respond to solar energy changes at the middle latitudes in the Northern Hemisphere.

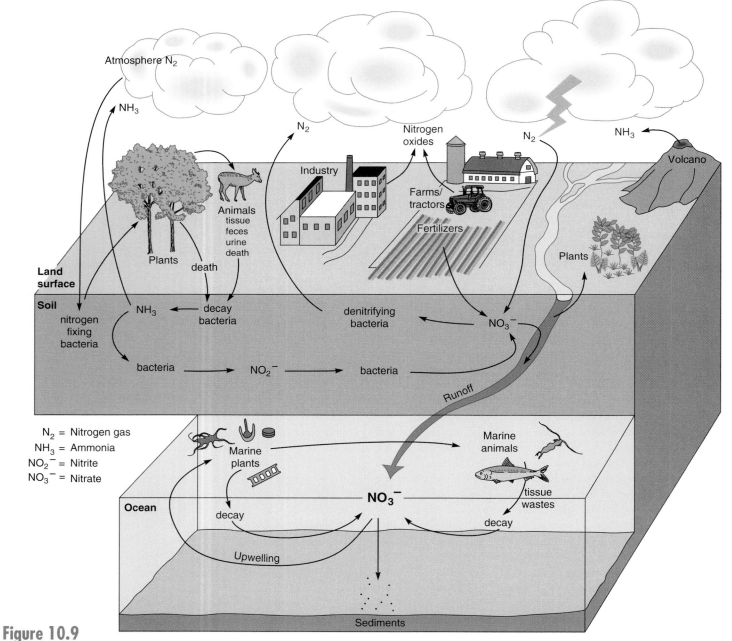

N₂ = Nitrogen gas
NH₃ = Ammonia
NO₂⁻ = Nitrite
NO₃⁻ = Nitrate

Figure 10.9

The nitrogen cycle. Although nitrogen makes up 78% of the earth's atmosphere, only a few microorganisms are able to change nitrogen gas to the nitrate that is used by land and sea plants, which are in turn eaten and recycled by animals.

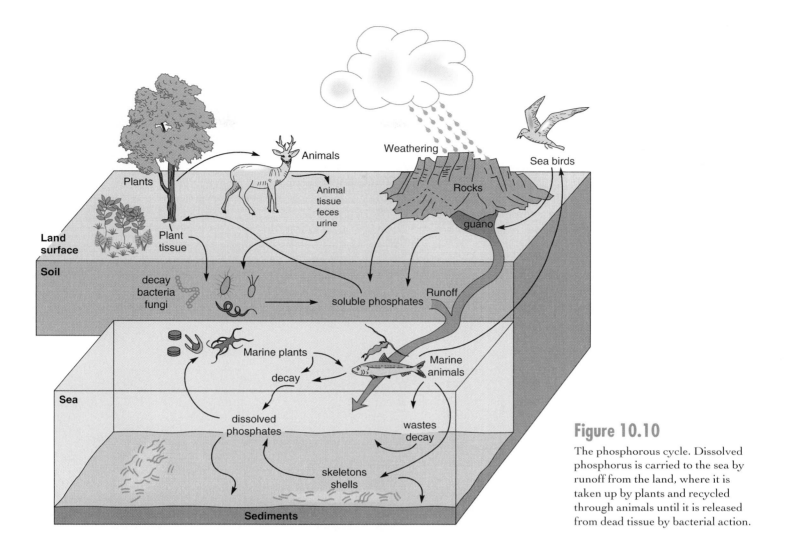

Figure 10.10

The phosphorous cycle. Dissolved phosphorus is carried to the sea by runoff from the land, where it is taken up by plants and recycled through animals until it is released from dead tissue by bacterial action.

ammonia, other species that convert ammonia to nitrite, and still other species that change nitrite to nitrate, the form most easily absorbed by the plants.

While nitrogen in the form of nitrate is required for life, the excess nitrogen from fertilizers and the combustion of fossil fuels that is entering the sea is degrading near shore areas. Nitrates increase phytoplankton blooms, but the decomposition of the resulting overabundant organic material decreases the oxygen in the water. Recently such problems have been reported from the Baltic and North seas, the western Mediterranean, and Narragansett and Chesapeake bays along the U.S. Atlantic coast. In the Gulf of Mexico an oversupply of nitrates delivered by the Mississippi River has been linked to a low-oxygen zone along the Louisiana coast; see Dead Zones in section 11, chapter 9.

There are many other inorganic nutrients that are related to phytoplankton growth. Some areas of the oceans have abundant nitrates and phosphates but show puzzlingly low levels of primary production. Experiments in the equatorial Pacific in 1994 and 1997 proved that it is possible to cause a surge in phy-

toplankton production by adding soluble iron to the surface water. However, after an initial spurt, the growth slows and levels off as the iron reacts with other dissolved substances and sinks. Nineteen ninety-five studies of primary production in Antarctic waters that are high in nutrients but low in production showed similar results.

It is possible to follow the cyclic path of any molecule required for life. Water is a requirement for life; it is part of all life processes and is cycled over and over through life-forms, time, and space; refer to the hydrologic cycle in chapter 2.

What role does the nutrient supply play in the primary production of polar, midlatitude, and equatorial oceans?

Explain the concept of nutrient cycles.

What part do decomposers play in nutrient cycles and primary production levels?

Measuring Primary Production   10.6

To determine the amount of plant material present in a water sample one can filter out the phytoplankton, count the cells, and multiply by the average mass per individual cell. Other less direct but less tedious methods are also available. Biomass may be estimated by extracting the chlorophyll from a sample and determining the concentration of pigment present. Another method exposes the chlorophyll in the phytoplankton cells to certain wavelengths of light that cause the chlorophyll pigment to fluoresce. The intensity of the fluorescence gives a measure of the chlorophyll, which is converted to phytoplankton biomass present in a given volume of water.

A simple technique to measure rate of biomass change collects seawater samples with their natural phytoplankton populations at selected depths and light levels. The amount of oxygen present in the seawater at the time of collection is measured. Samples are incubated in sealed containers for several hours at the light levels of the sampled depths, as well as in the dark. The oxygen in the samples is measured again at the end of the incubation period. From these data it is possible to calculate rates of gross primary production, net primary production, and respiration. Similar experiments add radioactive isotope carbon-14 to each sample and, after a period of time, the phytoplankton are filtered from the water and the amount of carbon-14 incorporated into their cells is measured to determine primary production.

Direct measurement of the rate of primary production (known as primary productivity) is expensive and limited to the sampling areas of oceanographic and fisheries research vessels. The map in figure 10.11 was compiled from many years of such sampling. More recently satellites have been recording sea surface chlorophyll measurements. Figure 10.12a–d are images of global distribution of chlorophyll concentrations or phytoplankton biomass prepared from satellite data collected between 1978 and 1986. Each image shows chlorophyll concentration averaged over three months. Changes in chlorophyll

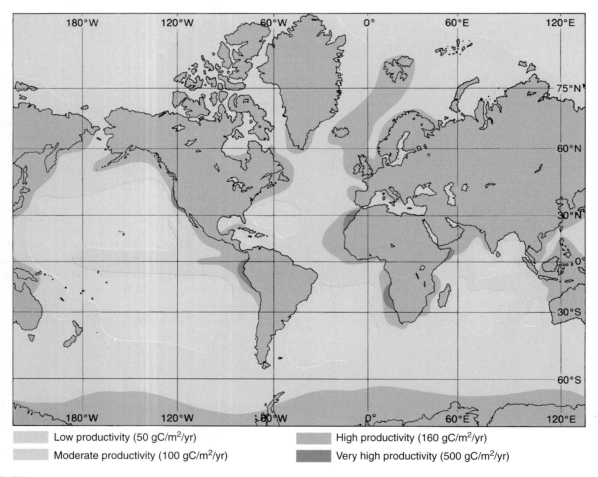

Low productivity (50 gC/m²/yr)	High productivity (160 gC/m²/yr)
Moderate productivity (100 gC/m²/yr)	Very high productivity (500 gC/m²/yr)

Figure 10.11

The distribution of primary productivity in the world's oceans.

More Information: opp.gsfc.nasa.gov/index.html

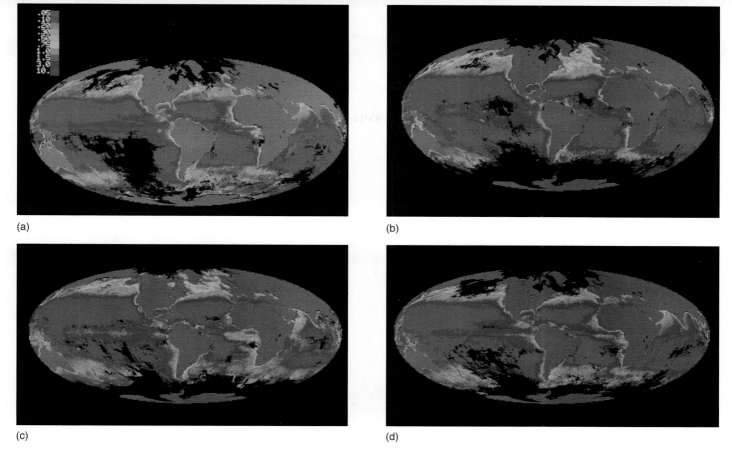

(a)

(b)

(c)

(d)

Figure 10.12

Satellite observations of global chlorophyll concentration. (a) January, February, March. (b) April, May, June. (c) July, August, September. (d) October, November, December. Color is equivalent to milligrams of chlorophyll pigment per cubic meter of surface water; see color bar in (a).

More Information: seawifs.gsfc.nasa.gov/SEAWIFS/CZCS_DATA/DOC/CZCS_BACKGROUND.html

concentration from one image to the next document standing crop changes which are governed by the combined effects of seasonal cycles of light intensity, depth of light penetration, nutrients, water temperature, currents, grazing by herbivores, reproduction, and population die-off. In these satellite images phytoplankton abundance is shown to be low in the central ocean gyres (magenta to deep blue) and higher along coastal and upwelling areas (yellow, orange, and red); areas of moderate phytoplankton population are green-blue. Black ocean areas indicate insufficient data.

> Explain one large-scale and one small-scale method of measuring primary production.

Global Primary Production   10.7

Refer to figure 10.12 while reading the following discussion. Coastal areas are generally more productive than open ocean, because along the coasts and in the estuaries, rivers and land runoff supply nutrients. Also, currents cause mixing and turbu-lence that supply nutrients from shallow depths to surface lay-ers. The fresh water helps to create a stable, low-density surface layer, which keeps the plant cells up where sunlight is plentiful.

Narrow areas of very high productivity are found against the west coasts of North and South America, the west coast of Africa, and along the west side of the Indian Ocean. These are major up-welling zones on the sheltered sides of landmasses in the trade wind belts and in areas of abundant sunlight. These are also the areas of the world's greatest fish catches. In these nutrient-rich zones, the popula-tions of phytoplankton form the first step in the feeding relationships that have produced great schools of commercially valuable fish.

The equatorial Pacific demonstrates the influence of up-welling on primary production caused by the equatorial diver-gence, as does the surface divergence surrounding Antarctica. In contrast, the downwelling centers of the large ocean gyres, where surface convergences occur, are areas of low nutrient availability that depresses primary production. Return to chap-ter 7, section 8 and figure 7.17. Compare areas of surface diver-gence and convergence, upwelling and downwelling, with the distribution of primary production rates in figure 10.11.

On the average, upwelling areas are four times more pro-ductive than coastal areas and five times more productive than the open ocean for the same units of area and time. This

table 10.1   World Ocean Primary Production

Area	Primary Production (gC/m^2/yr)	World Ocean Area (km^2)	(%)	Total Primary Production (metric tons of carbon/year)
Upwellings	640	0.36×10^6	0.1	0.23×10^9
Coasts	160	54×10^6	15.0	8.6×10^9
Open oceans	130	307×10^6	85.0	39.9×10^9

Source: Data from S. Smith and J. Hollibaugh, "Coastal Metabolism and the Oceanic Organic Carbon Balance" in Review of Geophysics *31 (1):75–89, 1993.*

table 10.2   Gross Primary Production: Land and Ocean

Ocean Area	Range (gC/m^2/yr)	Average (gC/m^2/yr)	Land Area	Amount (gC/m^2/yr)
Open ocean	50–160	130 ± 35	deserts, grasslands	50
Coastal ocean	100–250	160 ± 40	forests, common crops, pastures	25–150
Estuaries	200–500	300 ± 100	rain forests, moist crops, intensive agriculture	150–500
Upwelling zones	300–800	640 ± 150	sugarcane and sorghum	500–1500
Salt marshes	1000–4000	2471		

Source: Data from S. Smith and J. Hollibaugh, "Coastal Metabolism and the Oceanic Organic Carbon Balance" in Review of Geophysics *31 (1):75-89, 1993.*

relationship is shown in table 10.1. Notice that there is one hundred and fifty times more coastal area than upwelling area, and nearly six times more open-ocean area than coastal area. As a result, most of the carbon compounds produced by the phytoplankton are formed at low rates over large areas of the open sea. Smaller total amounts of organic carbon are produced along the coasts and in upwelling regions, but these amounts are concentrated in smaller areas, making these coastal and upwelling areas very high in rates of primary production. To compare primary production rates on land with those at sea, study table 10.2. Primary production in the open sea is about the same as that of the deserts on land, while coastal oceans are comparable to pastureland and lush forests, and certain upwelling zones and estuaries approach the most heavily cultivated land.

Shallow coastal areas and estuaries in which the sea bottom is exposed to sunlight support masses of phytoplankton and attached seaweeds. These provide a large amount of organic material to the animal population. Primary production occurs at all depths in shallow estuaries, and organic material is also produced in the bordering marshlands, resulting in an extraordinary rate of primary production. In addition when these organisms die and begin to decay, they form organic litter known as **detritus** which provides another source of nutrition. The estuary is able to support large and varied populations of animals as well as contributing organic matter that is exported to coastal waters where it nourishes still other populations.

Where are the areas of highest and lowest marine primary production? What are their equivalents in land vegetation?

What phenomena increase the primary production of an area and what phenomena decrease it?

Why is the total primary production of the open ocean higher than that of upwellings and coastal areas?

Food Chains, Food Webs, and Trophic Levels   10.8

Primary production forms the first link in the **food chain** that connects plants and animals. Where there is high primary production by phytoplankton, seaweeds, or other marine plants, there are large populations of animals. Animals are consumers; they feed on primary producers or other consumers. The **herbivores** eat plants directly; the **carnivores** feed on the herbivores or other carnivores. The most numerous (and the greatest biomass) of herbivores are the plant-eating animal plankton, or **zooplankton.** These are the primary consumers that convert plant tissue to animal tissue, and then become the food for other zooplankton, the flesh-eating carnivores or secondary consumers. Food chains may be long or short, but they are rarely simple and linear. Food chains are more likely to show complex interrelationships among organisms, in which case it is more appropriate to call the pattern a **food web.** Species may change levels in the food chain or web at different stages of their life cycle or consumers may feed at more than one level. The herring food web in figure 10.13 and the overall ocean food web in figure 10.14 are simplified diagrams that only begin to indicate the complexity of these systems.

Food chains and food webs represent the pathways followed by nutrients and food energy as they move through the succession of plants, grazing herbivores, and carnivorous predators. These relationships are often simplified and demonstrated in the form of a pyramid made up of **trophic levels** that represent links in a food chain (fig. 10.15). Trophic levels are numbered from the bottom of the pyramid to the top; the primary producers are always the first trophic level. The herbivorous zooplankton are the second trophic level, and the carnivores form the upper levels up to the top carnivore, a predator on which no other marine organism preys (for example, sharks and killer whales).

In general, moving upward from the first trophic level, the size of the organisms increases and the numbers and biomass of organisms decrease. The larger numbers of small organisms at the lower trophic levels collectively have a much larger biomass than the smaller numbers of large organisms at the upper levels. Figure 10.15 relates trophic levels, the primary energy source of the Sun, decomposition, recycling of nutrients, and energy loss.

The overall efficiency of energy transfer up each layer of an open-ocean trophic pyramid is estimated at about 10 percent. If, in order to add 1 kilogram (2.2 lbs) of weight, a person ate 10 kg of salmon, to attain that weight the salmon had to consume 100 kg of small fish, and the fish needed to consume 1000 kg of carnivorous zooplankton, which in turn required 10,000 kg of herbivorous zooplankton, needing 100,000 kg of phytoplankton to supply the eventual 1 kilogram gain at the top of the pyramid. The 90% energy loss at each trophic level goes to the metabolic needs of the organisms (feeding, breathing, moving, reproducing, and heat loss) and includes organic material that is not ingested but decays in the water column or on the sea floor. An organism feeding on 100 units from the level directly below converts only 10 units to body tissue that is

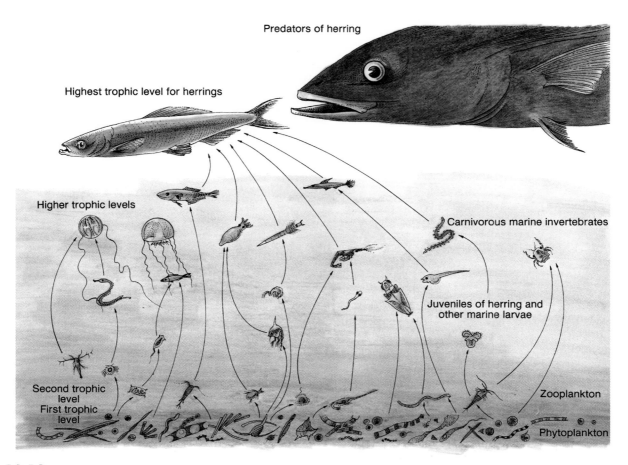

Figure 10.13

The food web of the herring at various stages in the herring's life. Juvenile herring feed at lower trophic levels than adult herring.

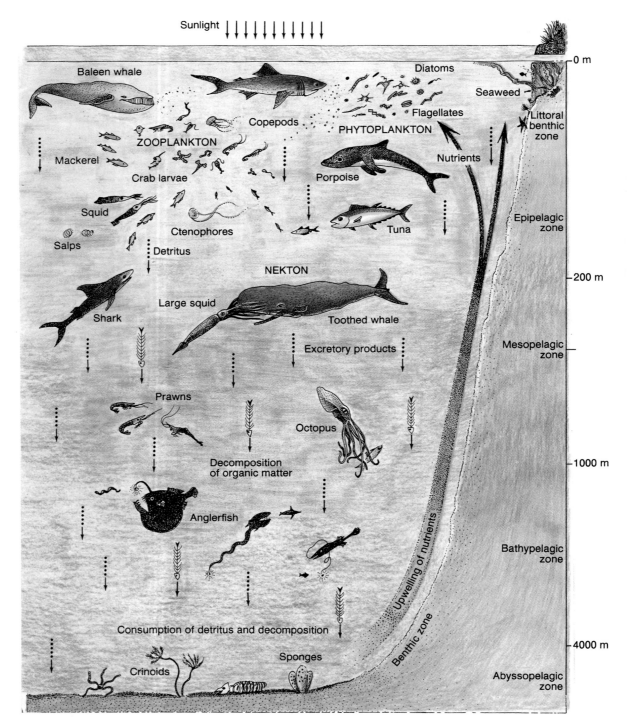

Figure 10.14

The ocean food web. Plant, animal, and bacterial populations are dependent on the flow of energy and the recycling of nutrients through the food web. The initial energy source is the Sun, which fuels the primary production in the surface layers. Herbivores graze the phytoplankton and benthic algae and are in turn consumed by the carnivores. Animals at lower depths depend on organic matter from above. Upwelling recycles nutrients to the surface, where they are used in photosynthesis.

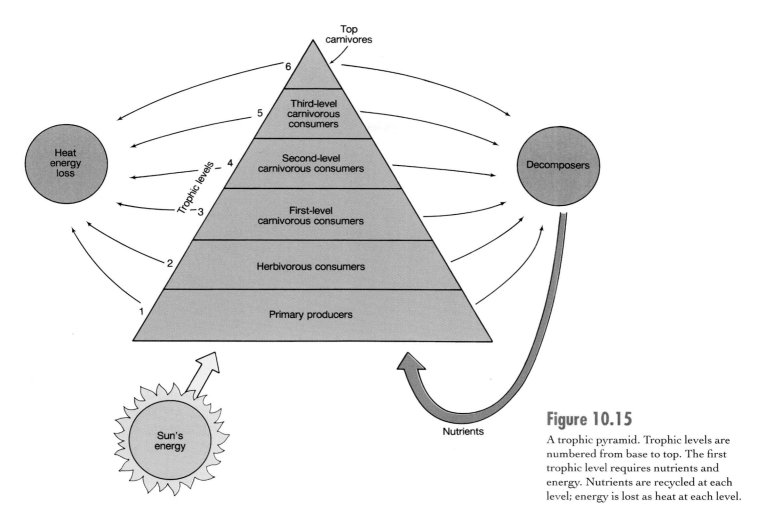

Figure 10.15

A trophic pyramid. Trophic levels are numbered from base to top. The first trophic level requires nutrients and energy. Nutrients are recycled at each level; energy is lost as heat at each level.

then available to predators in the level above. In areas with abundant phytoplankton populations food is obtained with less effort and the energy transfer up the pyramid increases.

But trophic efficiency does not necessarily remain the same in all environments. It may be influenced by various environmental phenomena such as El Niño (ch. 6), episodes of coastal pollution (ch. 9), and coral bleaching (ch. 12). A food web that is dependent on a specific trophic level may adjust as populations change with time. For example, the Antarctic food web, fish, seals, whales, and birds, has its base in the Antarctic **krill** *(Euphausia superba)*; see figure 10.16. Krill are shrimp-like, 6 cm (2.4 in) long, herbivorous organisms that feed on plantlike cells that multiply each summer as the Antarctic ice pack melts. Some species of Antarctic fish stay in the Southern Ocean year-round, feeding on krill; other species migrate into Antarctic waters each summer to feed on krill. All Antarctic fish combined are estimated to consume as much as 100 million tons of krill each year. One hundred fifteen million tons are eaten by Antarctic birds every year, including penguins (except emperors and kings that feed on fish and squid only). The crabeater seal is the most abundant seal in the world and its diet is about 94% krill; the leopard seal's diet is 37% krill. Stocks of Antarctic seals consume over 130 million tons of krill (two or three

times the current consumption by whales). Krill is the major food of the blue, fin, sei, minke, humpback, and right whales of the Southern Ocean. The total quantity of krill taken by all predators in the Southern Ocean may be about 500 million tons per year. At one time it was thought that the loss of the great whales would produce an overabundance of krill, but the smaller whales, the birds, and the seals now play larger roles in krill predation than they did a century ago, and there is no oversupply of krill. The organisms have struck a new balance, but still directly or indirectly, the Antarctic food web depends on krill.

> Compare a food chain and a food web.
>
> Sketch a trophic pyramid of four levels; label primary producers, herbivores, and carnivores. How are energy and nutrients related to these trophic levels?
>
> Why does energy lessen when moving upward from one trophic level to the next?
>
> Explain the importance of krill in the Antarctic food web.

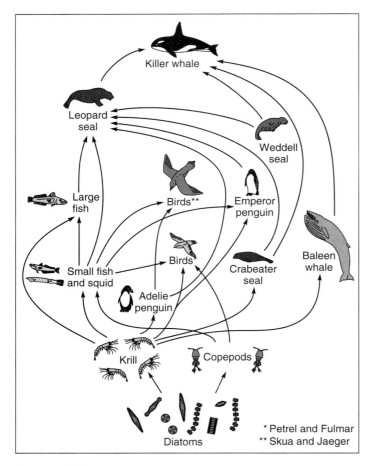

Figure 10.16

The Antarctic food web. The shrimp-like krill feed on the diatoms and all other members of the food web depend on the krill.

More Information: www.antdiv.au/science/bio/issues_krill/krill_features.html

Harvesting Efficiency   10.9

Custom, culture, economics, and availability influence the harvesting of the oceans by humans. Commercial harvests are made from both the higher trophic levels (for example, salmon, tuna, halibut, and swordfish), and the lower levels (for example, herring, shellfish, and anchovy). Harvesting high in the trophic pyramid is less energy efficient than at low

trophic levels. Overharvesting any level endangers levels both above and below it by removing food resources from higher levels and preventing recycling of nutrients from higher to lower levels.

The yields of food resources for humans are highest when the harvest is conducted at the lowest possible trophic level. The relationship between total plant production and theoretical production of fish at its harvested trophic level is shown in table 10.3. The third column in this table gives the average efficiency of energy conversion between the trophic levels, and the fourth column shows the trophic level at which humans usually harvest fish. The well-mixed, nutrient-rich waters of the coasts and estuaries support short, efficient food chains (fig. 10.17a,b). Because of the high rates of plant production in these areas, food is easily obtained and energy is transferred with greater efficiency to the next higher trophic level; consumers obtain food with less effort, and fish production is high. Compare the efficiency of these food chains with the open-ocean food chain in figure 10.17c and the open-ocean fish production in table 10.3.

> Why is it more efficient to harvest anchovies than tuna?
>
> Why is the efficiency of energy transfer per trophic level less in the open ocean than in upwelling zones?

Chemosynthetic Communities   10.10

In March 1977, an expedition from Woods Hole Oceanographic Institution using the research submersible *Alvin* discovered communities of animals living around hot-water, hydrothermal vents of the Galápagos Rift (a branch from the East Pacific Rise) at depths of 2500 to 2600 meters (8000–8500 ft). Since that first discovery vent communities have been found scattered throughout the world at seafloor spreading centers. These communities are known between 9° and 21°N latitude along the East Pacific Rise, in the Marina and Okinawa troughs and North Fiji Basin of the western Pacific, along the Mid-Atlantic Ridge, in the East Pacific from 49° to 22°S, and on the Gorda and Juan de Fuca Ridges of the north-

table 10.3   Oceanic Food Production

Area	Plant Production (metric tons of carbon/year)	Efficiency of Energy Transfer per Trophic Level	Trophic Level Harvested	Fish Production (metric tons/year)
Open ocean	39.9×10^9	10%	5	4.0×10^6
Coastal regions	8.6×10^9	15%	4	29.0×10^6
Upwelling areas	0.23×10^9	20%	2.5	46.0×10^6

Source: Data from S. Smith and J. Hollibaugh, "Coastal Metabolism and the Oceanic Organ's Carbon Balance" in Review of Geophysics 31 (1):75-89, 1993.

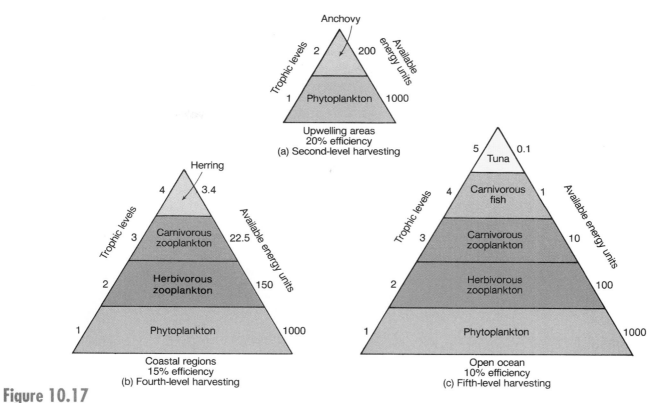

Figure 10.17

Trophic level efficiency varies among (a) upwelling areas, (b) coastal regions, and (c) the open ocean. The number of trophic levels and the level at which humans harvest fish differ with location.

east Pacific. Animals in these areas include filter-feeding clams and mussels, in addition to anemones, worms, barnacles, limpets, crabs, and fish. The clams are very large and fast growing, up to 4 cm (2 in) per year. The tube worms are startling in size, up to 3 m (10 ft) long; they are the fastest growing of all known marine invertebrates—nearly 1 m (3.3 ft) per year. These crowded communities have a biomass per area of seafloor that is five hundred to one thousand times greater than the biomass typical of the deep seafloor.

Dense clouds of bacteria are the base of the food pyramid for these communities. These bacteria form organic compounds with energy derived from dissolved chemicals in the seawater, a process known as **chemosynthesis,** which is another form of primary production. Instead of being dependent on sunlight for energy to produce organic matter by photosynthesis, these ocean-bottom bacterial populations rely on dissolved hydrogen sulfide in the hot vent water. These bacteria are able to do in the dark what the phytoplankton do in the sunlight: they fix carbon dioxide into organic molecules. The vent animals feed on the bacteria; no sunlight is necessary, and no food is needed from the sea surface. These self-contained vent communities are among the most productive in the world. In areas where the vents have become inactive, the animal communities have died when their energy source has been removed.

Small animals, such as small shrimplike organisms, may graze the bacteria directly. Soft-bodied organisms may absorb dissolved organic molecules released by dead bacteria. The

tube worms (fig. 10.18) have no mouth and no digestive system; the soft tissue mass of their internal body cavity is filled with bacteria. All synthesis of organic molecules is done within the bacteria. Despite their size and abundance these tube worms are completely dependent on the bacteria for their

Figure 10.18

Tube worms crowd around deep-sea hydrothermal vents. The internal body cavity of each worm is filled with bacteria that synthesize organic molecules by chemosynthesis.

Photo by R. A. Lutz

nutrition. The mussels have only a rudimentary gut, and the clams and mussels have large numbers of bacteria in their gills. Both the clams and the tube worms have red flesh and red blood based on the oxygen-binding molecule hemoglobin.

During studies of the large shrimp populations surrounding some vent areas, researchers noticed a reflective spot just behind the head of the shrimp. The reflective spot connects to the shrimp's brain, and light-sensitive pigment related to visual pigment is also present. This system is thought to allow the shrimp to detect radiation associated with the hot vents (refer to fig. 3.32) and may enable the shrimp to orient themselves with the vents and their food supply.

Off the coasts of Florida, Oregon, and Japan, communities based on chemosynthesis have been found that are associated with cold seepage areas. Along the continental slope of Louisiana and Texas, oil and gas seep up to the surface of the sea floor. In 1985, clams, mussels, and large tube worms (fig. 10.19) were collected from these sites at depths between 500 and 900 m (1600–3000 ft). The mussels' gills contain bacteria that use methane gas as a carbon source; the mussels are able to use organic carbon compounds produced by the bacteria. The tube worms and clams acquire their carbon from methane, biologically degraded oil, and the decay of other organisms. Carbon compounds from methane have been identified in animals preying on the shellfish, which demonstrates a pathway for chemosynthetic carbon to enter the general deep-sea animal community.

In 1990, salt seeps were reported at 650 m (2000 ft) on the floor of the Gulf of Mexico. Depressions are filled with salt brine more than 3.5 times the usual salinity of seawater, and the brine also contains methane gas. These extremely dense brine lakes are surrounded by large communities of chemosynthetic mussels.

Ninety-five percent of the nearly three hundred species found at these vents and seeps are new to science, and many are only distantly related to other earth creatures, appearing more closely related to ancient species. The presence of a plentiful food supply allows their rapid growth, despite the surrounding low temperatures and their remoteness from the photosynthetic layer at the sea's surface. These discoveries have changed our view of ocean food webs and the deep sea floor; scientists will never again consider it to be sparsely populated or inhospitable to life.

> How is it possible for large animal populations to live without a photosynthetic base?
>
> Draw a three-level trophic pyramid for a chemosynthetic community.

Extremophiles   10.11

Microorganisms that thrive under conditions that would be disastrous for other life forms (extreme temperatures, high levels of acid or salt, lack of oxygen) are called **extremophiles.**

Heat-loving extremophiles in the oceans require temperatures in excess of 80°C (176°F) for maximum growth. *Pyrolobus fumarii* is an extreme example; it was found at a depth of 3650 m (12,000 ft) in a hot vent in the mid-Atlantic Ridge. Its name means "fire lobe of the chimney" from its shape and the black smoker vent where it was found. *P. fumarii* stops growing below 90°C (194°F) and reproduces at temperatures up to 113°C (235°F). It uses hydrogen and sulfur compounds as sources of energy, can utilize nitrogen gas, and is able to live with or without oxygen. Other extremophiles are commonly found in the plumes of hot water that occur after an undersea eruption. It is unclear how deep into the Earth's crust these microorganisms can exist. See figure 10.20a,b.

Cold water microorganisms have been found living at all latitudes in water below 100 m (330 ft), in deep-sea sediments, and in the guts of deep-sea cucumbers. A microorganism found in Antarctic sea ice grows best in 4°C (39°F) and does not reproduce at temperatures above 12°C (54°F). Other extremophiles have been found living in the salt ponds constructed for the evaporation of seawater.

> What is an extremophile?
>
> Where are extremophiles found?

Figure 10.19
Mussels and tube worms at a Gulf of Mexico oil and gas seep.

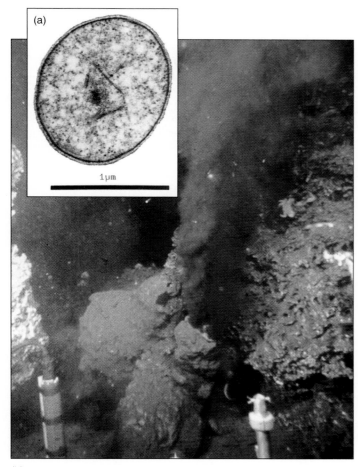

Figure 10.20ab

The extremophile *Pyrococcus endeavorii* (a) was isolated from an East Pacific Rise black smoker (b). This microorganism grows at temperatures that exceed 100°C (212°F).

More Information:
www-archbac.u-psud.fr/Projects/Biotech/BoEFacePlate.html

Animals use color for concealment and camouflage, and to warn predators of poisonous flesh and bitter taste. Barriers for marine organisms include water properties, light intensity, and seafloor topography.

Plants use photosynthesis to produce organic compounds from carbon dioxide and water in the presence of sunlight and chlorophyll; the new plant material is called primary production. Respiration breaks down organic compounds with the addition of oxygen to yield energy, water, and carbon dioxide.

Gross primary production is the total amount of organic material produced by photosynthesis per volume per unit of time; net primary production is the gain in organic material after deducting the amount used in plant respiration. Biomass is the amount of organic matter produced expressed as number or weight of organisms. The biomass available at a location at a specific time is the standing crop.

Primary production is controlled by the interaction of sunlight, nutrients, and the stability of the surface water in an area. At polar latitudes, the availability of light controls phytoplankton growth; in the tropics, sunlight is available year-round, but the nutrient supply is poor and limits production. At midlatitudes, light, nutrients, and water column stability vary with the seasons. Nutrients cycle through the land and sea, the plants and animals, and are returned to the water by death and bacterial decay.

Primary production is estimated by counting plant cells, measuring chlorophyll concentration, measuring the rate at which oxygen is produced or the rate of carbon-14 incorporation into plant cells. Satellite images provide evidence of standing crop changes over the yearly cycle.

In general, coastal waters are more productive than the open ocean. Good mixing, nutrients in the land runoff, and density-stable water combine to make shallow coastal areas very productive. Upwelling areas are four times more productive than coastal water and five times more productive than the open ocean, but there is six times more open-ocean area than coastal water area and one hundred fifty times more coastal water area than upwelling area.

The phytoplankton are preyed upon by herbivorous zooplankton that are preyed upon in turn by carnivorous zooplankton. The term "food web" is more appropriate than "food chain" to describe interconnecting feeding patterns. Trophic pyramids present the relationship between primary producers, herbivores, and carnivores in terms of the transfer of biomass and energy.

In the open ocean, the transfer efficiency between trophic levels is approximately 10 percent. Efficiency is higher in coastal and upwelling areas.

Self-contained deep-ocean animal communities depend on chemosynthetic bacteria for the first step in their food chains. Communities have been found in the vicinity of hydrothermal vents, around cold seeps of oil and methane gas, and around salt seeps. Extremophiles exist under conditions of extreme temperature, lack of oxygen, and high levels of acid or salt.

Summary

The marine environment is subdivided into the benthic and pelagic zones; there are subdivisions of each of these zones. The lighted surface is also known as the photic zone and all depths without sunlight form the aphotic zone.

Marine organisms are buoyed and supported by the seawater. Adaptations for staying afloat include floats, swim bladders, oil and fat storage, and extended surface-area appendages.

Most marine fish lose water to their environments. They drink continually and excrete salt to prevent dehydration. Salinity is a barrier to some organisms; others can adapt to large salinity changes.

Key Terms

aphotic zone, 240
benthic zone, 240
bioluminescence, 242
biomass, 244
buoyancy, 241
carnivore, 250
chemosynthesis, 255
chlorophyll, 244
decomposer, 245
detritus, 250
ecology, 240
extremophile, 256
food chain, 250
food web, 251
herbivore, 250
intertidal zone, 240
krill, 253
littoral zone, 240

neritic zone, 240
oceanic zone, 240
osmosis, 242
pelagic zone, 240
photic zone, 240
photosynthesis, 244
phytoplankton, 243
plankton, 243
primary production, 244
primary productivity, 244
respiration, 244
standing crop, 244
sublittoral zone, 240
subtidal zone, 240
supralittoral zone, 240
trophic level, 251
zooplankton, 250

Critical Thinking

1. How does the role of bioluminescence differ from the role of sunlight in the sea?
2. Many ocean organisms are delicate and fragile; how do they exist without damage in the oceans?
3. Primary production rates of estuaries approximately equal the most intensively cultivated land: why are estuaries less important than cultivated land as producers of human foods?
4. Why is a shallow-water estuary more productive than either a deep-water estuary or an upwelling zone?
5. How are photosynthesis and chemosynthesis similar; how are they different?

Suggested Readings

Life and the Ocean Environment

Levinton, J. S. 1995. *Marine Biology: Function, Diversity, Biodiversity, Ecology.* Oxford University Press, New York.

Life in the Sea, readings from *Scientific American.* 1982. Freeman, San Francisco. 248 pp. (Articles on marine organisms, the conditions in which they live, and their food resources.)

Marshall, J. 1998. Why Are Reef Fish so Colorful? *Scientific American Quarterly* 9(3):54–57.

Nelson, K., and C. Arneson. 1985. Marine Bioluminescence: About to See the Light. *Oceanus* 28(3):13–18.

Nybakken, J., and S. Webster. 1998. Life in the Ocean. *Scientific American Quarterly* 9(3):74–87.

Seizen, R. 1986. Cuttlebone: The Buoyant Skeleton. *Sea Frontiers* 32(2):115–22.

Sumich, J. L. 1996. *An Introduction to the Biology of Marine Life,* 6th ed. Wm. C. Brown, Dubuque, Ia. 461 pp.

Svitil, L. 1995. Collapse of a Food Chain. *Discover* 16(7):36–76.

Ward, P. L. Greenwald, and O. E. Greenwald. 1980. The Buoyancy of the Chambered Nautilus. *Scientific American* 243(4):190–203.

Chemosynthetic Communities

Alper, J. 1990. The Methane Eaters. *Sea Frontiers* 36(6):22–29. (Life at petroleum seeps.)

Ballard, R. D., and J. F. Grassle. 1979. Return to Oases of the Deep. *National Geographic* 156(5):689–705.

Grassle, J. F. 1987/88. A Plethora of Unexpected Life. *Oceanus* 31(4):41–46.

Lutz, R. A. 1991/92. The Biology of Deep-Sea Vents. *Oceanus* 34(4):75–83.

Lutz, R. A., and R. Haymon. 1994. Rebirth of a Deep-Sea Vent. *National Geographic* 186(5):114–26.

Menon, S. 1997. Deep Sea Rebirth. *Discovery* 18(7):34.

Tunnicliffe, V. 1992. Hydrothermal-Vent Communities of the Deep Sea. *American Scientist* 80(4):336–49.

Van Dover, C. L. 1987/88. Do "Eyeless" Shrimp See the Light of the Glowing Deep-Sea Vents? *Oceanus* 31(4):47–52.

Extremophiles

DeLong, E. 1998. Archaeal Means and Extremes. *Science* 280(5363):542–43.

Madigan, M., and G. Marrs. 1997. Extremophiles. *Scientific American* 276(4):82–87.

internet references

worldwide websites

Ocean Productivity

SeaWiFs—Home Page
    seawifs.gsfc.nasa.gov/SEAWIFS.html

SeaWiFs Project
    seawifs.gsfc.nasa.gov/SEAWIFS/CZCS_DATA/
    DOC/CZCS_BACKGROUND.html

Satellite Ocean Color

Color from Space—Primary Productivity
    xtreme.gsfc.nasa.gov/CAMPAIGN_DOCS/
    OCDST/ocdst_primary_productivity.html

Primary Production Science Computing Facility—
Home Page
    opp.gsfc.nasa.gov/index.html

Comparing Rates of Primary Production Science
Computing Facility
    sfbay.wr.usgs.gov/access/ColeCloern/Rates.html

Antarctic Food Web

Krill Features
    www.antdiv.gov.au/science/bio/issues_krill/
    krill_features.html

Antarctic Phytoplankton Food of Antarctic Krill
(Euphausia supberba)
    www.ecoscope.com/phtopla.htm

Krill Concentrations
    www.abdn.ac.uk/~nhi600/temp/tsld018.htm

Antarctic Krill and Ocean Currents
    www.antdiv.gov.au/news/news%5Freleases/2000/
    030800.html

Antarctic Marine Treaties
    www.antcrc.utas.edu.au/opor/Treaties/ccamlr.html

Antarctic Penguins Disrupted
    www.greenpeace.org/~climate/database/records/
    zgpz0288.html

Extremophiles

Biology of Extremophiles
    www-archbac.u-psud.fr/Projects/Biotech/
    BoEFacePlate.html

The Extremophiles
    www.accessexcellence.com/BF/bf03/somero/
    som_imgs/somTBL8.html

When settlers and traders came to the North American continent more than three and a half centuries ago they brought with them many organisms, plant and animal, intentional and unintentional. Among the unintentional were large communities of widely varied organisms that attached to and bored into the wooden hulls of their ships in the harbor or bay where their journeys began. Many of the

Biological Invaders

organisms transported across the oceans in this way would have been swept from the ships' sides by the waves and currents, but a few invaders would have survived when the ships anchored at their journey's ends. Hundreds of years later there is no way of knowing when the first European barnacles or periwinkle snails began to colonize the east coasts of the United States and Canada. Today these two organisms are considered typical of this region.

The Ballast Water Problem

Today's ships do not carry such communities, for their steel hulls are protected by antifouling paints, and they move through the water at speeds sufficient to sweep away many of the organisms that wooden hulls carried with them. However, today's ships do carry ballast water that is loaded and unloaded to preserve the stability of the ship as it unloads and loads cargo. Tens of thousands of vessels with ballast tanks ranging in capacity from hundreds to thousands of gallons of water move across our oceans. These vessels rapidly transport this ballast water and its population of organisms across natural oceanic barriers; the release of ballast water and organisms may occur days or weeks later, thousands of kilometers from their point of origin.

Biologists sampled ballast water from 159 cargo ships from various Japanese ports that docked in Coos Bay, Oregon.* The ballast water contained members of all the major groups of floating or planktonic organisms: copepods, marine worms, barnacles and barnacle larvae, flatworms, jellyfish, and shellfish. The copepod density was estimated to be greater than 1500 per cubic meter; the larvae of marine worms, barnacles, and shellfish were greater than 200 organisms per cubic meter.

Some Examples

Whether organisms arrive in ballast water, in the packing of commercially harvested fish and shellfish, attached to the floats of seaplanes, or are released from home aquariums, they all leave behind the natural controls of predators and disease found in their native environments. If these invaders are introduced into a hospitable new environment, they may flourish and severely disrupt its biological relationships, forcing out some species and destroying others. The results may be ecological catastrophes.

In 1985 an Asian clam (Potamocorbula amurensis) (fig.1) was discovered in a northern arm of San Francisco Bay. This clam was previously unknown in the area and appears to have arrived in its juvenile or larval form in the ballast water of a cargo ship from China. During the next six years the clam spread southward into the bay and formed dense colonies, as many as 10,000 clams per square meter. The Asian clam feeds on microscopic plants and the larvae of crus-

*Source: J. T. Carlton and J. B. Geller, "Ecological Roulette: The Global Transport of Nonindigenous Marine Organisms" in Science 261 (5117):78–82, 1993.

taceans, and the food requirements of these huge populations reduce amounts available for native species, stressing the system's species balance and food chains.

In 1990, a second intruder, the European green crab (Carcinus maenas) (fig. 2), was recognized in the southern part of San Francisco Bay. This crab moved from Europe to the East Coast of the United States in the 1820s and to Australia in the 1950s. It may have hitchhiked to the bay in ballast water or perhaps in seaweed wrapped around Maine lobsters. This crab is less than 6.5 cm broad (3 in), voracious and belligerent. It feeds on clams and mussels and may spawn several times from a single mating. It has populated most of the bay in the last two to three years and feeds enthusiastically on the Asian clam. The green crab may clean the bay of the Asian clam, but it can also feed on native species and outcompete them for food. The green crab has moved northward and has been found in Willapa Bay and

Figure 1

Asian clam (Potamocorbula amurensis).

Figure 2

Green crab (Carcinus maenas).

Grays Harbor, Washington. Some shellfish exports from Willapa Bay has been temporarily banned by state agencies attempting to prevent the crabs from moving farther north into Puget Sound.

In 1982 the ballast water of a ship from America carried a jellyfish-like organism known as a comb jelly *(Mnemiopsis leidyi)* into the Black Sea. From the Black Sea it spread to the Azov Sea and has recently moved into the Mediterranean. It has no predator in these areas and devours huge quantities of animal plankton, small crustaceans, fish eggs, and fish larvae. Russian scientists claim this small organism now dominates the Black Sea, devastating the fish populations and catches of the last five years. The cost to the Black Sea fisheries is an estimated $250 million and the fisheries of the Azov Sea shut down after catches dropped an estimated 200,000 tons.

Ballast water carrying the resting cells of Japanese red-tide-causing organisms was responsible for shutting down the natural and aquaculture shellfish harvests of Tasmania and southern Australia in the 1980s. Fish can also be transported this way; for example, Japanese sea bass were introduced into the Sydney harbor region in 1982–83.

Not all invaders are animals. In 1985 an attractive, fast-growing bright green, tropical seaweed *(Caulerpa taxifolia)* escaped into the Mediterranean Sea during a routine tank-cleaning at Monaco's aquarium. Since then it has spread over more than 4,600 hectares (11,000 acres) along the coasts of France, Spain, Italy, and Croatia. It smothers seafloor life and is toxic to many fish. No effective method of removing it has been found.

Cordgrass *(Spartina alterniflora)* (fig. 3) is a deciduous, perennial, flowering plant native to salt marshes of the U.S. Atlantic and Gulf coasts. It spreads by underground stems, seeds, and colonies that form when pieces of the root system or whole plants float into an area and take root. Once established it begins to trap sediment, eventually producing a high marsh system. In its native estuaries cordgrass prevents erosion and marshland deterioration, but along the coasts of England, France, New Zealand, China, and the western U.S., it has crowded out the native plants and turned tidal mudflats into high marshes that are inhospitable to fish and waterfowl dependent on the mudflats. *Spartina* also hampers shellfish growth and harvesting and interferes with recreational use of beaches. Efforts to control it have included burning, flooding, shading plants with plastic, smothering plants with mud or clay, applying herbicides, and re-peated mowing. Little success has been reported; mowing small accessible patches and biological controls for long-term regulation may be the most realistic approach.

Solutions

In the case of ballast water, the introduction of zebra mussels into the Great Lakes and their subsequent invasion of the rivers and streams of the central United States sounded the alarm for the freshwater environment. In 1990 the Nonindigenous Aquatic Nuisance Prevention and Control Act (NANPCA) was enacted in the United States. Under this law the United States adopted voluntary regulations for exchanging ballast water on the high seas for vessels bound for the Great Lakes; this provision became law in 1993. In 1999 a presidential order was given to federal agencies to fight invasive organisms, but little has been done. Proposals have included exchanging of ballast water at sea (not all organisms may be flushed out), treating the ballast water by adding poisonous chemicals, heating the water, filtering the water, and exposing it to ultraviolet radiation. But, none of the world's cargo vessels are designed for ballast management, and all these proposals require some redesign or refit of vessels.

Controlling ballast water will not close all doors to invading marine species, and it will be costly, but it will lead to fewer foreign invasions. However, it is well to keep in mind that according to James T. Carlton of the Maritime Studies Program at Williams College, "No introduced marine organism, once established, has ever been successfully removed or contained."

Readings

Carlton, J. T. 1993. Exotic Invaders. *The World & I* 8 (12):220–25.

Carlton, J. T., and J. B. Geller. 1993. Ecological Roulette: The Global Transport of Nonindigenous Marine Organisms. *Science* 261 (5117):78–82.

Cohen, A., and J. Carlton. 1995. Accelerating Invasion Rate in a Highly Invaded Estuary. *Science* 279 (5350):555–57.

Hedgpeth, J. W. 1993. Foreign Invaders. *Science* 261 (5117):34–35.

Meinesz, A. 1999. *Killer Algae, The True Tale of a Biological Invasion.* University of Chicago Press.

National Oceanic and Atmospheric Administration. 1994. *Nonindigenous Estuarine and Marine Organisms (NEMO).* United States Department of Commerce. (Proceedings of conference and workshop.)

Travis, J. 1993. Invader Threatens Black, Azov Seas. *Science* 262 (5138):366–67.

Internet Resources

Spartina Lab at Bodega Marine Laboratory
    www-bml.ucdavis.edu/spartina/home.html

Nonindigenous Aquatic Species
    nas.er.usgs.gov/

National Biological Information Infrastructure
    www.nbii.gov/

San Francisco Bay Program—USGS
    h2o.usgs.gov/public/wid/html/sfb.html

Phytoplankton from Ships' Ballast Waters
    www.maritimes.dfo.ca/science/mesd/he/ballast.html

Introduction of Nonindigenous Species to Chesapeake Bay via Ballast
    nas.er.usgs.gov/ballast.html

Figure 3

Colonies of *Spartina* proliferate over the mudflats of Willapa Bay, Washington.

chapter eleven

Life in the Water

School of barracuda in the Coral Sea near Australia.

Learning Objectives

After reading this chapter, you should be able to

- Separate marine organisms into three categories and give examples of each type.

- Distinguish between phytoplankton and zooplankton and appreciate the diversity of planktonic organisms.

- Explain the phenomenon known as "red tide" and its consequences.

- Understand the role of bacteria in the sea.

- Explain how plankton are sampled at sea.

- Recognize and appreciate the diversity of animals making up the nekton.

- Understand how sharks and rays differ from other fish and how deep-sea bony fish differ from commercial fish species.

- Describe the two categories of whales and discuss the current situation of whale populations.

- Explain the effect of the Marine Mammal Protection Act in U.S. waters.

- Discuss the present threat to sea turtle populations.

- Explain why populations of sea birds are in decline.

- Relate increased world fish catches to the decline of specific fisheries.

- Explain present trends in world fish catches.

- Discuss the role of fish farming as a means of ocean harvest.

ife in the water ranges from microscopic, single-celled organisms to the largest fish and the greatest whales. This chapter discusses the plankton, the wanderers and drifters of the oceans, existing in great swarms, and moving with the currents. Sharing the vast oceans above the sea floor are the nekton; these are the free swimmers of the oceans, moving through the water independent of the motion of currents and waves. The nekton include marine mammals, seagoing reptiles, and squid, but most of the nekton are fish. The plankton include the unicellular plantlike organisms at the beginning of the ocean food chains, and the nekton provide much of the commercial harvest taken from higher levels in the ocean food web. Information on world fish catches is included with quantities taken, methods used, and current population status. Red tides and toxic blooms due to increases in populations of marine microorganisms are also discussed in this chapter.

Classification of Organisms 11.1

A wide variety of organisms inhabit the environments of the oceans. Scientists classify all organisms, placing them in taxonomic groups to promote identification and to increase understanding of the relationships that exist among them. The most widely used classification system today places organisms in primary categories beginning with the kingdoms Monera (bacteria, single-celled organisms without nuclei), Protista (single-celled organisms with nuclei), Fungi, Plantae, and Animalia. See table 11.1 for the classification of some marine organisms and Appendix B for a more detailed classification of marine organisms using this system.

A new system based on recent biochemical and genetic research proposes three domains above the kingdom level. The Monera are placed in the Bacteria and Archaea domains, and all organisms with nuclei are placed in the Eukarya domain; see Appendix B, figure 1.

A much-used and practical method divides all marine organisms into three groups, based on how and where they live. Organisms that float or drift with the movements of the water are the **plankton.** Plantlike plankton are called **phytoplankton** and animal plankton are known as **zooplankton.** Organisms that live attached to or on the sea floor are the **benthos,** and the animals that swim freely and purposefully in the sea are the **nekton.**

Name and define the three commonly used categories of marine organisms.

The Plankton 11.2

Although many plankton have a limited ability to move toward and away from the sea surface, they make no purposeful motion against the ocean's currents and are carried from place to place suspended in the seawater. Some plankton are quite large; jellyfish may be the size of a large washtub, trailing 15-m (50-ft) tentacles. But the phytoplankton and many zooplankton must be observed under a microscope. Bacteria and the smallest phytoplankton cells may be less than 0.005 mm in diameter; these are collected using special filtering systems. The zooplankton and phytoplankton between 0.07 and 1 millimeter are known as "net plankton," because they are often captured in tow nets made of very fine mesh nylon.

How do plankton move from place to place?

What is the size range of planktonic organisms?

Phytoplankton 11.3

The phytoplankton (fig. 11.1) are mainly unicellular (single-celled) plantlike organisms known as **algae.** Each phytoplankton cell is an independent, photosynthesizing individual, and even in species in which the cells attach together in long chains or other aggregations, there is no division of labor between the cells. There is only one large, multicellular planktonic alga, the seaweed *Sargassum* that is found floating in the area of the North Atlantic known as the Sargasso Sea. *Sargassum* reproduces vegetatively by

table 11.1   Taxonomic Categories of Some Marine Organisms

	Killer Whale	Northern Fur Seal	Pacific (Japanese) Oyster	Giant Octopus	Sea Lettuce	Giant Kelp
Kingdom	Animalia	Animalia	Animalia	Animalia	Plantae	Plantae
Phylum	Chordata	Chordata	Mollusca	Mollusca	Chlorophyta	Phaeophyta
Class	Mammalia	Mammalia	Bivalvia (Pelecypoda)	Cephalopoda	Chlorophycae	Phaeophycae
Order	Cetacea	Carnivora (Pinnipedia)	Anisomyaria	Octopoda	Ulvales	Laminariales
Family	Delphinidae	Otariidae	Ostreidae	Octopodidae	Ulvaceae	Lessoniaceae
Genus	Orcinus	Callorhinus	Crassostrea	Octopus	Ulva	Macrocystis
Species	Orcinus orca	Callorhinus ursinus	Crassostrea gigas	Octopus dofleini	Ulva lactuca	Macrocystis pyrifera

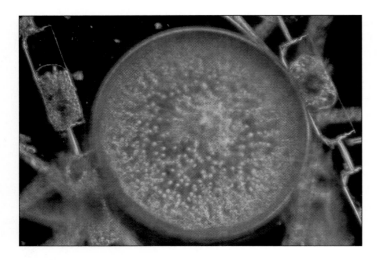

Figure 11.1

Phytoplankton. The large diatom, *Coscinodiscus*, between strands of *Ditylum*, another diatom. (×100)

fragmentation to form large mats, which provide shelter and food for a wide variety of organisms, including specialized fish and crabs found nowhere else.

Diatoms and **dinoflagellates** are two major groups of organisms belonging to the phytoplankton. Diatoms are sometimes called golden algae, because their characteristic yellow-brown pigment masks their green chlorophyll. Some diatoms have radial symmetry; they are round and shaped like pillboxes, while others have bilateral symmetry and are elongate. The round diatoms float better than the elongate forms; therefore the elongate diatoms are often found on the shallow sea floor or attached to floating objects, while the round diatoms are more truly planktonic. All are found in areas of cold, nutrient-rich water. Some common diatoms are shown in figure 11.2.

A hard, rigid, transparent cell wall impregnated with silica surrounds each diatom. Pores connect the living portion of the cell inside to its outside environment (fig. 11.3). The buoyancy of the diatoms is increased by the low density of the interior of the cells and the production of oil as a storage product.

Figure 11.2

Diatoms exist as single cells or in chains.

More Information:
www.ucmp.berkeley.edu/help/taxaform.html

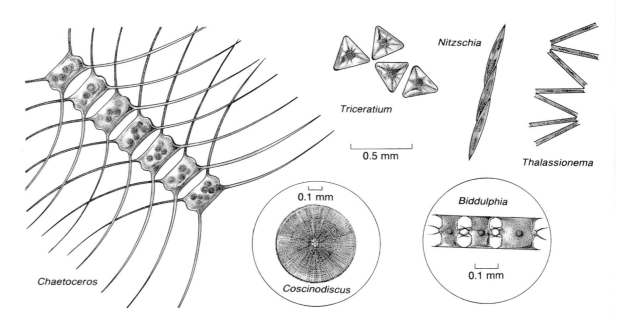

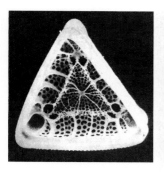
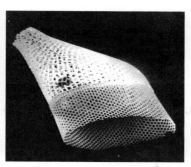
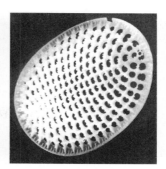
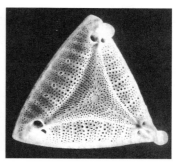

Figure 11.3

Stereoscan micrographs of diatom frustules show the pores through which the interior living cell is connected to its outside environment.

More Information: www.bgsu.edu/departments/biology/algae/index.html

Their small volume and accompanying relatively large surface area help these cells to stay afloat. Their large surface area also provides them with greater exposure to sunlight and water containing the gases and nutrients for photosynthesis and growth. In addition, some diatoms have spines or other projections that increase their ability to float.

Diatoms reproduce very rapidly by cell division; when their populations discolor the water, it is known as a **bloom.** Diatoms are most important as primary producers. Those that are not consumed by herbivores eventually die and sink to the ocean floor. In shallow areas, the cells reach the sea floor with some organic matter still locked inside, and they may, over a very long period of time, form oil deposits. The cell walls of diatoms that sink to greater depths build up siliceous sediments, the siliceous ooze discussed in chapter 4.

Coccolithophores are relatives of the diatoms; they are photosynthetic with outer calcareous plates that form sediments on the sea floor. During the late summer–early fall of 1997 the water over most of the continental shelf in the eastern Bering Sea was colored by a massive bloom of coccolitho-

phores that was clearly visible from space (fig. 11.4). A bloom of this type had never before been recorded in this region and was attributed to unusual climatic conditions, reduced cloud cover, and few storms resulting in abundant sunlight and above-average surface-water temperature.

Dinoflagellates are red to green in color and can exist at lower light levels than diatoms, because they can both photosynthesize like a plant and ingest organic material like an animal. Their external walls do not contain silica; some are smooth and flexible, but others are armored with plates of cellulose. Dinoflagellates usually have two whiplike appendages or **flagella** that beat within grooves in the cell wall, giving the cells limited motility. Under favorable conditions, they multiply even more rapidly than diatoms to form blooms, but they are not as important as the diatoms as a primary ocean food source. Some dinoflagellates are called fire algae, because they glow with bioluminescence at night. Representative dinoflagellates are shown in figure 11.5.

Figure 11.4

A true color image showing the coccolithophore bloom over the continental shelf of the eastern Bering Sea in September 1997.

Image provided by the SeaWiFs Project.

What is *Sargassum* and where is it found?

Distinguish between diatoms and dinoflagellates.

What allows diatoms to stay in the photic zone?

What is a bloom?

Why are some dinoflagellates called fire algae?

Red Tides and Toxic Blooms 11.4

Certain species of dinoflagellates produce **red tides** (fig. 11.6). These blooms may or may not be poisonous to fish and other organisms, and may or may not produce symptoms of paralytic shellfish poisoning (PSP), neurotoxic shellfish poisoning (NSP), or diarrhetic shellfish poisoning (DSP) in humans. In North American waters several different dinoflagellates produce red tides: *Gonyaulax, Alexandrium,* and *Gymnodinium* are

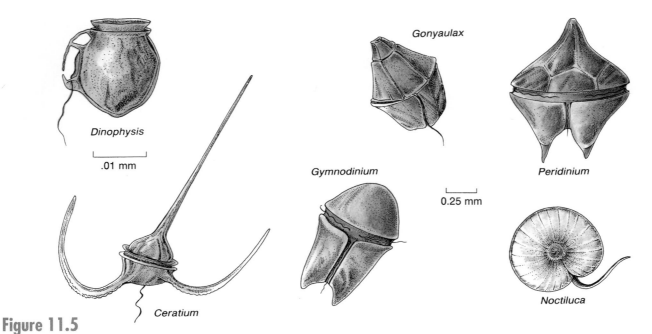

Figure 11.5

Dinoflagellates. *Noctiluca, Gymnodinium,* and *Gonyaulax* produce red tides. *Noctiluca* is a bioluminescent, nontoxic dinoflagellate. *Gymnodinium* and *Gonyaulax* produce toxic red tides and paralytic shellfish poisoning.

More Information: www.geo.ucalgary.ca/~macrae/palynology/dinoflagellates/dinoflagellates.html

Figure 11.6

This nontoxic red tide bloom of the dinoflagellate *Noctiluca* occurred in Puget Sound, Washington in 1996. The bloom, with its characteristic tomato soup color, extended for six miles along the shore and persisted for a week.

**More Information:
habserv1.whoi.edu/hab/nationplan/ECOHAB/ECOHABhtml.htm**

toxic; *Noctiluca* is not (figs. 11.5 and 11.6). Toxic species are generally not poisonous to shellfish feeding on a bloom, but the toxins are concentrated in the tissues of the shellfish and produce PSP, NSP, or DSP in humans eating the affected shellfish. *Gymnodinium* and a species of *Gonyaulax* kill fish in the Gulf of Mexico; Gonyaulax also kills shrimp and crab.

The toxins that produce PSP are powerful nerve poisons that can cause paralysis and death if the breathing centers are affected. One species of *Gymnodinium* can produce at least five different toxins, and the saxitoxins, which are found in butter clams *(Saxidomus)* but are produced by *Gonyaulax,* are fifty times more lethal than strychnine. The toxin is not affected by heat, so cooking the shellfish does not neutralize the poison. Even after the visible signs of red water due to a dinoflagellate bloom have disappeared, the shellfish can retain the toxin in their tissues for long periods, and so the beaches are kept closed to shellfish harvesting. NSP symptoms are similar to basic food poisoning; unlike PSP, few cases have been reported, and no known deaths have occurred. Only recently have researchers linked DSP outbreaks to the presence of a dino-flagellate; DSP outbreaks may have been occurring throughout history and been attributed to bacterial contamination. Scientists believe that the toxins are a defense against predators.

No one knows precisely what triggers the sudden blooms of these organisms, but it appears that red tides and toxic blooms are increasing in frequency and severity around the world (fig. 11.7). The blooms often happen in spring or summer after heavy rains have produced land runoff from sewage and agriculture that is high in nitrates and phosphates. Extra nutrients mean extra growth and this may explain why many of the world's coastal areas are experiencing such massive outbreaks. The dumping of ballast water by commercial vessels and the transfer of commercial shellfish stocks are considered an additional factor in the increasing number and severity of red tide episodes (see "Biological Invaders," end of chapter 10). High salinities are thought by some to trigger blooms in the

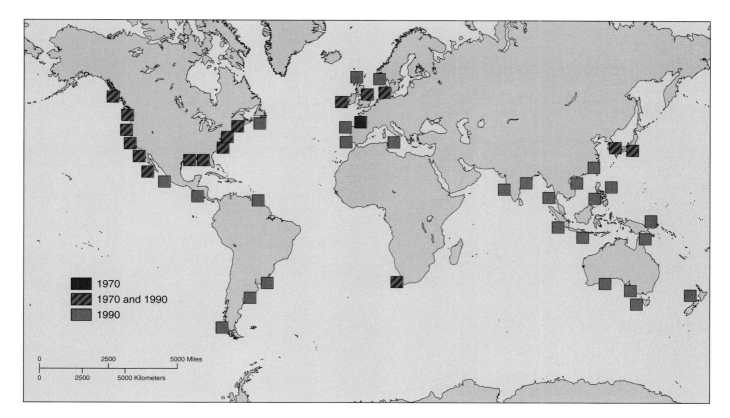

Figure 11.7

Global distribution of paralytic shellfish poisoning events observed in 1970 and in 1990. Eighteen events were recorded in 1970 versus forty-four in 1990.

Source: Data from D. M. Anderson, Scientific American, *Vo. 271, #2, August 1994.*

More Information: www.mdsg.umd.edu/seagrantmediacenter/news/whoi.html

Gulf of Mexico; along the New England coast, temperature and light may play major roles in activating blooms. Another suggestion is that cysts or thick-walled resting cells formed by the dinoflagellates settle to the bottom and may be brought to the surface by upwelling water. Cysts in the shallow waters of bays and harbors may be activated by sudden disturbance of the bottom; in deeper coastal water, the cysts appear to have an annual cycle.

The dinoflagellate *Pfiesteria piscicida,* popularly known as "the cell from hell," was first identified in North Carolina in 1993. Since then *P. piscicida* and similar organisms have caused major fish kills in estuaries and coastal waters of the mid-Atlantic and southeastern United States. These organisms have extraordinarily complex life histories (as many as twenty-four transformations in form) that are controlled by the availability of fish secretions, blood, and tissue. *P. piscicida* has been linked to human symptoms including sores, memory loss, nausea, respiratory distress, and vertigo.

Just as not all red tides are toxic, not all red water is caused by dinoflagellates. The Red Sea received its name because of dense blooms of nontoxic photosynthetic bacteria with large amounts of red pigment. The Indian Ocean has red tides due to a toxic photosynthetic bacterium, not a dinoflagellate.

An estimated 10,000 to 50,000 people eating fish in tropical regions of the world are affected by ciguatera poisoning each year. There is no way to prepare affected fish to make it safe to eat. Symptoms of ciguatera poisoning are extremely variable and may occur in various combinations; they include headache, nausea, vomiting, abdominal cramps, reduced blood pressure, and in severe cases, convulsions, muscular paralysis, hallucinations, and death. Several dinoflagellates are associated with ciguatera, but *Gambierdiscus toxicus* is most often the cause. Researchers in the Gilbert Islands found that dinoflagellates are eaten by herbivores, and the ciguatoxins then move through the food web in a cycle that appears to take at least eight years. Ciguatera is an international problem; it has delayed development of fisheries in the Red Sea, Sri Lanka, New Guinea, and Puerto Rico; the loss to the Floridean/Caribbean/Hawaiian seafood industry is estimated at $10 million yearly.

Shellfish contaminated with domoic acid, traced to a bloom of the diatom *Pseudonitzschia pungens,* caused three deaths in eastern Canada in 1987. Prior to that time diatoms were not known to produce toxins. Shellfish and crab fisheries along the entire U.S. West Coast were closed in 1991 due to blooms from another species of this diatom. Domoic acid poisoning in humans is known as amnesic shellfish poisoning (ASP) and may cause short-term memory loss as well as nausea, disorientation, and muscle weakness.

How are red tides, dinoflagellates, and paralytic shell-fish poisoning related?

What environmental conditions are thought to trigger red tides?

Why are shellfish unsafe to eat during and after red tide episodes?

Distinguish between ciguatera and domoic acid poisoning.

Where have *Pfiesteria* outbreaks occurred?

Zooplankton   11.5

The zooplankton are either herbivores, grazing on the phytoplankton, or carnivores feeding on other members of the zooplankton. Many of the zooplankton have some ability to swim and can even dart rapidly over short distances in pursuit of prey or to flee predators. They may move vertically in the water column, but, like the phytoplankton, they are also transported by the currents. A sample of zooplankton is shown in figure 11.8.

The life histories of zooplankton types are varied and show different strategies for survival in a world where reproduction rates are high and life spans are short. These animals may produce three to five generations a year in warm water,

Figure 11.8

This zooplankton sample shows an octopus larva (upper right), an arrowworm (center), a euphausiid (lower left), and a copepod (lower right). ×3

where food supplies are abundant and higher temperatures accelerate life processes. At high latitudes, where the season for phytoplankton growth is brief, the zooplankton may produce only a single generation in a year. Rapid growth rates and short life spans of the carnivorous zooplankton are responsible for the rapid liberation of nutrients to be recycled by the phytoplankton.

Zooplankton exist in patches of high population density between areas that are much less heavily populated. High population patches attract predators, and the more sparse populations between the denser patches preserve the stock. Turbulence and eddies disperse individuals from the densely populated patches to the intervening sparse areas. Convergence zones and boundaries between water types concentrate zooplankton populations, which attract predators.

Some zooplankton migrate toward the sea surface each night and return to the depths each day, either in an attempt to maintain their light level or in response to the movement of their food resource. The amount of daily migration varies between 10 m (30 ft) and 500 m (1500 ft). Accumulations of these organisms reflect sound waves from depth sounders and are seen on a depth recording as a false bottom or **deep scattering layer,** known as the DSL.

Foraminiferans and **radiolarians** are single-celled microscopic members of the zooplankton; both are shown in figure 11.9. Foraminiferans are encased in compartmented calcareous coverings, while the radiolarians are surrounded by silica walls. The radiolarian coverings are ornately sculptured and covered with delicate spines. Protrusions or pseudopodia (false feet), many with skeletal elements, radiate from the cell and catch diatoms and small protozoa. Both foraminiferans and radiolarians are found in the warmer regions of the oceans. After death, their coverings accumulate on the ocean floor contributing to the ocean sediments, as discussed in chapter 4.

Among the most common and widespread zooplankton types worldwide are the small **crustaceans** (shrimplike animals): **copepods** and **euphausiids** (figs. 11.8 and 11.10). These animals are basically herbivorous and consume more than half their body weight daily. Copepods are smaller than euphausiids; euphausiids move more slowly and live longer than copepods. Both reproduce more slowly than the diatoms, doubling their populations only three to four times a year. They may make up more than 60% of the zooplankton in any of the oceans and are an important food source for small fish. There are more copepods than any other kind of zooplankton in the oceans; copepods are a major link between phytoplankton (producers) and first level carnivores (consumers). In the Arctic and Antarctic, the euphausiids are the **krill,** occurring in such quantities that they provide the main food for the baleen whales (see section 11.11; also information on the Antarctic food web in section 10.7). The first commercial krill harvests were made by the former USSR in the 1960s, and in later years, Japan, Korea, Poland, and Chile participated. The largest harvests (529,000 metric tons) were taken in the Antarctic summer of 1980–81. In 1996 with only Japan, Poland, and Ukraine participating, the catch was 102,000 metric tons, dropping further in 1997 to

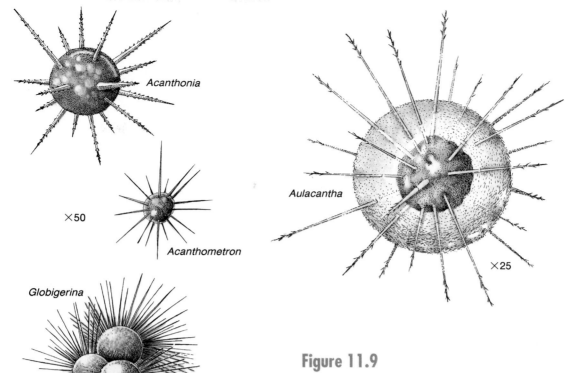

Figure 11.9

Selected members of the radiolaria (*Acanthonia, Acanthometron,* and *Aulacantha*) and a foraminifera (*Globigerina*).

More Information:
www.ucmp.berkeley.edu/help/taxaform.html

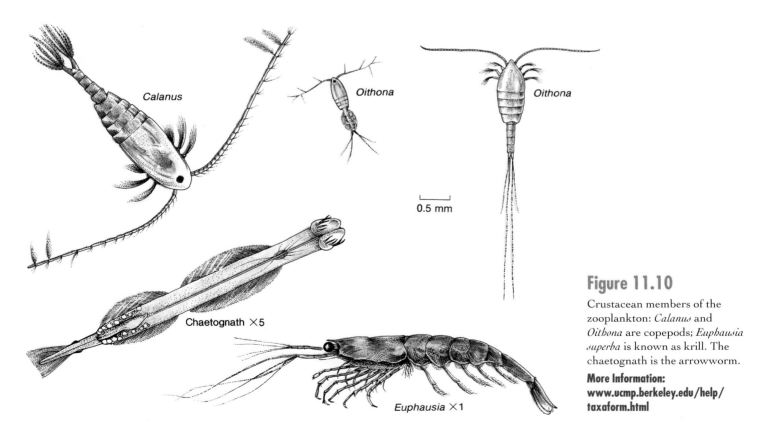

Figure 11.10

Crustacean members of the zooplankton: *Calanus* and *Oithona* are copepods; *Euphausia superba* is known as krill. The chaetognath is the arrowworm.

More Information:
www.ucmp.berkeley.edu/help/taxaform.html

85,000 metric tons. Krill is used as livestock and poultry feed in eastern Europe and as fish feed by the Japanese. Marketing krill for human consumption has not been very successful. Because the krill deteriorate rapidly, krill used for human consumption must be processed within three hours, limiting the daily harvest. Also, the distance to the fishing ground is long and the costs of the vessels and fuel are high.

Arrowworms (figs. 11.8 and 11.10) are common, carnivorous members of the zooplankton. They are abundant in ocean waters from the surface to the great depths. Arrowworms are 2 to 3 cm (1 in) long, nearly transparent, and feed on other zooplankton. There are several species found in seawater, and in some cases a particular species is limited to a certain type of water.

Other zooplankton include swimming snails, comb jellies, and true jellyfish. The swimming snails are **pteropods,** modified **mollusks** that may or may not have a small shell, but they all have a transparent, gracefully undulating "wing" (fig. 11.11). Some pteropods are herbivores, and some are carnivores. Comb jellies or **ctenophores** (fig. 11.12) float in the surface waters. Some have trailing tentacles; all are propelled slowly by eight rows of beating **cilia** or hairlike cell extensions. The round sea gooseberries or sea walnuts are small, but the beautiful, tropical Venus' girdle may be 30 cm (12 in) or more in length. All ctenophores are carnivores, feeding on other zooplankton.

True jellyfish (fig. 11.13) resemble comb jellies, but they come from another unrelated group of animals, the Cnidaria or **coelenterates.** Some jellyfish spend their entire lives as drifters; others are planktonic for only a portion of their lives, as they eventually settle and change to a bottom-dwelling form similar to a sea anemone. Another group of unusual jellyfish include the Portuguese man-of-war, *Physalia,* and the small by-the-wind-sailor, *Velella* (fig. 11.13). Both the Portuguese man-of-war and the by-the-wind-sailor are **colonial** animals; they are collections of individual but specialized organisms, some of which have the task of gathering food, reproducing, or protecting the colony with stinging cells, while some form the float.

All of the zooplankton discussed to this point, with the exception of certain jellyfish, spend their entire lives as plankton and are called **holoplankton.** However, an important portion of the zooplankton spends only part of its life as plankton; these are members of the **meroplankton.** The eggs and **larvae,** or juvenile forms, of oysters, clams, barnacles, crabs, worms, snails, starfish, fish, and many other organisms are a part of the zooplankton for a few weeks. The currents carry these larvae to new locations, where they find food sources and areas to settle. In this way, areas where a species may have died out are repopulated, and overcrowding in the home area is made less likely. Sea animals produce larvae in enormous numbers, and these meroplankton are an important food source for the zooplankton and other animals. The parent animals may produce millions of spawn, but only small numbers of males and females need survive to adulthood in order to guarantee survival of the stock.

Larvae often look very unlike the adult forms into which they will develop (fig. 11.14). Early scientists who found and described these larvae gave each a name, thinking they had discovered a new type of animal. We keep some of these names today, referring, for example, to the trochophore larvae of worms, the veliger larvae of sea snails, the zoea larvae of crabs, and the nauplius larvae of barnacles. Other members of the meroplankton include fish eggs and juvenile fish. Some large

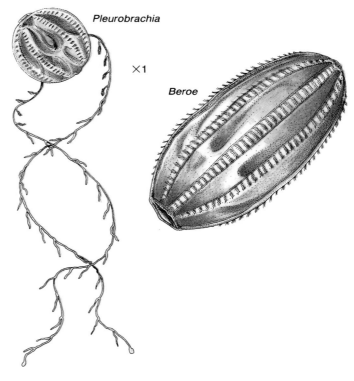

Pleurobrachia

×1

Beroe

Figure 11.12

The comb jellies (ctenophores) *Pleurobrachia* and *Beroe. Pleurobrachia* is often called a sea gooseberry.

More Information: www.ucmp.berkeley.edu/help/taxaform.html

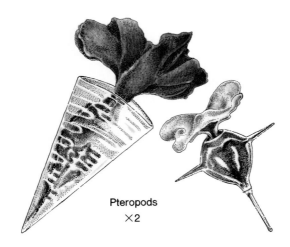

Pteropods
×2

Figure 11.11

The pteropods are planktonic snails.

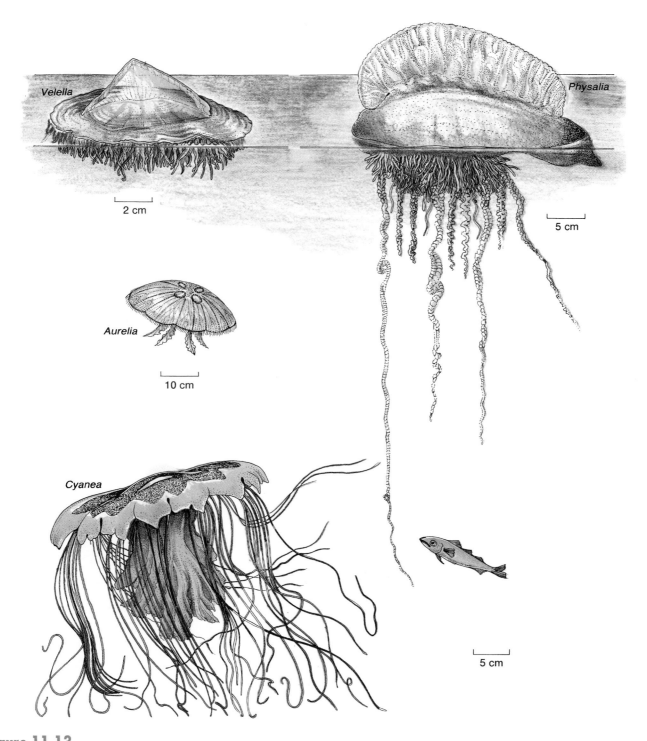

Velella

Physalia

2 cm

5 cm

Aurelia

10 cm

Cyanea

5 cm

Figure 11.13

Jellyfish belong to the Coelenterata or Cnidaria. *Velella,* the by-the-wind-sailor, and *Physalia,* the Portuguese man-of-war, are colonial forms.

More Information: www.ucmp.berkeley.edu/help/taxaform.html

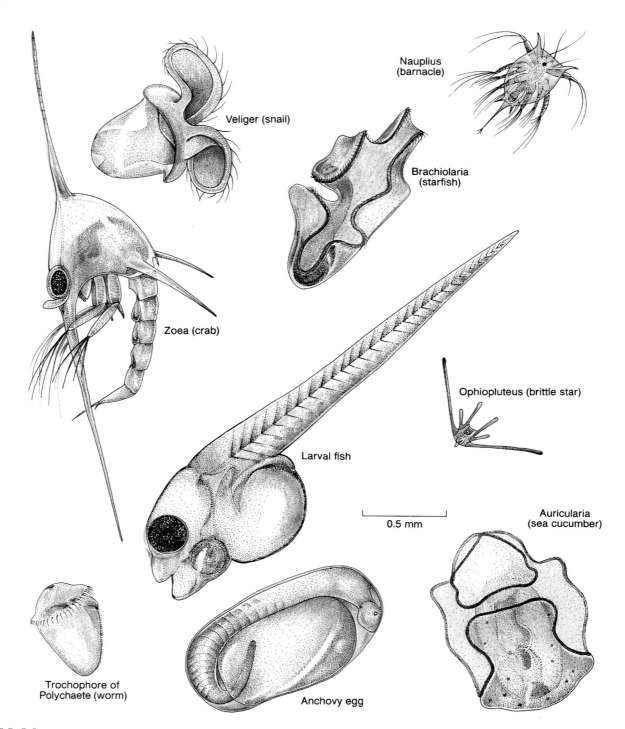

Veliger (snail)

Nauplius
(barnacle)

Brachiolaria
(starfish)

Zoea (crab)

Ophiopluteus (brittle star)

Larval fish

0.5 mm

Auricularia
(sea cucumber)

Trochophore of
Polychaete (worm)

Anchovy egg

Figure 11.14

Members of the meroplankton. All are larval forms of nonplanktonic adults.

seaweeds release reproductive cells that drift in the plankton until they are consumed or settle out to grow attached to the sea floor.

Where would you expect to find large quantities of zooplankton?

What causes the "deep scattering layer"?

Give an example of a single-celled member of the zooplankton.

What kinds of animals form the majority of the zooplankton?

Why are krill important in ocean food webs?

Describe several larger members of the zooplankton.

What is a colonial animal? Give an example.

Distinguish between holoplankton and meroplankton.

Bacteria   11.6

Bacteria are the smallest living organisms; they are microscopic, single cells and many have the ability to reproduce by cell division every few minutes when they inhabit a favorable environment. Bacteria are the most numerous organisms in the oceans; 1×10^{29} bacterial cells are estimated to inhabit marine environments. Bacteria live free in the seawater and exist on every available surface within it, for example decaying material, the surfaces of organisms, the sea floor, and pieces of floating wood. They produce dense blooms in estuaries and warm-water regions; certain bacteria are associated with the hydrothermal vents of the deep sea floor, where they are the primary producers for the animal communities surrounding the vents (refer to section 10.10).

There are about 100 million bacteria in every liter of saltwater at all depths and at all latitudes. Bacteria play an important role in the decay and breakdown of organic matter, returning it to the sea as basic chemicals and compounds to be used again by new generations of plants and animals (refer to section 10.5). Bacteria also serve as a major protein source. A film of bacteria is found on minute particles of floating organic material; these are an ideal food for planktonic larvae and a variety of single-celled protozoans (single-celled animals), which in turn serve as food for tiny worms, clams, and crustaceans.

Where are the roles of bacteria in the sea?

Where are marine bacteria found?

Sampling the Plankton   11.7

Traditionally, plankton are sampled by towing conical nets of fine mesh material through the sea behind a vessel or by dropping a net straight down the side of a stationary vessel and pulling it straight up like a bucket (fig. 11.15). After the net is returned to the deck, it is rinsed carefully and the "catch" is collected. To measure the volume of water that has flowed through the net, a flow meter is placed in the mouth of the net. The speed at which the net is towed is important; the tow must be rapid enough to catch the organisms but slow enough to let the water pass through the net. Fine nets are used for sampling phytoplankton while coarser mesh nets are reserved for the larger zooplankton. Plankton may also be sampled by using a water bottle or a submersible pump to collect water that is then filtered to remove the plankton.

Today's oceanographers sample zooplankton with multiple-net systems mounted on a single frame; the nets are opened and closed on command from the ship. The frame also carries electronic sensors that relay data on salinity, temperature, water flow, light level, net depth, and the angle of the tow to the ship's computer.

Whatever the sampling method, the number and kinds of plankton must be determined. A subsample of the catch is inspected directly under the microscope. If the amount of

Figure 11.15

After a plankton tow the net is rinsed to wash the plankton down into the sample cup at the narrow end of the net.

plankton is of greater interest than the kinds of organisms, an electronic particle counter may be used to find the total count of organisms, or a subsample may be dried and weighed. The biomass of phytoplankton can also be determined by dissolving out the chlorophyll from a sample and measuring the pigment's concentration. New optical instruments measure the natural fluorescent signal coming from chlorophyll pigment in phytoplankton cells. These measurements are made directly and instantaneously at the desired depth and then related to phytoplankton abundance. Instruments carried by satellites can also measure phytoplankton populations near the sea surface; see figures 10.12 and 1.19. In order to learn how planktonic organisms behave, researchers at the Woods Hole Oceanographic Institution use a strobe light with four video cameras at four different magnifications to photograph the plankton. Eventually they hope to program the system to recognize different kinds of plankton, allowing the researchers to acquire species and population information.

> **What different methods are used to sample plankton?**

Table 11.2 is a summary of the plankton, their distribution, and their characteristics.

The Nekton   11.8

Approximately 5000 species of nekton swim freely through the pelagic and neritic regions of the oceans. The only **invertebrate** animals (animals without backbones) among these are the squid and a few species of shrimp. The other members of the nekton are **vertebrates** (animals with backbones); these are the fish, the reptiles, and the mammals.

Squid (fig. 11.16) are elusive, abundant, and until recently, little known inhabitants at all ocean depths. The present use of video cameras in submersibles and seagoing robots is providing researchers with a record of rarely-seen deep-water squids, many never before seen alive. They swim rapidly, and their wide range of bioluminescence and coloration allows them to change color and disappear swiftly. They range from a few centimeters in length to the giant squid *(Architeuthis)* that may be 20 m (65 ft) long. Scientists know little about these huge creatures but suspect that there may be large numbers of them in upper aphotic zone.

The fish dominate the nekton. Fish are found at all depths and in all the oceans, but their distribution is determined directly or indirectly by their dependence on primary producers. Fish are concentrated in upwelling areas, shallow coastal areas, and estuaries. The surface waters support much greater populations per unit of water volume than the deeper zones, where food resources are sparser.

Fish come in a wide variety of shapes related to their environment and behavior. Some are streamlined, designed to move rapidly through the water (tuna and mackerel); others are flattened for life on the sea floor (sole and halibut); while still others are elongate for living in soft sediments and under rocks (some eels). Fins provide the push or thrust for locomotion and are found in a variety of shapes and sizes. Fish use their fins to change direction, turn, balance, and brake. A sample of the variety found among marine fish is discussed in the next two sections.

> **What invertebrate animals are part of the nekton?**

table 11.2	Plankton Summary	
Type	**Distribution**	**Characteristics**
Plantlike Organisms		
Phytoplankton		
Diatoms	Cool, nutrient-rich surface waters, inshore and offshore.	Single photosynthetic cells or chains of cells with silica covering, radial or bilateral symmetry.
Dinoflagellates	All surface and near-surface waters, inshore and offshore.	Single cells, two flagella, cellulose covering. Some photosynthetic; some ingest organics. Bioluminescence common; some toxic. Red tides formed by some.
Animals		
Zooplankton		
Holoplankton (permanent plankton)	Worldwide polar to tropic, surface and deeper waters, inshore and offshore.	Many varieties, large range in size, migrate vertically, vary from single cells to complex multicellular and colonial forms. Herbivores and carnivores. Copepods and krill especially important.
Meroplankton (part-time plankton)	Worldwide polar to tropic, surface and deeper waters, inshore and offshore.	Developmental stages (eggs and larvae) of nonplanktonic animals. Important to distribution and food production.

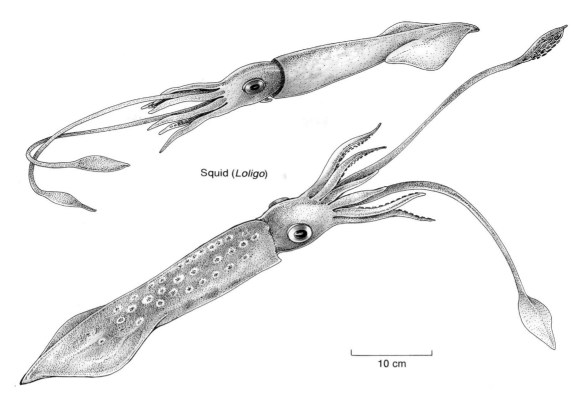

Squid (*Loligo*)

10 cm

Figure 11.16

The squid is a swimming mollusk considered to be a member of the nekton.
More Information: is.dal.ca/~ceph/TCP/index.html

Sharks and Rays   11.9

Sharks differ from other fish by their skeletons of cartilage rather than bone and by their toothlike scales. Shark scales have a covering of dentine similar to vertebrate teeth and are extremely abrasive; sharkskin has been used as a sandpaper and as a polishing material. The shark's teeth are modified scales; they are replaced rapidly if they are lost, and they occur in as many as seven overlapping rows. Sharks are acutely aware of their environment through good eyesight and excellent senses of smell, hearing, mechanical reception, and electrical sense. They are most sensitive to chemicals associated with their feeding and are able to detect them in amounts as dilute as one part per billion. Sharks can detect the movement of water from currents or from injured or distressed animals; they use their electroreception sense to locate prey and recognize food. As a shark swims through the earth's magnetic field, an electric field is produced that varies with direction, giving the shark its own compass.

There are more than 300 known species of sharks widely spread through the oceans; some are shown in figure 11.17. The whale shark is the world's largest fish, reaching lengths of more than 15 m (50 ft). This graceful and passive animal feeds on plankton and is harmless to other fish and mammals. The docile basking shark, 5 to 12 m (15–40 ft) long, is another plankton feeder.

Many sharks are swift and active predators, attacking quickly and efficiently using their rows of sharp serrated teeth. They also play an important role as scavengers by eliminating the diseased and aged animals. Sharks can and do attack humans, although the reasons for these attacks and the frenzied feeding sometimes observed in groups of sharks are not understood. A human swimming inefficiently at the surface may look like a struggling, ailing animal and be attacked; a diver swimming completely submerged may appear as a more natural part of the environment and be ignored.

Skates and rays, also shown in figure 11.17, are flattened, sharklike fish that live near the sea floor. They move by undulating their large side fins, which gives them the appearance of flying through the water. The large manta rays are plankton feeders, but most rays and skates are carnivorous, preferring crustaceans, mollusks, and other benthic organisms. Their tails are usually thin and whiplike, and that of the stingray carries a poisonous barb at the base. Some of the skates and a few of the rays have shock-producing electric organs; they are along the sides of the tail in skates and on the wings of the rays. Their purpose appears to be mainly defensive. Like the sharks, most rays bear their young live. Skates enclose the fertilized eggs in a leathery capsule, called a mermaid's purse or a sea purse, that is ejected into the sea and from which the fully formed young emerge in a few months (fig. 11.17).

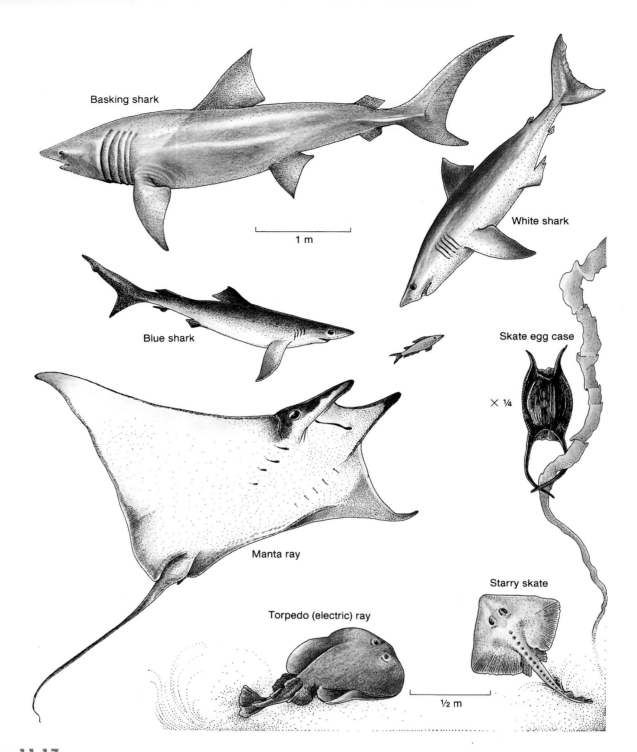

Basking shark

White shark

Blue shark

1 m

Skate egg case

× ¼

Manta ray

Starry skate

Torpedo (electric) ray

½ m

Figure 11.17

Representative members of the cartilaginous fish. The leathery egg case, or mermaid's purse, is that of a skate.

More Information: www.flmnh.ufl.edu/fish/sharks/sharks.htm

How do sharks and rays differ from other fish?

At what trophic level do the largest sharks feed?

What roles do sharks play in the ocean food web?

How does a skate or a ray differ from a shark?

Bony Fish 11.10

Most commercially valuable fish are found between the ocean's surface layers and about 200 m (600 ft) deep, and most of these fish are streamlined, active, predatory, and capable of high-speed, long-distance travel. Among the most important species fished commercially are the enormously abundant small herring-type fish, such as sardine, anchovy, menhaden, and herring. These fish feed directly on plankton and are found in large schools in areas of high primary production.

Fish schools may consist of a few fish in a small area or thousands of fish covering several square kilometers. Usually the fish are all of the same species and similar in size. Fish schools have no definite leaders, and the fish change position continually. Most schooling fish have wide-angled eyes and the ability to sense changes in water displacement, allowing them to keep their relationship to one another constant as the school moves or changes direction. Schooling probably developed as a means of protection, since each fish in the school has less chance of being eaten than it does alone. The school may also keep reproductive members of the population together.

Other valuable fish that are harvested commercially include mackerel, pompano, swordfish, and tuna. These fish are caught at sea, out of sight of land. Fish that live on or near the bottom do not swim as rapidly as those that swim freely in the water. The flounder, halibut, turbot, and sole are commercially valuable bottom fish. Perch and snapper tend to congregate along the sea floor in the shallower, nearshore areas. They are often called rock fish because they hide among the rocks and live in cracks of reefs and rocky areas.

Representative bony fish that are fished commercially are shown in figure 11.18. The commercial exploitation and economic significance of food fish are discussed in section 11.14.

The fish of the deep sea are not well known, for the depths at which they live are difficult and expensive to sample. A variety of deep-sea types is shown in figure 11.19.

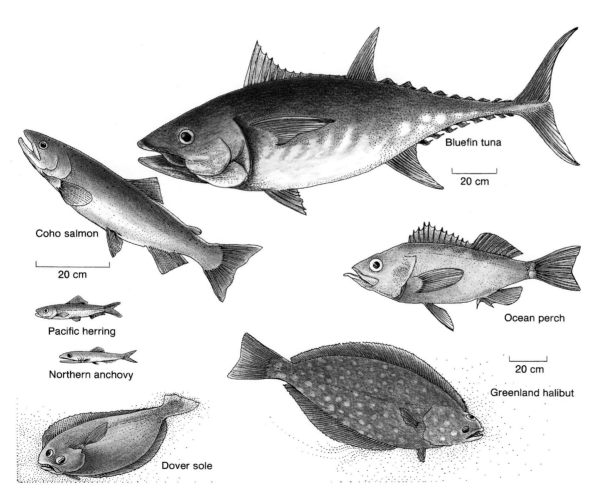

Figure 11.18

Commercially harvested members of the bony fish.

More Information:
www.fao.org/default.htm

Bluefin tuna

20 cm

Coho salmon

20 cm

Pacific herring

Northern anchovy

Dover sole

Ocean perch

20 cm

Greenland halibut

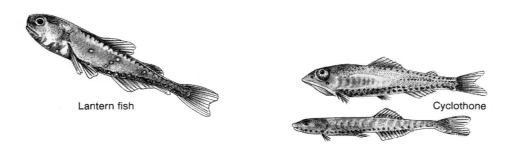

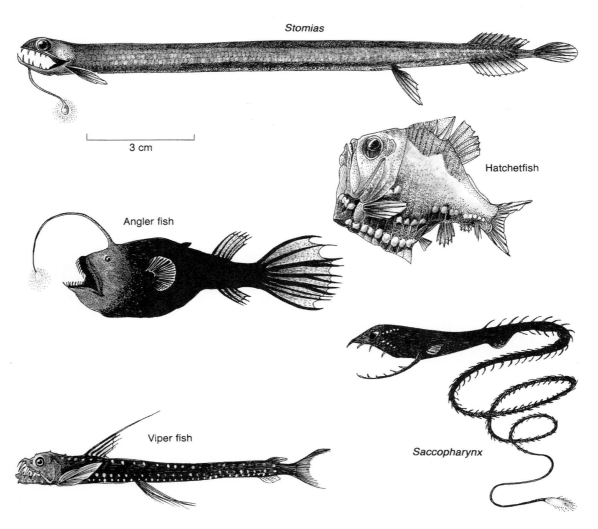

Lantern fish

Cyclothone

Stomias

3 cm

Hatchetfish

Angler fish

Saccopharynx

Viper fish

Figure 11.19

Fish from the deep sea.

In the dim, transitional layer between 200 and 1000 m (600–3000 ft), the waters support vast schools of small luminous fish. The many species of *Cyclothone* are believed to be the most common fish in the sea. Deeper-living species are black, while the shallower-living species are silvery, to blend with the dim light. The lantern fish has a worldwide distribution; some 200 species are distinguished by the pattern of light organs along their sides. They are a major item in the diet of tuna, squid, and porpoises. *Stomias*, a fish with a huge mouth, long, pointed teeth, and light organs along its sides, is also found here, as is the large-eyed hatchetfish. These fish prey on the great clouds of euphausiids and copepods found at this depth.

In the perpetually dark zones are predators with highly specialized equipment for catching their prey. These fish are equipped with light-producing organs used as lures, large teeth that in some species fold backward toward the gullet so their prey cannot escape, and gaping mouths with jaws that unhinge to accommodate large fish. Among the most famous of these predators is *Saccopharynx* with its funnel-like throat and tapering body. When the stomach is empty the fish appears slender, but it expands to accept anything the great mouth can swallow. The female angler fish has a dorsal fin modified into a fishing rod with an illuminated lure dangling just above her jaws.

Although fierce and monstrous in appearance, most of these deep-water fish are small, between 2 and 30 centimeters (1–12 in) in length. They breathe slowly, and the tissues of their small bodies have a high water and low protein content. These fish go for long periods between feedings and use their food for energy rather than for increased tissue production.

> At what depths are most commercial fish found?
>
> How do fish benefit from schooling?
>
> List several characteristics of deep-water fish.

Mammals   11.11

Marine mammals are warm-blooded air breathers. They may spend all of their lives at sea, or they may return to land to mate and give birth. In either case, the young are born live and are nursed by their mothers. Included in the group are large and small whales (including porpoises and dolphins), seals, sea lions, walruses, sea otters, and sea cows. See figure 11.20 and table 11.3 for representative whales.

table 11.3   **Principal Characteristics of the Great Whales**

	Distribution	Breeding Grounds	Average Weight (tons)	Greatest Length (m)	Food
Toothed Whales					
Sperm	Worldwide; breeding herds in tropic and temperate regions	Oceanic	35	18	Squid, fish
Baleen Whales					
Blue	Worldwide; large north-south migrations	Oceanic	84	30	Krill
Finback	Worldwide; large north-south migrations	Oceanic	50	25	Krill and other plankton, fish
Humpback	Worldwide; large north-south migrations along coasts	Coastal	33	15	Krill, fish
Right	Worldwide; cool temperate	Coastal	(50)	17	Copepods and other plankton
Sei	Worldwide; large north-south migrations	Oceanic	17	15	Copepods and other plankton, fish
Gray	North Pacific; large north-south migrations along coasts	Coastal	20	12	Benthic invertebrates
Bowhead	Arctic; close to edge of ice	Unknown	(50)	18	Krill
Bryde's	Worldwide; tropic and warm temperate regions	Oceanic	17	15	Krill
Minke	Worldwide; north-south migrations	Oceanic	10	9	Krill

Based on K. R. Allen, Conservation and Management of Whales, *Washington Sea Grant Program, 1980. Reprinted by permission.*
Key: ( ) = estimate.

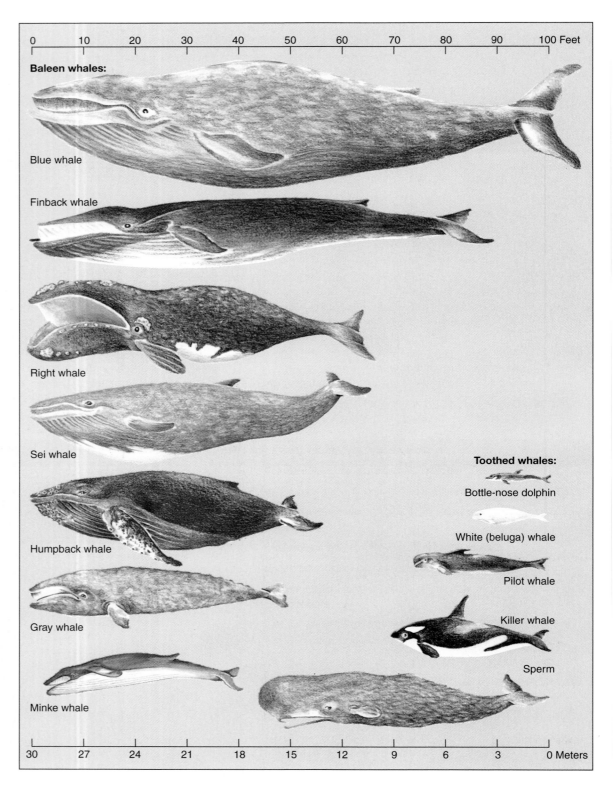

Figure 11.20

Relative sizes of baleen and toothed whales.

More Information:
www.webmedia.com.au/
whales/whales3.html

Whales belong to the mammal group called **cetaceans.** Some cetaceans are toothed, pursuing and catching their prey with their teeth and jaws (for example, the killer whale, the sperm whale, and the small whales known as dolphins and porpoises); others have mouths fitted with strainers of **baleen,** or whalebone. Baleen whales gulp water and plankton, then expel the water through the baleen, leaving the krill behind.

The mouths of toothed and baleen whales are compared in figure 11.21. The blue, finback, right, sei, gray, and humpback whales are baleen whales. The gray whale feeds mainly on bottom crustaceans and worms.

The "great whales" that have been the focus of the whaling industry are the blue, sperm, humpback, finback, sei, and right whales. The earliest known European whaling was by the

(a)   (b)

Figure 11.21

(a) The killer whale *(Orcinus orca)* is a toothed whale. (b) Bowhead whale *(Balaena mysticetus)* baleen. This dead whale has been hauled onto the ice and is shown lying on its back.

Norse between A.D. 800 and 1000 and continued locally, for the next 400 years. In the 1500s, the Basque people of France and Spain crossed the Atlantic to Labrador to hunt whales for their oil, and by 1600 whaling had become a major commercial activity among the Dutch and the British. At about the same time the Japanese began harvesting whales, and in the 1700s and 1800s, whalers from the United States, Great Britain, the Scandinavian countries, and other northern Europeans pursued them far from shore. By the end of the nineteenth century, whales were being hunted with explosive harpoons and high-speed hunting vessels that brought the dead whales to factory ships for conversion to oil. These methods continued into the twentieth century greatly increasing the efficiency of the hunt and rapidly depleting whale stocks.

In the 1930s, the blue whale harvest reduced the population to less than 4% of its original numbers, threatening the species with extinction. In 1946, representatives from whaling nations established the International Whaling Commission (IWC) to set catch quotas and self-regulate the industry. The annual harvest of whales between 1910 and 1980 is given in figure 11.22, and world population estimates are given in table 11.4.

In order to study whale populations and assess their ability to recover, an IWC moratorium on commercial harvesting of whales (except dolphin and porpoise), began in 1985–86. To assess stocks, Japan has practiced "scientific" or "research" whaling, harvesting 330 minke whales each year under IWC direction. Norway chose to set its own national catch limits and resumed whaling in 1993 (289 whales). In 1994 the IWC voted, by a margin of 23 to 1 (Japan) and with several abstentions, to establish a whale sanctuary in Antarctic waters below 55°S. In 2000 Japan expanded its "scientific" hunt; six vessels began hunting 160 minke whales, 50 Bryde's whales, and 10 sperm whales. This is the first time Japan has hunted Bryde's or sperm whales since 1987. Iceland left the IWC in 1999 with plans to resume whaling in 2001.

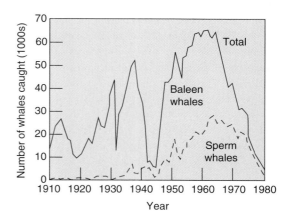

Figure 11.22

Total annual catches of baleen and sperm whales in all oceans from 1910 to 1980. A moratorium on commercial whaling began in 1985–86.

table 11.4   World Population Estimates of Great Whales

Species	Initial[1]	Present[2]	Status[1]
Right whale	100,000	8600	Severely depleted
Blue whale	>196,000	10,320	Severely depleted
Finback whale	>464,000	118,320	Severely depleted
Bowhead whale	>55,000	8000	Severely depleted
Humpback whale	>120,000	39,000	Severely depleted
Sei whale	>105,000	27,640	Depleted
Sperm whale	277,000	234,246	At or above optimum sustainable population
Gray whale	<20,000	26,900	Recovered
Minke whale	900,000	900,000	Sustained

[1]From Endangered Whales: Status Update, 1994. *National Marine Mammal Laboratory, National Marine Fisheries Service, National Oceanic and Atmospheric Administration.*

[2]Perry, S. L., D. P. DeMaster, G. K. Silber. 1999. The Great Whales: History and Status of Six Species Listed as Endangered Under the U.S. Endangered Species Act of 1973. *Marine Fisheries Review, Special Issue* 61 (1):1–74.

The IWC permits whaling by native peoples (in Alaska, Denmark [Greenland], and the former Soviet Union) who depend on limited whaling for subsistence, cultural, and nutritional needs. In 1997 the Native American Makah tribe of Washington State, citing an 1895 treaty with the United States, received permission to harvest five gray whales from the annual limit of 140 gray whales harvested in the North Pacific for traditional aboriginal subsistence needs. One whale was harvested by the Makahs in 1999.

Representative **pinnipeds** ("feather-footed"), named for their four characteristic swimming flippers, are shown in figure 11.23. These animals still retain ties to land, spending considerable time ashore on rocky beaches, ice floes, or in shore caves. They are found from the tropics to the polar seas, ranging from the nearly extinct monk seal of the western Hawaiian Islands and Mediterranean area to the fur seals of the Arctic. The common harbor seal, the harp seal of the northwest Atlantic, the leopard seal of the Antarctic, and the elephant seal, the male with its great pendulous snout, are all true seals, or seals without external ears and with torpedo-shaped bodies that require them to use a wriggling motion on land. The northern fur seal and the sea lion are eared seals with longer necks and supple forelimbs tipped with broad flippers, on which the animal can walk and hold its body in a partially erect position when on land.

Seals and sea lions are currently enjoying a period of relative peace, compared to the sealing days of the nineteenth and early twentieth century, but other problems still confront them. The northern fur seal's 1867 population was 2.5 to 3 million; heavy hunting reduced it to 200,000 to 300,000 by 1910. Although the population recovered from these excessive harvests and the present population is approximately 1.1 million, there has been a long-term downward trend since the mid-1950s. This has been blamed on entanglement with nets, lines, plastic strapping lines and other debris, but more recent studies suggest that the decline is related to a reduced prey base caused by the large commercial fish catches in the region.

Manatees and **dugongs** or **sea cows** are members of the sirenia (fig. 11.24). These slow-moving animals are the world's only herbivorous marine mammals. Manatees are found in the brackish coastal bays and waterways of the warm southern Atlantic coasts and in the Caribbean. Dugongs are found in waters around Southeast Asia, Africa, and Australia. At present, the growth of human populations is putting increasing hunting pressure on the dugong in the southern Pacific. Manatees in the coastal waters of the Caribbean and tropical Atlantic are frequently injured and killed by collisions with the propellers of large and small vessels. In this way 67 manateees were killed in Florida in 1998 and 82 in 1999. Periodically the Florida coastal habitat produces a toxic red tide that produces a neurotoxin *(Gymnodinium)*; during 1996 at least 158 manatees died from its effects. Levels of this neurotoxin were fifty to one hundred times normal in the tissues of dead manatees. Also destruction of manatee habitat in Florida continues to accelerate as salt marshes, sea grass beds, and mangrove areas are drained, reclaimed, and otherwise destroyed.

In 1972 the U.S. Congress established the Marine Mammal Protection Act. The act includes a ban on the taking or importing of any marine mammals or marine mammal product. "Taking" is defined by the act as harvesting, hunting, capturing, or killing any marine mammal or attempting to do so. The act covers all U.S. territorial waters and fishery zones. The act effectively removed these animals and their products from commercial trade in the United States. Only under strict permit procedures and with the approval of the Marine Mammal Commission can a few individual marine mammals be caught for scientific research and public display. Since passage of the act, the numbers of some species have increased, and the inter-

Marine Birds   11.13

About 3% of the Earth's 8600 species of birds are considered marine species. Marine birds may spend their lives almost continuously at sea or their association with the sea may be only periodic. Some are so well adapted to oceanic life that they rarely come ashore; others move daily into coastal waters to feed, but all return to shore to nest.

The wandering albatross of the southern oceans is the most truly oceanic marine bird, spending four to five years at sea before returning to nesting sites ashore. Wilson's petrel, the smallest of the oceanic birds breeds in Antarctica and flies 16,000 km (10,000 mi) to Labrador during the Southern Hemisphere winter. Penguins do not fly but swim, in underwater flocks, at almost 16 kilometers per hour (10 mph). Other large fishing birds are pelicans and cormorants that fish mostly in coastal areas. Gulls and terns (fig. 11.26) are common all over the world; both are strong flyers. Puffins, murres, and auks are limited to the North Atlantic, North Pacific, and Arctic; these are heavy-bodied, short-winged, and short-legged diving birds.

Shorebirds are usually migratory and arrive on intertidal sands and mud flats in very large numbers (thousands and hundreds of thousands) at certain times of the year (fig. 11.27). Food is not depleted because the different kinds of birds arrive at different times of the year. Also different birds select different shore lands, and when the shorebirds feed, each bird exploits a different food resource related to the length of its neck, bill, and legs. Herons and egrets wade into the water; sandpipers probe the sand and mud; oystercatchers feed in the rocky intertidal zone, and gulls and terns swim and dive in the estuaries.

All over the world increasing numbers of people choose to live or vacation along coasts and beaches. Wetlands are diked and filled, removing food sources and nesting sites. Vacation homes, harbors, and resorts are built along once-lonely beaches, and runoffs from town, industry and agriculture contaminate estuaries, while accidental spills and industrial dumping pollute inshore waters. The intensity of fishing has increased in all the oceans, and there is less fish for both humans and birds. Most recreational uses of shore areas are in the late spring and early summer and this is also the time that most marine birds look for isolation to nest and rear their young.

Natural disasters such as landslides disturb nesting areas and events such as El Niño (chapter 6) depress the great bird populations. However, these disruptions are usually cyclical; a population decrease in one year builds up again over the next series of unaffected years. Human changes are generally more permanent, and the flocks of migratory seabirds that were described by naturalists of the seventeenth, eighteenth, and nineteenth centuries may never be seen again.

> Compare the wandering albatross to a heron—how do they differ and how are they the same?
>
> Why are marine bird populations decreasing?

Commercial Fisheries   11.14

World Fish Catches

In 1950, the total world marine fish catch was approximately 21 million metric tons. During the next forty-seven years, as human populations exploded, the fishing effort by all nations intensified, and the technology and gear used to hunt and catch the fish improved dramatically. The UN Food and Agriculture Organization (FAO) lists 76 million metric tons caught in 1985 and 86 million metric tons in 1996 and 1997.

Figure 11.26

Crested terns face the wind on the beach of Heron Island, a part of the Great Barrier Reef, Australia.

Figure 11.27

A mixed flock of dunlins and sanderlings takes flight along the water's edge, Long Beach, Washington.

Fisheries and Overfishing

Around the world too many fishing boats are taking too many fish, too fast. Diminishing fish populations are the victims of relentless overfishing and management that fails to acknowledge declining stocks. The 1997 report from FAO lists 44% of world fish stocks as being fully exploited, 16% as overfished, 6% as depleted, and 3% recovering slightly. According to the U.S. Office of Fisheries Conservation Management, 41% of the species in U.S. waters are overfished.

Canada closed its two hundred-year-old Newfoundland cod fishery in 1992, resulting in the displacement of many fisheries workers. In 1993 the National Marine Fisheries Service (NMFS) closed large parts of the U.S. cod fishery, and Iceland warned its fishing industry that their cod stock could collapse unless the annual catch was cut back by 40%. North Atlantic swordfish landings in the United States have declined 70% from 1980 to 1990, and the average weight of the swordfish has fallen from 115 to 60 pounds. The western Atlantic adult bluefin tuna population dropped 80% between 1970 and 1993. The Gulf of Mexico population has dropped 70% since 1975 and the Mediterranean population of bluefin has declined 50%. Many believe this fish is doomed for it is the world's most valuable fish, selling in Japan at $200/kg (nearly $91 per pound). These data mean trouble for people who depend on fish as their source of protein as well as for the hundred million people that make their living by catching, processing, or selling fish.

Today, as overfishing takes its toll of familiar species, the pressure is placed on other species. Until recently, sharks were fished for sport and for local markets. The Asian demand for shark fin soup supports an expanding shark fishery that amputates the shark fins and throws the carcasses back into the ocean to die, a process known as "finning," but all sharks are now fished heavily around the world. The 1990 world catch was 690,000 metric tons; the 1997 catch had climbed to 790,000 metric tons. Annual U.S. sales of shark steaks and fillets have increased from 3000 metric tons in 1985 to an average of 6000 metric tons per year in the 1990s.

Overfishing appears to be decimating shark populations in all the world's oceans. Since 1970 there has been been a 40% drop in shark caught in the northwestern Atlantic and a 50% drop in the southeast Atlantic. When U.S. commercial shark fishing began in the late 1970s, 21 metric tons of thresher sharks were brought to California docks; by 1982 the thresher catch was 1100 metric tons, and in 1989 the fishery collapsed when the catch fell to 300 metric tons. The National Marine Fisheries Service (fig. 11.28) has worked out a management proposal for U.S. shark fishing that reduces commercial fishing by 30% and bans finning. Population problems are intensified by sharks' low reproductive rate, their slow growth, and the long period they require for maturation. Concern over the loss of the ocean's most widespread predator mounts; no one knows how many sharks are in the world's oceans and no one knows how their loss will affect the ocean food web.

Overfishing is sometimes combined with environmental degradation or with environmental change that intensifies its

Figure 11.28

National Marine Fisheries Service scientists sample shark populations to evaluate commercial fisheries impacts.

More Information: www.nmfs.noaa.gov/

effects. The anchovy fishery concentrated in the upwelling zone off the coast of Peru has produced the world's greatest fish catches for any single species. Anchovies are small, fast-growing fish that feed directly on the phytoplankton in the upwelling and travel in dense schools that are easy to net in large quantities. These fish are converted to fish meal that is exported as feed for domestic animals.

The anchovy fishery began in 1950 with a harvest of 7000 metric tons; by 1970 about 12.3 million metric tons were taken. As the fishing increased the average size of the fish taken decreased and it took more and more fish (smaller and younger) to make up the same catch. After 1970 the Peruvian coast was visited three times (1972, 1976, 1982–83) by El Niño (refer to chapter 6); see figure 11.29. Between 1985 and 1992 catches along the west coast of South America stabilized between 3 and 5 million metric tons, rose to 8 million in 1993, and 12.5 million metric tons in 1994. Catches dropped in 1995 (8.6 million metric tons), rose slightly in 1996 (8.8 million metric tons), and El Niño struck again in 1997–98. The impact of this El Niño reduced the area's catch to 7.7 million metric tons. In this fishery overfishing combined with environmental changes intensifies the effect of overharvesting.

The wild salmon runs of the Pacific Northwest and Alaska are becoming smaller, and the fishing is now tightly controlled. The fishing season is shortened and the catches decrease, causing profits for those who fish to fall; management of fish stocks requires more and more regulation in attempts to ensure a returning population for the next year. Returning stocks of coho, chinook, and sockeye salmon along the coasts of British Columbia, Washington, and Oregon have diminished greatly in the last decade. Because fish that spawn in the rivers under one governmental jurisdiction are caught as they move

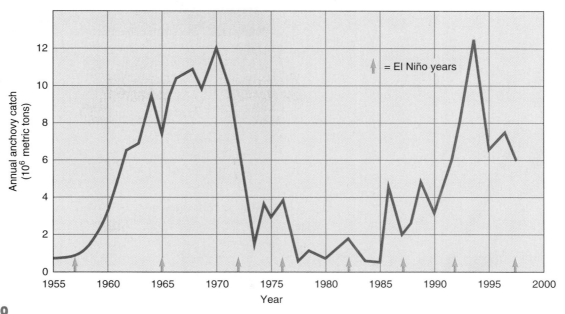

Figure 11.29

The anchovy catch of Peru and Ecuador by calendar year, given in millions of metric tons. Years of El Niño are 1957, 1965, 1972, 1976, 1982–83, 1992–93, and 1997–98.

More Information: seawifs.gsfc.nasa.gov/OCEAN_PLANET/HTML/peril_fishing.html

through the coastal waters of other governmental jurisdictions, much friction occurs among those negotiating the length of fishing periods and the number of fish to be caught. At the present time there appears to be more concern for the "right to fish" than there is with assuring the survival of the fishery.

The degradation of the freshwater environment to which the wild fish must return for spawning adds to their difficulties. Salmon need high-quality fresh water and clean, gravel-bottomed, shady, cool streams to spawn. But rivers are routinely dammed for power and flood control, cutting off salmon from their home streams. Trees are harvested to the stream edges, removing shade and allowing mud and silt eroded from the exposed land to cover the streambed. Wastes in the stream degrade the water quality and damage the juvenile fish and the returning runs. Although these activities are being corrected in many areas, the populations of naturally spawning fish continue to decline. However, variations in the return of combined wild and hatchery salmon that mature in one, two, or three years can still produce high fishing yields. For example, the 1991 Alaska pink salmon harvest set a record of 175 million fish, surpassing the 1990 record of 155 million fish.

Even with increased management and regulation of fish stocks, the general trend of ocean fishing is down and the costs are up. One way to increase the fish harvest is by developing new fisheries such as the North Pacific bottom fishery and the South African pilchard fishery, and another is to foster an increased consumer demand for new fish and fish products. For example, Alaskan pollock, a bottom fish, is processed to remove the fats and oils that give the fish its flavor, and a highly refined fish protein called **surimi** is processed and flavored to

form artificial crab, shrimp, and scallops. Surimi is a major fish product in Japan and has a growing market in the United States.

In all cases the fishing boat and gear, the crew's wages, and the fuel required are increasing the cost of commercial fishing (fig. 11.30). In the United States, those who fish are usually independent operators who sell their catch directly to the processor. The U.S. consumer prefers fish fillets and fish steaks, not low-cost minced fish products; therefore, the fishing is for high-cost fish such as salmon, swordfish, and halibut.

Incidental Catch

In every fishery large numbers of unwanted fish are caught incidentally while fishing for other species. These fish are referred to as **incidental catch** or **by-catch** and represent a tremendous waste of marine resources. The UN Food and Agricultural Organization estimates that each year 27 million tons of fish, about 25% of all reported commercial marine landings, are caught as by-catch and discarded. The world's shrimp fishery is estimated to have an annual catch of 1.8 million tons; the associated discarded by-catch is 9.5 million tons. Each year, worldwide, ocean longline fisheries report an incidental catch of 40,000 sea turtles with an estimated 42% mortality, and shrimp trawlers catch more than 45,000 sea turtles with an estimated mortality rate of 12,000. Alaskan trawlers for pollock and cod throw back to the sea some 25 million pounds of halibut, worth about $30 million, as well as salmon and king crab because they are prohibited from keeping or selling this by-catch. Another 550 million pounds of groundfish are discarded

Figure 11.30

A salmon purse seiner hauling in its net.

Fish type	Percent Discard of Landed Weight	Landed Weight (millions of metric tons)
Shrimp and prawns	520	1.83
Crabs	249	1.12
Flounder, halibut, sole	75	1.26
Redfish, bass, conger eel	63	5.74
Lobster, spiny-rock lobster	55	0.21

table 11.6 Global Incidental Catch

Data from D. M. Alverson et al., "A Global Assessment of Fisheries Bycatch and Discards," 1994. FAO Technical Paper No. 339.

by fishers in Alaskan waters to save space for larger or more valuable fish. See table 11.6 for global incidental catch rates for the most severely affected fisheries.

The discard rate of incidental catch varies from place to place; if the incidental catch does not bring a high enough price, and if processors or a market are not available, these "trash fish" will be returned to the sea, usually dead. In the Gulf of Mexico 1 pound of shrimp results in 10.3 pounds of by-catch, which is nearly all discarded, but in Southeast Asia and other areas with local fisheries and fresh fish markets, much of the by-catch is used.

Fish Farming

An alternative way to increase the fish harvest is by fish farming. Farming the water, known as **mariculture** or **aquaculture,** began in China some 4000 years ago. In China, Southeast Asia,

and Japan, fish farming in fresh and salt water has continued to the present as a practical and productive method of raising large quantities of fish. The methods are labor intensive, and most fish farms are small, family-run operations. In **monoculture,** fish are raised as a single species; in **polyculture,** surface feeders and bottom feeders are raised together, making use of the total volume of the pond.

Fish farming in the United States produces only 2% of our fishery products, but that amount includes 50% of the catfish and nearly all of the trout. To succeed in the United States, fish farming requires a proven market and a large-scale operation to keep costs down. To be economically productive, a farmed species must have juvenile forms that survive well under controlled conditions, gain weight rapidly, eat a cheap and available food, and have adults that fetch a high market price.

Aquaculture of marine species has nearly tripled its output in the last decade. In 1987 marine aquaculture provided 4 million metric tons of fish and shellfish; in 1996 the harvest was 11 million metric tons. Farming of Atlantic and Pacific salmon raises more than 650,000 metric tons of fish each year. The world's largest salmon producers are Norway (46%), Chile (22%), Scotland (13%), and Canada (7%). One-third of the salmon consumed worldwide now come from salmon pens. Nearly 2 million metric tons of wild fish are required by the farmed salmon's diet of 45% fish meal and 25% fish oil, thus increasing this pressure on world fisheries rather than decreasing it.

Marine fish are grown in floating fish cages or pens (fig. 11.31) kept in shallow coastal areas, which are becoming less and less available as shorelines are increasingly built over and used for recreation. There are also problems with the impact of fish pens on water quality (increasing nutrients, decreasing oxygen, releasing antibiotics into the marine environment), and possible transmission of disease to wild populations. These concerns are pushing the development of offshore submersible systems that will distribute waste and excess food over larger areas with better circulation of seawater.

Aquaculture research is focusing on several other fish species for marketing by the year 2010. In Hawaii pen-reared

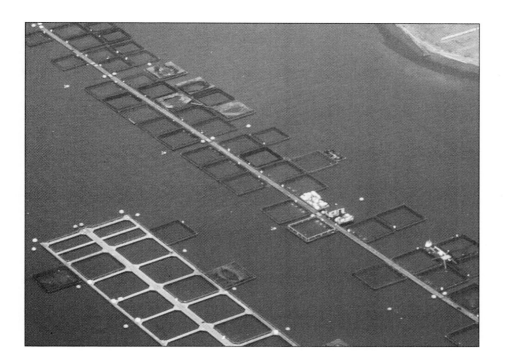

Figure 11.31

An aerial view of salmon pens in which fish are raised to market size by Sea Farm Washington, Inc.

More Information:
www.fao.org/waicent/faoinfo/fishery/publ/sofia/sofiaef.htm

mahi mahi now grow to 1.5 kg (3 lbs) size in 150 days, and are expected to be the first to enter the market. Norway has the technology to grow Atlantic cod, but the cost is still too high for the commercial market. The United States and Canada are experimenting with the Atlantic halibut, and in Texas the craze for the cajun dish, blackened redfish, has opened the way to redfish farming.

How has the world fish catch changed in the last half of this century?

What has happened to the ocean's fish stocks during the same period?

What factors influence the Peruvian anchovy harvest?

What is incidental catch and what is its effect on the environment?

Compare the positive and negative aspects of fish farming.

Summary

Marine organisms are grouped as plankton, nekton, and benthos. Phytoplankton are plantlike and zooplankton are animals.

The plankton are the drifting organisms of the oceans. Phytoplankton are single cells or chains of cells. *Sargassum* is the only large planktonic seaweed. Diatoms are round or elongate golden-brown algae with silica coverings. They reproduce rapidly by cell division and make up the first trophic level of the open ocean.

Heavy blooms of some dinoflagellates produce red tides; some are toxic and others are not. Dinoflagellate toxin is concentrated in shellfish and produces several types of poisoning in humans. Red tides are apparently triggered by certain combinations of environmental factors and the disturbance of dormant dinoflagellates. Domoic acid, produced by diatoms, is a newly discovered toxin. Ciguatoxin is another dinoflagellate toxin that affects humans and hampers fishery development.

Concentrations of zooplankton are found at convergence zones and along density boundaries. Some zooplankton migrate toward the sea surface at night and away from it during the day.

The calcareous-shelled foraminiferans and the silica-shelled radiolarians are single cells. The copepods and euphausiids are the most abundant zooplankton. Euphausiids, also known as krill, are the basic food of the baleen whales. Other small members of the plankton are the carnivorous arrowworms and the swimming snails or pteropods. Large zooplankton include the nearly transparent comb jellies and jellyfish.

Zooplankton that spend their entire lives in the plankton are holoplankton. Meroplankton are juvenile (or larval) stages of nonplanktonic adults including fish eggs, very young fish, and the larvae of barnacles, snails, crabs, and starfish. Marine bacteria are also planktonic and exist on any available surface.

Plankton are sampled with nets and pumps. Various techniques are used to determine the kinds and numbers of organisms.

Nekton swim freely. Squid are fast-swimming invertebrate nekton, but fish dominate the nekton. Sharks, rays, and skates are fish with cartilaginous skeletons. All other fish have bony skeletons, including the commercially fished species and the highly specialized types of the deeper sea.

Marine mammals include whales, seals, sea lions, and sea cows. The whalers hunted the great baleen whales and the toothed sperm whale. As hunting techniques changed, harvesting became more intensive, and some species of whales have been threatened with extinction. The International Whaling Commission regulates whaling on a voluntary basis; its moratorium on commercial whaling continues. Japan takes a research harvest of whales under the IWC; Norway and Iceland set their own catch limits. The IWC has established a whale sanctuary in Antarctica.

Pinnipeds are widespread; fur seals were hunted heavily during the nineteenth and early twentieth centuries but are now protected, and their populations are increasing. Manatees and dugongs are found in the warm waters of the Indian and Atlantic Oceans. The Marine Mammal Protection Act protects all marine mammals in U.S. waters and prohibits U.S. commercial trade in marine mammal products.

Sea snakes, the marine iguana, marine crocodiles, and sea turtles are reptile members of the nekton. Sea turtle populations are under great pressure from hunters and poachers.

World fish catches increased dramatically between 1950 and the 1980s due to increased fishing efforts and improved technology. Catches declined in the early 1990s but soon reached new highs. Overfishing has been relentless. Atlantic cod fisheries and North Pacific coastal salmon fisheries have been cut back and some closed in the 1990s. The Peruvian anchovy fishery nearly collapsed with overfishing and El Niño events. World shark fisheries are threatened with collapse due to its recent explosive growth. Declining fish catches, increased regulation, incidental catch, and increased costs are worldwide problems.

Aquaculture of marine species has tripled in ten years. It has proved to be practical and productive in Asia, but in the United States, it is a small industry that requires large-scale operations, proven markets, and cost-cutting technology. Salmon farming has increased, especially in Norway, Chile, and Canada. Offshore systems are under development and new species are being farmed.

Key Terms

algae, 263	dinoflagellate, 264
aquaculture, 290	dugong, 282
baleen, 280	euphausiid, 268
benthos, 263	flagella, 265
bloom, 265	foraminifera, 268
by-catch, 289	holoplankton, 270
cetaceans, 280	incidental catch, 289
cilia, 270	invertebrate, 274
coccolithophore, 265	krill, 268
coelenterate, 270	larvae, 270
colonial, 270	manatee, 282
copepod, 268	mariculture, 290
crustacean, 268	meroplankton, 270
ctenophore, 270	mollusk, 270
deep scattering layer (DSL), 268	monoculture, 290
diatom, 264	nekton, 263

phytoplankton, 263	red tide, 265
pinniped, 282	sea cow, 282
plankton, 263	surimi, 289
polyculture, 290	vertebrate, 274
pteropod, 270	zooplankton, 263
radiolarian, 268	

Critical Thinking

1. Why is a juvenile planktonic stage important to a nonplanktonic adult?

2. Episodes of toxic phytoplankton blooms appear to be increasing along the world's coasts. What are some possible reasons for this increase?

3. Some marine mammal populations have increased in U.S. waters since the passage of the Marine Mammal Protein Act. These mammals now compete with humans for food resources of economic value. Should any adjustments be made to the Marine Mammal Protection Act? Give reasons for your answer.

4. The world's marine fish catch has increased dramatically in the last twenty-five years, but at the same time, the catch from many large fisheries has declined. How do you explain this contradictory situation?

5. Why are bottom fish (rather than fish in the water column) more likely to show the first effects of a degraded nearshore environment?

Suggested Readings

Levinton, J. S. 1995. *Marine Biology: Function, Biodiversity, Ecology.* Oxford University Press, New York.

Plankton
Coniff, R. 2000. Jellyfish. *National Geographic* 197(6):82–101.
Mistry, R. 1992. Lilliputian World of Plankton. *Sea Frontiers* 38(1):42–47.
Nelson, K., and C. Arneson. 1985. Marine Bioluminescence: About to See the Light. *Oceanus* 28(3):13–18.
Stevens, J. 1995. The Secret Lives of Krill. *Sea Frontiers* 41(2):26–31.
Wrobel, D. J. 1990. Transient Jewels. *Sea Frontiers* 40(3):14–17, 60.

Red Tides
Anderson, D. M. 1994. Red Tides. *Scientific American* 271(2):62–68.
Baden, D. G. 1990. Toxic Fish: Why They Make Us Sick. *Sea Frontiers* 36(3):8–14. (Ciguatera poisoning)
Cherfas, J. 1990. The Fringe of the Ocean—Under Siege from Land. *Science* 248(4952):163–65. (Red tides and toxic blooms)
Culotta, E. 1992. Red Menace in the World's Oceans. *Science* 257(5076):1476–77.
Epstein, P., T. Ford, and R. Colwell. 1992. Marine Ecosystems. *The Lancet* 342(1881):1216–19. (Algal toxins and human health)
The Ecology and Oceanography of Harmful Algal Blooms. 1997. *Limnology and Oceanography* 42(5 part 2):1009–1305. (A collection of papers)

Nekton—Squid
Conniff, R. 1996. In Search of the "White Ghost." *Smithsonian* 27(2):126–37.
Roper, C. F. E., and K. J. Boss, 1982. The Giant Squid. *Scientific American* 246(4):96–105.

Nekton — Fish

Alverson, D. 1998. *Discarding Practices and Unobserved Fishing Mortality in Marine Fisheries: An Update.* Washington Sea Grant, University of Washington.

Conniff, R. 1993. Disappearing Shadows in the Surf. *Smithsonian* 24(2):32–43. (Sharks)

Curtsinger, B. 1995. Close Encounters with the Gray Reef Shark. *National Geographic* 187(1):45–65.

Klimley, A. P. 1994. The Predatory Behavior of the White Shark. *American Scientist* 82(2):122–33.

Pauly, D. et al. 2000. Fishing Down Aquatic Food Webs. *American Scientist* 88(1):46–51.

Robinson, B. H. 1995. Light in the Ocean's Midwaters. *Scientific American* 273(1):60–65.

Safina, C. 1997. *Song for the Blue Ocean.* Henry Holt, N.Y. (Fish stocks, methods, and culture explained)

Schmidt, K. 1998. Ecology's Catch of the Day. *Science* 281(5374):190–92. (Coastal fisheries)

VanDyk, J. 1990. The Long Journey of the Pacific Salmon. *National Geographic* 178(1):3–37.

Wu, N. 1990. Fangtooth, Viperfish and Black Swallower. *Sea Frontiers* 36(5):32–39. (Deep-sea fish)

Nekton — Mammals

Ackerman, D. 1992. Last Refuge of the Monk Seal. *National Geographic* 181(3):128–44.

Chadwick, D. 1998. Bottlenose Whales. *National Geographic* 194(2):78–79.

Chadwick, D. 1999. Listening to Humpbacks. *National Geographic* 196(1):110–129.

Gentry, R. 1987. Seals and Their Kin. *National Geographic* 171(4):475–501.

Gerber, L. R., D. P. DeMaster, and S. P. Perry. 2000. Measuring Success in Conservation. *American Scientist* 88(4):316–324. (Present whale population status)

Kovacs, K. 1997. Bearded Seals. *National Geographic* 191(3):124–37.

Nicklin, F. 1995. Bowhead Whales. *National Geographic* 188(2):114–29.

O'Shea, R. J. 1994. Manatees. *Scientific American* 271(1):66–72.

Perry, S. L., D. P. DeMaster, and G. K. Silber. 1999. The Great Whales: History and Status of Six Species Listed as Endangered Under the U.S. Endangered Species Act of 1973. *Marine Fisheries Review* 61(1):1–74.

Trites, A. 1992. Northern Fur Seals: Why They Declined. *Aquatic Mammals* 18(1):3–18.

Whitehead, H. 1995. The Realm of the Elusive Sperm Whale. *National Geographic* 188(5):56–73.

Wolkomir, R. 1996. The Fragile Recovery of California Sea Otters. *National Geographic* 187(6):42–61.

Zimmer, C. 1992. Portrait in Blubber. *Discover* 13(3):86–89. (Elephant seals)

Nekton — Reptiles

Allen, L. 1994. Cast Ashore. *Nature Conservancy* 44(3):16–23. (Florida's sea turtles)

Lohmann, K. J. 1992. How Sea Turtles Navigate. *Scientific American* 266(1):100–106.

Minton, S. A., and H. Heatwole. 1978. Snakes and the Sea. *Oceanus* 11(2):53–56.

Ritchie, R. 1989. Marine Crocodiles. *Sea Frontiers* 35(4):212–19.

Rudloe, A., and J. Rudloe. 1994. Sea Turtles In A Race for Survival. *National Geographic* 185(2):94–121.

Nekton — Fisheries

Baldwin, R. F. 1990. Fundy Farming. *Sea Frontiers* 36(5):40–45. (Salmon farming)

Crowley, M. 1995. Trends and Technology. *National Fisherman* 76(3):18–19. (Fish farming)

Kane, A. P. 1994. Growing Fish in Fields. *World Watch* 6(5):20–27.

Kelley, K. 1995. Can Fishermen Become Fish Farmers? *National Fisherman* 76(3):20–22.

Kunzig, R. 1995. Twilight of the Cod. *Discover* 16(4):44–58.

Safina, C. 1995. The World's Imperiled Fish. *Scientific American* 273(5):46–53.

Wacker, R. 1994. Strip Mining the Seas. *Sea Frontiers* 40(3):14–17, 60.

internet references

worldwide websites

The Plankton

Web Lift to Information on Taxa
www.ucmp.berkeley.edu/help/taxaform.html
Go to Protists

Algal Microscopy and Image Digitization, Diatoms
www.bgsu.edu/departments/biology/algae/index.html
Go to Images and Scanning Electron Micrographs

Dinoflagellates
www.geo.ucalgary.ca/~macrae/palynology/dinoflagellates/dinoflagellates.html
Links to Red Tides, Dinoflagellate Toxins, Images

The Plankton Net
www.uoguelph.ca/zoology/ocean/index.htm

Red Tides and Toxic Blooms

Ecology and Oceanography of Harmful Algal Blooms
habserv1.whoi.edu/hab/nationplan/ECOHAB/ECOHABhtml.html

Bad Bug Book: Introduction to Foodborne Pathogenic Microorganisms and Natural Toxins
vm.cfsan.fda.gov/~mow/intro.html
Under Natural Toxins go to Ciguatera Poisoning and Shellfish Toxins

Algae Blooms—Red Tides
www.mdsg.umd.edu/seagrantmediacenter/news/whoi.html

Red Tides in Western Gulf of Maine
crusty.er.usgs.gov/wgulf/wgulf.html

Facts about *Pfiesteria*
www.ehnr.state.nc.us/EHNR/files/pfies.htm

Harmful Algal Blooms—Homepage
www.nwfsc.noaa.gov/hab/

The Nekton

Web Lift to Information on Taxa
www.ucmp.berkeley.edu/help/taxaform.html
Go to Metazoa for Basal Vertebrates, Fish, and
Mammals

The Cephalopod Page: Octopuses, Squid, Cuttlefish,
and Nautilus
is.dal.ca/~ceph/TCP/index.html

Shark Research Program, University of Florida Museum
of Natural History
www.flmnh.ufl.edu/fish/sharks/sharks.htm

Turtle Survival League/Caribbean Conservation
www.cccturtle.org/

Sea Turtles
www.seaworld.org/Sea_Turtle/seaturtle.html

Protected Marine Species
www.rtis.com/nat/user/elsberry/marspec.html

NOAA Office of Protected Resources
www.nmfs.noaa.gov/prot_res/prot_res.html

Marine Mammals

International Whaling Commission
ourworld.compuserve.com/homepages/iwcoffice/

SeaWorld Information Database
www.seaworld.org/
Go to Animal Resources

Whale Information Network
www.webmedia.com.au/whales/whales3.html

Cetacean Society International—Home Page
elfi.com/csihome.html

Marine Mammals—Hawaii Whale Research
Foundation
www.hwrf.org/hwrf/html/marmam.html

Marine Mammals Management, U.S. Fish and Wildlife
Service
www.r7.fws.gov/mmm/

Understanding the Seal Fishery
www.dfo-mpo.gc.ca/communic/seals/understa/
under_e.htm

Manatees
www.fmri.usf.edu/manatees.htm

Fish and Fisheries

Our Living Oceans
www.st.nmfs.gov/st2/tm_spo19/spo19.html

Food and Agriculture Organization of the United
Nations
www.fao.org/default.htm
Go to Fisheries

FAOSTAT Database Gateway
apps.fao.org/fishery/fprod1-e.htm

FAO—Fishery Department
www.fao.org/waicent/faoinfo/fishery/publ/sofia/
sofiaef.htm

National Coalition for Marine Conservation
www.savethefish.org/

Seafood Watch, Monterey Bay Aquarium
www.mbayaq.org/

FishBase: A Global Information System on Fishes
www.fishbase.org/

NOAA Fisheries
www.nmfs.noaa.gov/

Welcome to FIS—Fish Information Service
www.fish.com/

Ocean Planet Overfishing
seawifs.gsfc.nasa.gov/OCEAN_PLANET/HTML/
peril_fishing.html
Go to Atlantic Cod, Salmon, Bluefin Tuna,
bycatch, etc.

The Fishing Industry
www.gov.nf.ca/publicat/tags/text/page23.htm

Office of Protected Resources—List of Sites
www.nmfs.noaa.gov/prot_res/prot_res.html

Oceanlink
Oceanlink.island.net
Go to Aquafacts

Aquaculture Trends—FAO
www.fao.org/WAICENT/faoinfo/fishery/trends/
aqtrends/aqtrend.htm

Aquaculture Links
www.ag.auburn.edu/dept/faa/aquaculture.html

Another new kind of community has been discovered on the seafloor, a place once thought to be cold, dark, and unsuited to life. In 1987 Craig Smith and colleagues from the University of Hawaii accidentally discovered the carcass of a blue whale on the floor of the Santa Catalina Basin off California (fig. 1). Their studies show that a whale carcass or whale fall supports a large community of organisms. The

Whale Falls

scavengers such as hagfish, crabs, and sharks reduce the body to bones in as little as four months. Whale bones are rich in fats and oils that give the animal buoyancy in life; after death they provide nutrition for bacteria. The bacteria decompose the fatty substances and generate sulfides and other compounds that diffuse out through the bone. Bacterial mats then form over the skeleton and are grazed by worms, mollusks, crustaceans, and other organisms. Other animals may be attracted, getting their nutrition from the bacterial mats, the fatty substances in the bones, or other animals at the site.

More than five thousand animals from 178 species were isolated from five vertebral bones recovered from one whale. Small mussels and limpets are shown on recovered whale bone in figure 2. Ten of these species including worms and limpets have been found only associated with whale skeletons. By comparison the most fertile hydrothermal vent areas (chapter 10, figure 10.18) have

yielded only 121 species and hydrocarbon seeps just 36 (chapter 10, figure 10.19).

We do not know how abundant whale falls are on the sea floor, for they can be anywhere and are difficult to locate. In 1993 the U.S. Navy searched 20 km^2 (8 mi^2) of the Pacific Missile Range off California with side scan sonar while seeking a lost missile. Eight whale falls were videotaped and one of these was subsequently located and examined by Smith and his team using a submersible. With permission from the National Marine Fisheries Service, Smith has taken two dead stranded whales out to sea and sunk them to observe the colonizing of the carcasses and learn more about the diversity of the organisms associated with whale falls.

More Information

Deming, J., A. Reysenbach, S. Macko, and C. Smith. 1997. Evidence for the Microbial Basis of a Chemoautotrophic Invertebrate Community at a Whale Fall on the Deep Sea Floor: Bone-Colonizing Bacteria and Invertebrate Endosymbionts. *Microscopy Research and Technique* 37:162–70.

Life Among the Whale Bones. 1998. *Science* 279 (5355):1302.

Figure 1

The skull, jawbones, and vertebrae of a 21-meter (70 foot) blue whale on the floor of the Santa Catalina basin off southern California, depth 1240 m (4000 ft). The skull is about 1.5 m (5 ft) long, and each vertebra is about 40 cm (16 in) long.

Figure 2

Mussels (each about 1 cm [0.4 in] long) clustered on a recovered whale bone. Up to 178 species of animals have been found living on a single carcass.

chapter twelve

Life on the Sea Floor

Sea urchins and sea anemones are found in rocky habitats.

The exposed rock ... under the towering tide, but they were more than that, they were ferocious with life. There was ... exuberant fierceness in the littoral here, a vital competition for existence. Everything seemed speeded-up, starfish and urchins were more strongly attached than in other places, and many of the univalves were so tightly fixed that the shells broke before the ... hold. Perhaps the force of the ... shore has much to do with ... ere. It is noteworthy that the ... such beaten shores for the safe ... increase their toughness and ... rival. This ... and makes us feel ... resisting qualities of the animals, it almost seems that they are excited too.

John Steinbeck,
from The Log from the Sea of Cortez

Learning Objectives

After reading this chapter, you should be able to

- Describe a typical alga and explain its structure.
- List the major groups of seaweeds and explain where they are found.
- Understand how benthic algae modify the marine environment for other marine organisms.
- Recognize and appreciate the diversity of benthic animals.
- Discuss factors that modify the distribution of benthic animals.
- Compare intertidal zonation on rocky beaches with intertidal zonation on sandy or mud beaches.
- Give examples of adaptations that allow various benthic animals to colonize different seafloor environments.
- Explain symbiotic relationships and give examples of the different kinds.
- Discuss the conditions required for a stony coral reef and the interrelationships between its organisms.
- Describe the present state of commercial benthos harvests: wild and farmed.
- Explain how the new techniques of genetics are being used to improve farmed fish and shellfish.

Biodiversity in the Oceans

m ost marine animal species and nearly all of the larger marine algae are benthic. The algae are found along the sunlit, shallow coastal areas and in the intertidal zone, while the benthic animals are found at all depths on the sea floor or in the sediments. The benthos includes a remarkably rich and diverse group of organisms, among them the luxuriant, colorful tropical coral reefs, the great cold-water kelp forests, and the hidden life below the surface of the mudflat and sandy beach. Benthic organisms are important food resources and provide valuable commercial harvests (for example, clams, oysters, crabs, and lobsters).

The Seaweeds   12.1

Seaweeds are the large plantlike, benthic algae. They photosynthesize, but their body forms, reproduction, and biochemistry are very different from land plants. These large algae have simple tissues; they do not produce flowers or seeds; and their pigments and storage compounds vary from group to group.

Seaweeds grow attached to rocks, shells, or any solid object. Those found floating at the water's edge or thrown up on the shore have been dislodged from the bottom. Seaweeds are attached by a **holdfast** that anchors the plant firmly to a solid base or **substrate.** The holdfast is not a root; it does not absorb water or nutrients. Above the holdfast is a stemlike portion known as the **stipe.** The stipe may be so short that it is barely identifiable, or it may be up to 35 m (115 ft) in length. It acts as a flexible connection between the holdfast and the **blades,** the plant's photosynthetic organs. Seaweed blades serve the same purpose as a leaf but do not have the specialized tissues and veins of leaves. The blades may be flat, ruffled, feathery, or even encrusted with calcium carbonate.

Seaweeds grow attached to rocky substrates; they are not found in areas of mud or sand where their holdfasts have nothing to which they can attach. Because the benthic algae are dependent on sunlight, they are confined to the shallow, sunlit depths of the ocean where they are surrounded by water with its dissolved carbon dioxide and nutrients. They are efficient primary producers, exposing a large blade area to both the water and the Sun. The general characteristics of a benthic alga are shown in figure 12.1.

Algae can be divided into groups based on their pigments. Green algae are moderate in size and may form fine branched structures or thin, flat sheets. Brown algae range from microscopic chains of cells to the

kelps, which are the largest of the algae. Kelps have more structure than most algae, with strong stipes and holdfasts that allow them to grow in fast currents and heavy surf, or their holdfasts may attach below the depth of wave action and their blades float at the surface supported by gas-filled floats. Kelps are especially abundant along the coasts of Alaska, British Columbia, Washington, California, Chile, New Zealand, northern Europe, and Japan. Their brown color comes from a pigment that masks their green chlorophyll. Red algae are the most abundant and widespread of the large marine algae. Their body forms are varied, flat, ruffled, lacy, or intricately branched. They also contain pigments that mask the green chlorophyll.

Although the algae are commonly classified by color, the visible color can be misleading. Some red algae appear brown, green, or violet, and some brown algae appear black or greenish. Some representative algae are shown in figure 12.2.

The characteristic pattern for seaweed growing on a rocky shore is first the green algae, then the brown algae, and last the red algae as depth increases. Remember that the wavelength as well as the quantity of light changes with depth in the sea (see chapter 5). Green algae grow in shallow water because their chlorophyll absorbs light from the visible light spectrum available at the sea surface. Brown algae are found at moderate depths; their brown pigment is more efficient at trapping the shorter wavelengths of light available at these depths. At maximum growing depths, the algae are red, for the red pigment can best absorb the remaining blue-green light.

Seaweeds provide food and shelter for many animals. They act in the sea much as the trees and shrubs do on land. Some fish and other animals, such as sea urchins, limpets, and some snails, feed directly on the algae; other animals feed on shreds and pieces that settle to the bottom. Some organisms use large seaweeds as a

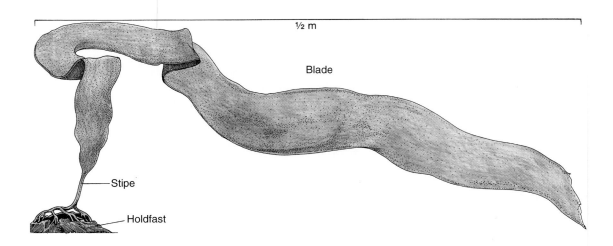

Figure 12.1

Benthic algae are attached to the sea floor by a holdfast. A stipe connects the holdfast to the blade. *Laminaria* is a genus of kelp and a member of the brown algae.

Blade

Stipe

Holdfast

½ m

place of attachment; some of the smaller algae grow on the large kelps. Algae that produce calcareous outer coverings are important in the building of tropical coral reefs, discussed in section 12.10.

Although the diatoms have been discussed as part of the plankton in the previous chapter, there are benthic diatoms as well. Benthic diatoms are usually elongate and grow on rocks, muds, and docks, where they produce a slippery brown coating.

> Sketch a large benthic alga (seaweed). Label its parts and give the function of each part.
>
> How do seaweeds differ from land plants?
>
> What are the three general classes of seaweeds?
>
> How are seaweeds distributed with depth and why are they distributed this way?
>
> What is a kelp? How are kelps modified for their habitat? (See figure 12.2, *Nereocystis, Macrocystis,* and *Postelsia.*)
>
> What roles are played by large algae in the marine environment?

Some Marine Plants   12.2

There are a few flowering plants with true roots, stems, and leaves that have made a home in the sea. Eelgrass, with its strap-shaped leaves, is found growing on mud and sand in the shallow, quiet waters of bays and estuaries along the Pacific and Atlantic coasts, and turtle grass is common along the Gulf Coast. Surf grass flourishes in more turbulent areas exposed to waves and surge action. These sea grasses are rich sources of food and shelter for animals and act as attachment surfaces for some algae.

Midlatitude salt marshes are dominated by marsh grasses able to tolerate the salty water (refer back to fig. 9.2g). Marsh grasses are partly consumed by marsh herbivores, but much of the grass breaks down in the marsh and is washed into the estuaries by the tidal creeks. There the plant remains are broken down further by bacteria that release nutrients to the water to be recycled by the marsh grasses and the algae.

In the tropics mangrove trees grow in swampy intertidal areas (refer back to fig. 9.2f and the discussion of wetlands in section 9.11). The intertwining roots of the mangrove trees provide shelter and also trap sediments and organic material, helping eventually to fill in the swamps and move the shore seaward.

> How are marine habitats modified by each of these plants: (a) sea grasses, (b) salt marsh grasses, (c) mangroves.

The Animals   12.3

Benthic animals are found at all depths and are associated with all substrates. There are more than 150,000 benthic species as opposed to about 3000 pelagic species. About 80% of the benthic animals belong to the **epifauna;** these are the animals that live on or are attached to the surface of rocky areas or firm sediments. Animals that live buried in the substrate belong to the **infauna** and are associated with soft sediments such as mud and sand.

Some animals of the sea floor are **sessile,** attached to the sea floor as adults (for example, barnacles, sea anemones, and oysters), while others are **motile,** or free-moving, all their lives (for example, crabs, starfish, and snails). Most benthic forms produce motile larvae that spend a few weeks of their lives as meroplankton (chapter 11), allowing the species to avoid over-crowding and to colonize new areas. Sessile adults must wait for their food to come to them, either under its own power or carried by waves, tides, and currents. Motile organisms are

Figure 12.2

Representative benthic algae. *Ulva* and *Codium* are green algae. *Postelsia, Nereocystis,* and *Macrocystis* are kelps. The kelps and *Fucus* are brown algae. The red algae are *Corallina, Porphyra,* and *Polyneura. Corallina* has a hard, calcareous covering.

More Information: seaweed.ucg.ie/seaweed.html

able to pursue their prey, scavenge over the bottom, or graze on the seaweed-covered rocks.

The distribution of benthic animals is controlled by a complex interaction of factors, creating living conditions that are extremely variable. The substrate may be solid rock, shifting sand, or soft mud. Temperature, salinity, pH, exposure to air, oxygen content of water, and water turbulence change abruptly in the shallow intertidal zone; the same factors are nearly constant in deep water. Benthic animals exist at all depths and are as diverse as the conditions under which they live.

Animals on Rocky Shores   12.4

The rocky coast is a region of rich and complex seaweed and animal communities living in an area of environmental extremes. At the top of the littoral zone, organisms must cope with long periods of exposure to air as well as heat, cold, rain, snow, and predation by land animals and seabirds, also the waves and turbulence of the returning tides. At the littoral zone's lowest reaches, the seaweeds and animals are rarely exposed but have their own problems of competition for space and predation by other organisms. The exposure to air endured by marine life at different levels in the littoral zone is shown in figure 12.3.

The distribution of the seaweeds and animals is governed by their ability to cope with the stresses that accompany exposure, turbulence, and loss of water. Along the rocky coasts of such areas as North America, Australia, and South Africa, biologists have noted that patterns form as the seaweeds and animals sort themselves out over the intertidal zone. This grouping is called **intertidal zonation** and is shown in figure 12.4. Zones vary with local conditions; they are generally narrow where the shore is steep or the tidal range is small, and wide where the beach is flat and the range of tides is large. The distribution of seaweeds, with the green algae in shallow water, the brown algae in the intertidal zone, and the red algae in the subtidal area, is an example of vertical zonation.

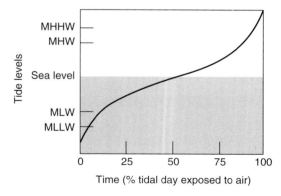

Figure 12.3

The time exposure to air for intertidal benthic organisms is determined by their location above and below mean sea level and by the tidal range. (MLLW = mean lower low water; MLW = mean low water; MHW = mean high water; MHHW = mean higher high water.

In the supralittoral (or splash) zone, which is above the high water level and covered with water only during storms and the highest tides, the animals and seaweeds occupy an area that is as nearly land as it is ocean bed. At the top of this area, patches of dark lichens and algae appear as crusts on rocks. Scattered tufts of algae provide grazing for small herbivorous snails and limpets (fig. 12.5). The small acorn barnacles filter food from the seawater and are able to survive even though they are covered with water only briefly during spring tides. The width of the supralittoral zone varies with the slope of the shore, variations in light and shade, exposure to waves and spray, tidal range, and the frequency of cool days and damp fogs.

Conspicuous members of the midlittoral (fig. 12.6) include several other species of barnacles, limpets, snails, mussels, and chitons. Chitons and limpets are grazers that scrape algae from hard surfaces, while mussels filter organic material from the water. A muscular foot anchors chitons and limpets to the rocks; strong cement secures barnacles; and special threads attach mussels. Tightly closed shells protect many of these organisms from drying out during low tide, and their rounded profiles present little resistance to the breaking waves. Species of brown algae found here have strong holdfasts and flexible stipes. Mussel beds provide shelter for less conspicuous animals such as sea worms and small crustaceans. Shore crabs of varied colors and patterns are found in the moist shelter of the rocks, and small sea anemones huddle together in large groups to conserve moisture. The area is crowded, and the competition for space is extreme. New space in an inhabited area becomes available by predation. Seasonal die-offs of algae and the battering action of strong seas and floating logs also clear space for newcomers.

A selection of organisms from the lower littoral is found in figure 12.7. The larger flowerlike anemones attach firmly to the rocks and spread their tentacles to grasp and paralyze any prey that touches them. Starfish of many colors and sizes make their home in this zone preying on shellfish, sea urchins, and limpets. The filter-feeding sponges encrust the rocks, and on a minus tide delicate, free-living flatworms and ribbon worms are found keeping moist under the mats of algae. Colorful sea slugs or nudibranchs are active predators, feeding on sponges, anemones, and the spawn of other organisms. Although soft-bodied, sea slugs have few if any enemies because they produce poisonous acid secretions. Species of chitons, limpets, and sea urchins graze on the algae covering the rocks. Sea cucumbers are found wedged in cracks and crevices. Tube worms secrete the leathery or calcareous tubes in which they live and extend only their feathery tentacles to strain their food from the water.

Octopuses are seen occasionally from shore on very low tides. They feed on crabs and shellfish, live in caves or dens, and are known for their ability to flash color changes and move gracefully and swiftly over the bottom and through the water. They are not aggressive, although they are curious and have been shown to have learning ability and memory. The world's largest octopus, found in the coastal waters of the eastern North Pacific, commonly measures up to 3 m (16.5 ft) from tip to tip

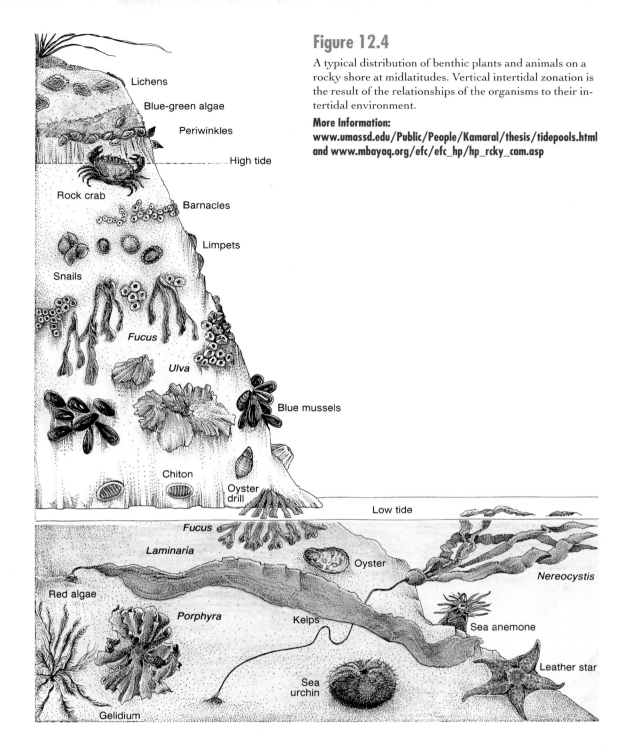

Figure 12.4

A typical distribution of benthic plants and animals on a rocky shore at midlatitudes. Vertical intertidal zonation is the result of the relationships of the organisms to their intertidal environment.

More Information:
www.umassd.edu/Public/People/Kamaral/thesis/tidepools.html
and www.mbayaq.org/efc/efc_hp/hp_rcky_cam.asp

and weighs 20 kg (45 lbs); specimens in excess of 7 m (23 ft) and 45 kg (100 lbs) have been observed.

The bottom of the littoral zone merges into the beginning of the sublittoral zone extending across the continental shelf. If the shallow areas of the subtidal zone are rocky, many of the same lower littoral zone organisms will be found. When soft sediments begin to collect in protected areas or deeper water, the population types change, and animals appear that are commonly found on mud and sand substrates.

Define intertidal zonation and explain what causes it.

Give examples of animals living in the supralittoral, the littoral, and the sublittoral zones along a rocky shore.

What are the survival strategies and adaptations of the animals at each level?

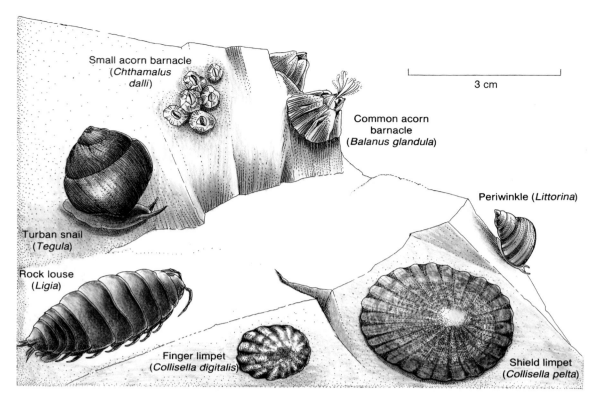

Figure 12.5

Organisms of the supralittoral zone. The limpets and the snail, *Littorina*, are herbivores. The barnacles feed on particulate matter in the water. *Ligia* is a scavenger.

Small acorn barnacle (*Chthamalus dalli*)

Common acorn barnacle (*Balanus glandula*)

Periwinkle (*Littorina*)

Turban snail (*Tegula*)

Rock louse (*Ligia*)

Finger limpet (*Collisella digitalis*)

Shield limpet (*Collisella pelta*)

3 cm

Tide Pools   12.5

A tide pool, left by the receding tide in a rock depression or basin, provides a habitat for animals common to the lower littoral zone. Some small, isolated tide pools provide a very specialized habitat of increased salinity and temperature due to solar heating and evaporation. Other tide pools act as catch basins for rainwater, lowering the salinity and temperature in fall and winter. Isolated tide pools often support blooms of microscopic algae that give the water the appearance of pea soup.

The deeper the tide pool and the greater the volume of water, the more stable its environment when isolated by a falling tide. The larger the tide pool, the more slowly it changes temperature, salinity, pH, and carbon dioxide-oxygen balance. Animals such as starfish, sea urchins, and sea cucumbers require large, deep pools. Fish species found in tide pools are patterned and colored to match the rocks and the algae within the pool and spend much of their time resting on the bottom, swimming in short spurts. Each tide pool is a specialized environment populated with organisms that are able to survive under the conditions established in that particular pool.

What happens to water isolated in a tide pool?

How does this affect the organisms that live in a tide pool?

Animals of the Sand and Mud   12.6

Along exposed gravel and sandy shores, waves produce an unstable benthic environment. Few seaweeds can attach to the shifting substrate, and therefore few grazing animals are found. When currents deposit sand and mud in quiet coves and bays, the habitat is more stable. Here the size and shape of the sediment particles and the organic content of the sediment determine the quality of the environment. The size of the spaces between particles regulates the flow of water and the availability of dissolved oxygen. Beach sand is fairly coarse and porous, gaining and losing water quickly, while fine particles of mud hold more water and replace the water more slowly. The finer the mud particles, the tighter they pack together and the slower the exchange of water; oxygen is not resupplied quickly, and wastes are removed slowly. Deeper into the sediments, the decomposition of organic material produces hydrogen sulfide with its rotten-egg odor, and lack of oxygen restricts the depth to which infauna can be found. Animals such as clams live in these low-oxygen environments by using their siphons to obtain food and oxygen from the water at the sediment surface. In locations protected from waves and currents, eelgrass and surf grass help stabilize the small-particle sediments and provide shelter, substrate, and food, creating a special community of plants and animals.

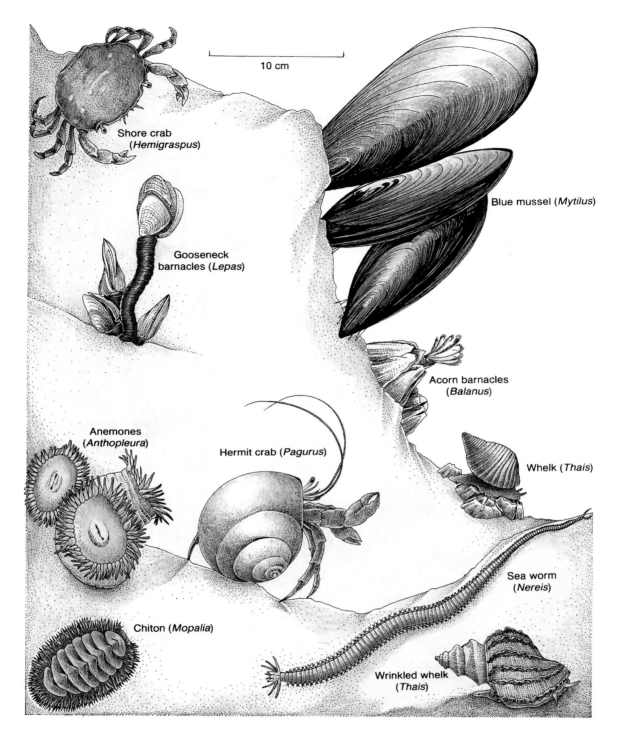

Figure 12.6

Representative organisms from the midlittoral zone. The mussels and barnacles filter their food from the water. The chitons graze on the algae covering the rocks. *Thais,* a snail, is a carnivore. Small shore crabs and hermit crabs are scavengers. *Balanus cariosus* is a larger and heavier barnacle than the barnacles of the supralittoral zone. *Nereis* is often found in the mussel beds. The anemones huddle together to keep moist when exposed.

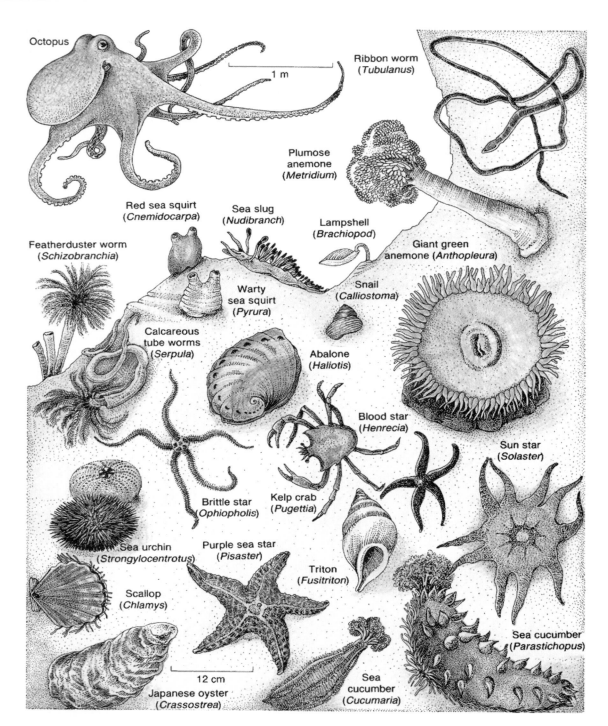

Figure 12.7

Lower littoral zone organisms. The starfish feed on the oysters; the sea urchins are herbivores; and the sea cucumbers feed on detritus suspended in the water. Among the mollusks are the oysters, scallops, snails, abalone, sea slugs, and octopuses. The oysters and scallops are filter feeders. *Calliostoma* and the abalone are grazers; the sea slugs, the octopus, and the triton snail are predators.

More Information: www.ucmp.berkeley.edu/help/taxaform.html

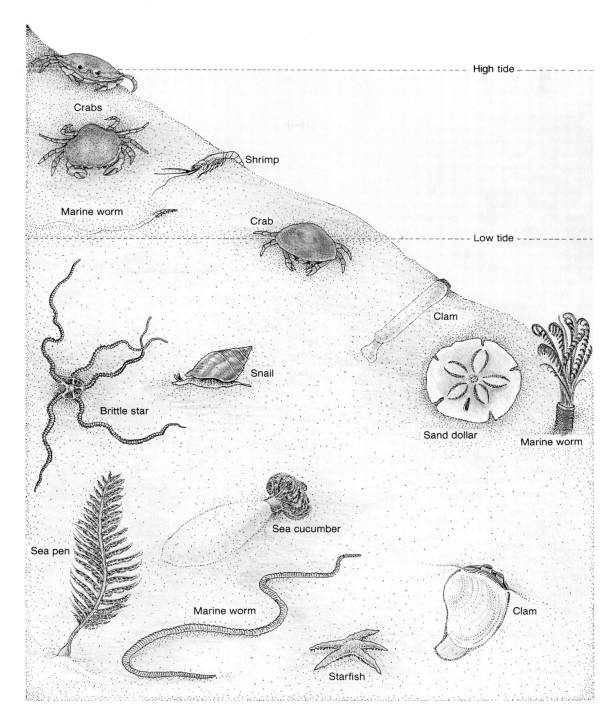

High tide

Crabs

Shrimp

Marine worm

Crab

Low tide

Clam

Snail

Brittle star

Sand dollar

Marine worm

Sea pen

Sea cucumber

Marine worm

Clam

Starfish

Figure 12.8

Zonation on a soft-sediment beach is less conspicuous than that found on a rocky beach. Animals living at the higher tide levels burrow to stay moist.

Most sand and mud animals are detritus feeders, and most detritus is former plant material that is degraded by bacteria and fungi. The sand dollar feeds on detritus particles found between the sand grains. Clams, cockles, and some worms are filter feeders, feeding on the detritus and microscopic organisms suspended in the water. Other animals are deposit feeders that engulf the sediments and process them in their guts to extract organic matter (for example, burrowing

sea cucumbers). Areas of mud that are high in organic detritus support large quantities of bacteria that are a food source for many small organisms.

The intertidal area of a soft-sediment beach shows some zonation of benthic organisms, but because of the low slope and overlapping boundaries between zones, zonation here is not nearly as clear-cut as it is along a steep, rocky shore. The distribution of life in soft sediments is shown in figure 12.8

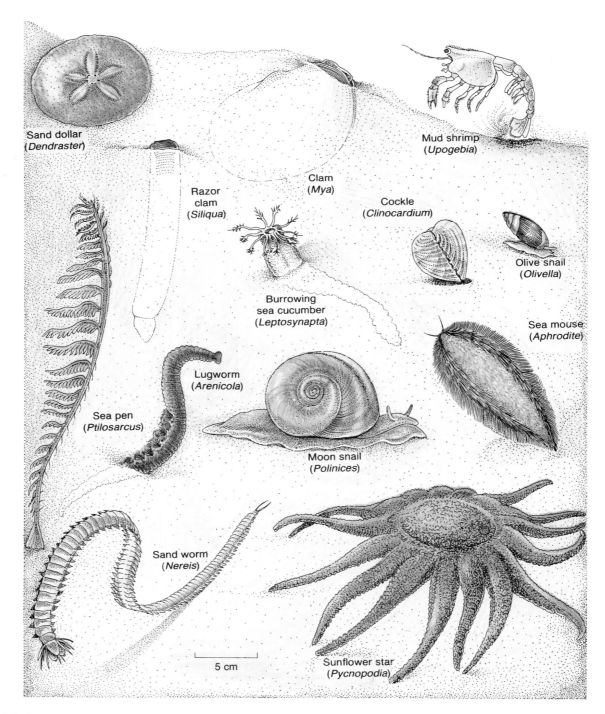

Figure 12.9

Organisms of the soft sediments. Infauna types include the mud shrimp, the lugworm, the clam, the cockle, and the burrowing sea cucumber. The sand dollar feeds on detritus; the moon snail drills its way into shellfish; the sea pens feed from the water above the soft bottom. The sea mouse, like *Nereis,* is a polychaete worm.

(a)

(b)

(c)

Figure 12.10

(a) A vase-shaped glass sponge 640 m (2100 ft) under the surface of the Brown Bear Sea Mount in the northeast Pacific Ocean. (b) A deep-sea crab photographed at a depth of 2000 m (6550 ft) on the Juan de Fuca Ridge, in an area that has very little life. (c) A group of deep-sea sponges and an anemone are seen at 684 m (2244 ft) on the Brown Bear Sea Mount.

Photos (a), (b), (c) by Dr. Verena Tunnicliffe, University of Victoria, B.C., Canada.

and a selection of animals from this region is found in figure 12.9.

Compare the habitats of a protected mudflat and an exposed sandy beach. What algae, plants, animals would you expect to find in each of these environments?

What algae and animals would you *not* expect to find in each environment?

What is detritus, and what is its importance in marine food webs?

Animals of the Deep Sea Floor   12.7

The deep sea floor includes the flat abyssal plains, the trenches, and the rocky slopes of seamounts and mid-ocean ridges. The seafloor sediments are more uniform and their particle size is smaller than those of the shallow regions close to land sources, and the environment is uniformly cold and dark.

The stable conditions of the sea floor appear to have favored deposit-feeding infaunal animals of many species. Many members of the deep-sea infauna are very small and measure 2 mm or less; dominant members include detritus-eating worms, burrowing crustaceans, and burrowing sea cucumbers. The deep sediments are continually disturbed and reworked by sea cucumbers and worms as they extract organic matter. This process, called **bioturbation,** results in a well-mixed, uniform sediment layer that covers vast areas of the ocean floor.

Single-celled protozoans are abundant and widely distributed. Glass sponges attach to the scattered rocks on oceanic ridges and seamounts; so do sea squirts and sea anemones (fig. 12.10). Their stalks lift them above the soft sediments into

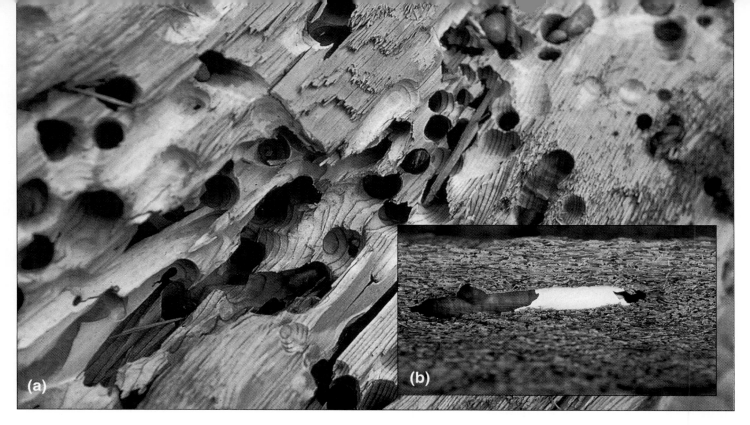

Figure 12.11

Unprotected wood, such as driftwood, is subject to destruction by marine borers. (a) This beach log has been riddled by shipworm. (b) The small holes surrounding a shipworm hole are bored by a crustacean known as gribble.

the water, where they feed by straining out organic matter. Stalked barnacles attach to glass sponges as well as to shells and boulders. Tube worms are common, ranging in size from a few millimeters to 20 cm (8 in), and sea spiders, brittle stars, and sea cucumbers are found at depths down to 7000 m (23,000 ft). Snails are found to the greatest depths; some in the deepest trenches have no eyes or eye stalks. The specialized communities of the hydrothermal vents are discussed in chapter 10.

> Describe the environment of deep-sea benthos. What kinds of organisms are found in this habitat?
>
> What produces bioturbation, and what is its effect?

Fouling and Boring Organisms   12.8

Organisms that settle and grow on pilings, docks, and boat hulls are said to foul these surfaces. Fouling organisms include barnacles, anemones, tube worms, mussels, and algae. When these organisms grow on a vessel's hull, they slow its movement through the water, adding to costs in shipping time and haul-outs for cleaning. Fouling organisms make it difficult to operate vessels and equipment in the ocean environment, and much research goes into experiments with paints and metal alloys to discourage and control these fouling organisms.

Other organisms naturally drill or bore their way into the substrate. Sponges bore into scallop and clam shells; snails bore into oysters; some clams bore into rock. Organisms that bore into wood are costly, for they destroy harbor and port structures as well as wooden boat hulls. The shipworm is a wormlike mollusk with one end covered by a two-valved shell (fig. 12.11a). Shipworms secrete enzymes that break down and partially digest the wood fibers, which are scraped away by the shell's rocking and turning movement. This motion allows the animal to bore deep and destructive holes all through a piece of wood. The gribble, a small crustacean, gnaws more superficial and smaller burrows but is also very destructive (fig. 12.11b).

> Give examples of fouling and boring organisms.
>
> Why are these organisms a problem for humans? How do humans combat them? What are the consequences of this human activity to the marine environment?

Intimate Relationships 12.9

Competition for food and space and predator-prey relationships are common in the marine environment, but relationships can also be cooperative. The beautiful, green sea anemone of the Pacific coast is green because a small, single-celled alga grows in the animal's cells. This mutually beneficial relationship or **mutualism,** in which the alga receives protection as well as carbon dioxide and nitrogenous compounds from the anemone, and the anemone acquires organic compounds and some oxygen from the alga, is a type of symbiotic relationship. **Symbiosis** refers to a state in which two dissimilar organisms live in a close, intimate relationship.

Another interesting symbiotic relationship is found in tropical waters between the sea anemone and the clownfish (fig. 12.12). The clownfish acquires protection by nestling among the anemone's stinging tentacles while acting to lure other fish within the anemone's grasp. Shellfish of all types play host to various worms and small crustaceans. Rather than being mutually beneficial, these relationships are often beneficial to one partner and of no harm to the other, a relationship known as **commensalism.**

A relationship in which one partner is harmed by the other is **parasitism.** Parasitic flatworms, roundworms, and bacteria infect marine animals in the same way that land animals are infected. Parasitic relationships are not confined to benthic organisms; both fish and sea mammals are often heavily parasitized.

> Compare the three types of symbiosis. Give an example of each.

Tropical Coral Reefs 12.10

Coral reefs are the most luxuriant and complex of all benthic communities (see fig. 12.12). The largest coral reef in the world, the Great Barrier Reef, stretches more than 2000 km (1250 mi) from New Guinea southward along the east coast of Australia. Corals are colonial animals, and individual coral animals are called **polyps** (fig. 12.13). A coral polyp is very similar to a tiny sea anemone with its stinging tentacles, but unlike the anemone, a coral polyp extracts calcium carbonate from the water and builds within its tissues a calcareous skeletal cup. Large numbers of polyps grow together in colonies of delicately branched forms or rounded masses.

Tropical reef-building corals have specialized requirements; they require warm, clear, shallow, clean water and a firm substrate to which they can attach. Because the water temperature must not go below 18°C and the optimum temperature is 23° to 25°C, their growth is restricted to tropical waters between 30°N and 30°S and away from cold-water currents.

Waters at depths greater than 50 to 100 m (150–300 ft) are also too cold for significant secretion of calcium carbonate.

Within the tissues of the coral polyps are masses of single-celled dinoflagellate algae called **zooxanthellae.** Polyps and zooxanthellae have a symbiotic relationship in which the coral provides the algal cells with a protected environment, carbon dioxide, and nitrate and phosphate nutrients, and the algal cells photosynthesize, return oxygen, remove waste, and produce carbon compounds which help to nourish the coral. Some coral species receive as much as 60% of their nutrition from their algae. Zooxanthellae also cause the coral to produce more calcium carbonate and increase the growth of their calcareous skeleton. The polyps feed actively at night, extending their tentacles to catch zooplankton, but during the day, their tentacles are contracted, exposing the outer layer of cells containing zooxanthellae to the sunlight. Most Caribbean corals are found in the upper 50 m of lighted water, whereas Indian and Pacific corals are found to depths of 150 m (400 ft) in the more transparent water of these oceans.

Corals are slow-growing organisms; some species grow less than 1 centimeter (0.4 in) in a year, and others add up to 5 cm (2 in) each year. The same coral may be found in different shapes and sizes, depending on the depth and the wave action of an area. Environmental conditions vary over a reef, forming both horizontal and vertical zonation patterns as shown in figure 12.14. On the sheltered (or lagoon) side of the reef, the shallow reef flat is covered with a large variety of branched corals and other organisms. Fine coral particles broken off from the reef top produce sand, which fills the sheltered lagoon floor. On the reef's windward side, the reef's highest point, or reef crest, may be exposed at low tide and is pounded by the breaking waves of the surf zone. Here the more massive rounded corals grow. Below the low tide line, to a depth of 10 to 20 m (30–60 ft) on the seaward side, is a zone of steep, rugged buttresses, which alternate with grooves in the reef face. Masses of large corals grow here, and many large fish frequent the area. The buttresses dissipate the wave energy, and the grooves drain off fine sand and debris, which would smother the coral colonies. At depths of 20 to 30 m (60–100 ft), there is little wave energy, and the light intensity is only about 25% of its surface value. The corals are less massive at this depth, and more delicately branched forms are found. Between 30 and 40 m (100 and 130 ft), the slope is gentle and the level of light is very reduced; sediments accumulate at this depth, and the coral growth becomes patchy. Below 50 m (150 ft), the slope drops off sharply into the deep water.

Coral reefs are complex assemblages of many different types of plants and animals, and competition for space and food is intense. It has been estimated that as many as 3000 animal species may live together on a single reef. The giant clam (*Tridacna*) measuring up to a meter in length and weighing over 150 kg was among the most conspicuous, but overharvesting and poaching have greatly reduced their numbers. These clams also possess zooxanthellae in large numbers in the colorful tissues that line the edges of the shell. Crabs, moray eels, colorful reef fish, poisonous stonefish, long-spined sea urchins, sea horses,

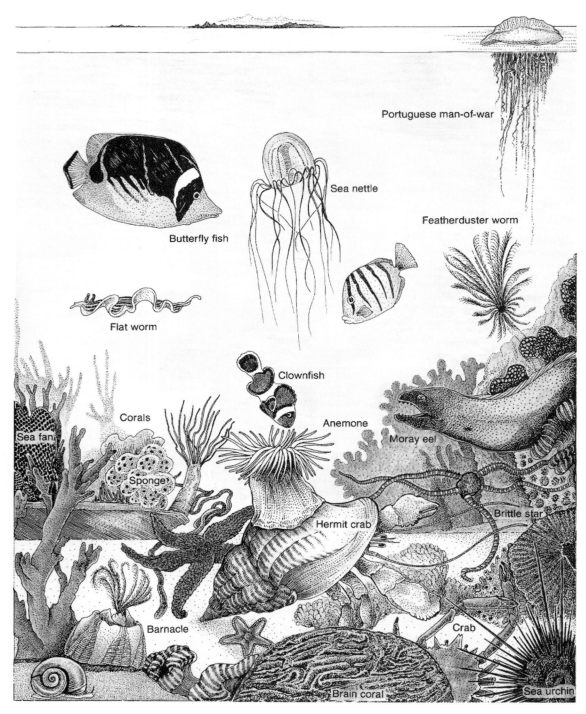

Figure 12.12

A coral reef is a complex, interdependent, but self-contained community. Members include corals, clams, sponges, sea urchins, anemones, tube worms, algae, and fish.

More Information: www.ust.hk/~webrc/ReefCheck/reef.html
coral.aoml.noaa.gov/

shrimp, lobsters, sponges, and many more organisms are all found living here together. The outer calcareous coverings of encrusting algae, shells, the tubes of worms, and the spines and plates of sea urchins are compressed and cemented together to form new places for more organisms to live. At the same time, some sponges, worms, and clams bore into the reef; some fish graze on the coral and the algae, and the sea cucumbers feed on the broken fragments, reducing them to sandy sediments.

Figure 12.13

Star coral polyps on a reef in the Caribbean Sea, Bonaire, Lesser Antilles.

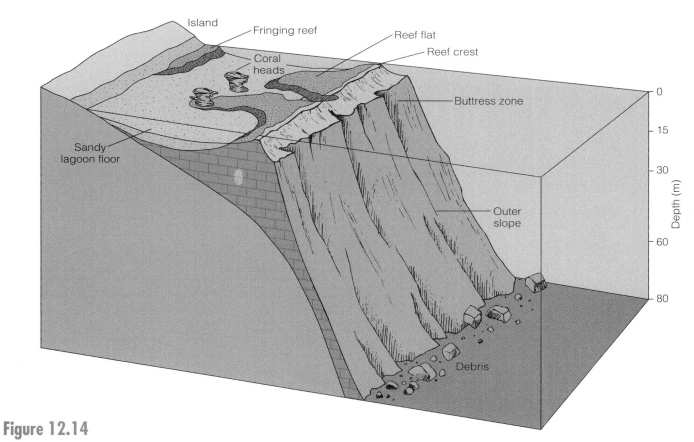

Figure 12.14

Coral reef zonation. Water depths and wave action vary over the reef. The reef crest may be exposed at low tide. Sand from coral debris fills the lagoon and drains off down the grooves of the outer slope, where buttresses dissipate wave energy from the open sea.

In the last ten to fifteen years, many of the world's coral reefs have been seriously damaged. From Florida to Fiji reefs are infested with algae, predators, and recurrent episodes of bleaching. Seventy percent of the Philippine coral reefs were reported in critical condition at the 1992 International Coral Reef Symposium.

Coral bleaching, in which the corals expel their zooxanthellae and turn white, has been occurring more frequently and with greater severity. In the tropical eastern Pacific in 1983 large amounts of bleaching, thought to have been associated with the 1982–83 El Niño event (see section 6.14), occurred in reefs off Costa Rica, Panama, and the Galápagos Islands. Between 70 and 90 percent of the corals in Panama and Costa Rica and more than 95% of the corals in the Galápagos were destroyed. Bleaching also occurred in the Society Islands and along the Great Barrier Reef. Shallow reefs of the Java Sea lost 80 to 90 percent of their coral cover, and five years later coral cover was only 50% of its former level. Mass bleaching in the Atlantic associated with elevated water temperatures began in 1987 in the Florida Keys and eventually stretched from Bermuda south to the Bahamas, affecting coral reefs throughout the Caribbean. In 1991 coral bleaching was reported again in the Caribbean and also in French Polynesia.

The 1997–98 El Niño, accompanied by record high ocean temperatures, moved through the Indian Ocean producing sudden and steep increases in coral bleaching. The Global Coral Reef Monitoring Network's report, given in 2000 at the 9th International Reef Symposium, reports 11% global reef destruction prior to 1998 and less than a year later, another 16% severely damaged reefs "from Brazil to the Indian Ocean." Some reefs have been recovering more quickly than expected, perhaps due to the survival of juvenile corals, but total recovery will require decades.

No one is sure why massive bleaching episodes occur. One possibility is that when the water warms it causes the algae to produce more oxygen (a by-product of photosynthesis) than the corals can stand; another theory is that stressed coral polyps provide fewer nutrients to the algae. Whatever the cause, when corals bleach the delicate balance between the zooxanthellae and the corals is disrupted, leaving them susceptible to disease and other stresses.

Reefs are also under attack by disease. Coralline lethal orange disease or CLOD is caused by a bright orange, bacterial pathogen that is lethal to the encrusting red algae that deposit calcium carbonate, which cements together sand, dead algae, and other debris to form a hard, stable substrate. CLOD was initially found in 1993 in the Cook Islands and Fiji. It spread to the Solomon Islands in 1994, New Guinea in 1995, and now has a 6000 km (3600 mi) range in the South Pacific. It is not known whether it has been introduced from some obscure location or whether it has been present on the reefs, but has developed into a more virulent form. Between 1996 and 1997 black band disease on Florida coral reefs increased by nearly 300%. Scientists do not know what triggered this change, but poor water quality and polluted runoff into Florida Bay are suspected.

Periodically a dramatic population increase of the crown-of-thorns sea star, which feeds on coral polyps, occurs. No one knows why the crown-of-thorns population increases so rapidly; evidence points to a correlation between rainy weather (low salinity) and runoff (increased nutrients), allowing large numbers of larvae to survive. The reefs have been able to regenerate in most cases when the crown-of-thorns starfish population dies back, but the outbreaks of starfish do appear to be occurring more frequently and sometimes become chronic.

In addition, coral reefs are mined for building materials, despoiled by shell and aquarium collectors, and dynamited and poisoned to harvest reef fish. Sodium cyanide is used to stun the fish, making them easy to capture for the restaurant and aquarium trades, and the poison also kills other reef organisms. The reefs are also damaged by careless sport divers trampling delicate corals and by boats grounding or dragging their anchors. The clearing of mangroves, and the expansion of citrus and banana plantations have brought silt, pesticides, and fertilizers into the reef environment. Sewage from new resorts and hotel complexes and the septic tanks from an increasing population provide nutrients for the algae. In Belize the largest barrier reef in the Western Hemisphere is being damaged by serious algal overgrowth fed by increased nutrients.

Whatever the cause, if the reef's chemical balance is upset the results can be disastrous; too little oxygen and the animals die, too many nitrates and the algae overgrow the reef. If the grazers decline, the predators decline, upsetting the balance again. Ensuring the continued existence of these beautiful and productive areas requires both an increased understanding of the complex nature of reef communities and the development of policies designed to protect them from human interference. To do so the United Nations Environment Program, the Intergovernmental Oceanographic Commission, and the World Meteorological Organization has launched a worldwide monitoring program, installing monitoring stations on the world's major reefs.

> What conditions are required for the growth of a tropical stony coral reef?
>
> Explain the symbiotic relationship between coral polyps and zoonxanthellae.
>
> How do reef growth and wave action combine to create basic reef features?
>
> Discuss the effect of the crown-of-thorns starfish and episodes of bleaching on coastal reefs.
>
> How do human activities impact these reefs?

Table 12.1 is a summary of the benthos, their distribution and characteristics.

table 12.1 Benthos

Type		Distribution	Characteristics
Algae			
	Diatoms	Cool waters, attached to rocks, mud, other surfaces. Photic zone.	Elongate diatoms with bilateral symmetry.
	Seaweeds	Polar to tropic waters; most abundant at midlatitudes. Limited to photic zone. Vertical zonation associated with pigments, light, and substrate.	Algae with holdfast, stipe, and blades; no flowers or seeds. Nutrients and gases absorbed from the water. Require hard substrate for attachment. Kelps reach 35 m in length.
Plants			
	Seed plants	Midlatitudes and tropics. Quiet shallow water; mud and sand substrates.	Plants with flowers and seeds, roots, stems, and leaves. Salt tolerant. Salt marshes and aquatic grass areas in bays. Special mangrove tree habitat in tropics.
Animals Littoral zone			
	Epifauna	Intertidal of all latitudes. Animals live on or attached to rocks or firm sediments. Vertical zonation governed by exposure, predation, competition, and substrate.	Carnivores and herbivores. Some motile, predators and scavengers. Some sessile, filter feeders or contact feeders. Adapted to withstand desiccation.
	Infauna	Intertidal of all latitudes. Animals live buried in substrate. Little zonation.	Animals buried in substrate. Some filter feeders; some process substrate to obtain organics.
Sublittoral	Epifauna	Animals live on or attached to rocks or firm sediments. No zonation; substrate controls distribution. Populations decrease with depth and availability of food.	Animals similar to lower littoral zone. Less environmental stress.
	Infauna	Animals live buried in substrate. No zonation; substrate controls distribution. Populations decrease with depth and availability of food.	Animals similar to lower littoral zone.
Coral reefs		Tropical. Warm, shallow, clean water between 30°N and 30°S. Vertical zonation and reef profile are products of wave action and water depth.	Luxuriant and complex benthic community. Includes reef-building corals, giant clams with internal photosynthetic zooxanthellae, and many other algae and animals.
Chemosynthetic communities (See section 10.10)		Isolated communities associated with deep-sea hydrothermal vents and continental shelf gas and oil seeps.	Life-forms feed on chemosynthetic bacteria or harbor bacteria within their tissues. Bacteria at vents use hydrogen sulfide gas as energy source. Hydrocarbons and methane used at seeps. Communities include large clams, large tube worms, crabs, mussels, and barnacles.

Harvesting the Benthos   12.11

Animals

Benthic animals are a valuable part of the seafood harvest that includes crabs, shrimps, prawns, and lobsters (crustaceans) and clams, mussels, and oysters (mollusks). World harvests for 1997 include catches of 6.8 million metric tons of mollusks and 5.3 million metric tons of crustaceans. Aquaculture added another 9.4 million tons of mollusks and 1.2 million tons of crustaceans. Because of the demand for mollusks and crustaceans, the catches are much more important in dollar value than the weight of the catches would suggest.

Many of the problems of the finfish fisheries are repeated in the benthic fisheries. For example, between 1975 and 1980,

Figure 12.15

Alaska king crab being unloaded from a crab pot.

More Information: seawifs.gsfc.nasa.gov/OCEAN_PLANET/HTML/peril_fishing.html

more and more boats entered the booming U.S. king crab fishery of the Bering Sea (fig. 12.15). The huge catches of 1979 (70,000 metric tons) and 1980 (84,000 metric tons) were followed by five years of declining catches dropping to 7,000 metric tons in 1985. The rapid decrease in catch was apparently due to overfishing combined with insufficient knowledge of the king crab's life history. Under tight controls the catch increased in the years following 1985, reaching 15,000 metric tons in 1990 but dropping to 8000 metric tons in 1997.

Bristol Bay is the most productive region of the Bering Sea. In 1993 it contributed 6500 metric tons to the king crab catch, but in 1994 the Bristol Bay king crab fishery was closed due to low numbers of female crabs in that area's population. In 1995 the catch rose again, 6600 metric tons, but in 1996 the king crab fishery was canceled and quotas for other types of crabs were also cut. Speculation on reasons for the reduced catches include historical overfishing, natural fluctuations of the environment, as well as by-catch and damage by other bottom fishers.

Attempts to increase the harvests of crustaceans and mollusks have focused on aquaculture, or mariculture. Oysters,

mussels, and clams are raised on aquaculture farms around the world (fig. 12.16). The world harvest of crustaceans, mainly shrimp produced on aquaculture farms, climbed from 200,000 metric tons in 1985 to 1.2 million metric tons in 1996. The value of the shrimp crop increased from $1.1 million to $7.0 million during this time.

Aquaculture projects for benthic species in the United States are faced with the same difficulties as those facing fish farms. High costs, strict licensing policies, the need for technology to replace hand labor, and the need for research to improve diet and disease control require considerable attention if we are to increase our seafood harvest in this way.

Algae

Seaweeds are gathered from the wild in northern Europe, Japan, China, and Southeast Asia; they are also an important part of the Japanese aquaculture industry. The 1994 estimate for total seaweed harvest was 1.1 million metric tons. Certain species of green algae, known as sea lettuces, are used in seaing, soup, pudding, cosmetics, and medicines. About one million pounds of agar and 10 million pounds of carrageenan are used in the United States each year.

Biomedical Products

Many benthic organisms produce biologically active compounds that have therapeutic use. Extracts of some sponges yield anti-inflammatory and antibiotic substances; some corals produce antimicrobial compounds; the sea anemone, *Anthopleura*, produces a cardiac stimulant and a muscle relaxant has been isolated from the marine snail, *Murex*. Thousands of active compounds from marine organisms have been screened during the last 10–15 years for their anticancer, anti-inflammatory, anti-tumor, immune suppressant and antiviral possibilities. Most of the substances studied come from soft-bodied organisms that rely on toxins for defense. These animals have developed an array of compounds that prevent predation and give them an advantage in the competition for space on a crowded coral reef or pier piling.

Collecting, extracting, identifying, testing, and evaluating active natural compounds from the world's benthos is time-consuming and expensive. From discovery to pharmacy takes ten to fifteen years and may cost hundreds of millions of dollars. If the process is successful, there is the additional problem of the demand for organisms exceeding their supply until the substance is successfully synthesized.

Assess the importance of benthic organisms in world food harvests.

Why have some harvests declined?

How are harvests being increased?

What are the uses of the seaweed harvest?

Why is there an interest in screening benthic organisms for useful biomedical products?

Figure 12.16

Raft culture of bay mussels in Puget Sound. Mussels adhere to ropes suspended from floating rafts.

More Information:
www.fao.org/WAICENT/FAOINFO/FISHERY/trends/aqtrends/aqtrend.asp

weed soups, in salad, and as a flavoring in other dishes. Kelp blades, fresh, dried, pickled, and salted, are used in soups and stews. Japanese nori is used in soups and stews and is rolled around portions of rice and fish for flavor. Historically in coastal areas, kelp was used as winter fodder for sheep and cattle and to mulch and fertilize the fields.

There are also important industrial uses for some algal products. **Algin** is extracted from brown algae, and **agar** and **carrageenan** are obtained from red algae. Algin derivatives are used as stabilizers in dairy products and candies as well as in paints, inks, and cosmetics. Agar is used as a medium for bacterial culture in laboratories and hospitals; in addition, it serves as an ingredient in desserts and in pharmaceutical products. Algae rich in carrageenan are gathered in the wild in New England and northeastern Canada; carrageenan is a stabilizer and emulsifier used to prevent separation in ice cream, salad dress-

Genetic Manipulation   12.12

Farm-raised fish and shellfish make up more and more of the world's food supply. Fisheries scientists in North America, Japan, and northern Europe are using gene transfer and chromosome manipulation techniques to keep fish farms stocked with species that will reach market weight faster and be more resistant to disease and freezing.

Aquaculturalists are transferring genes that boost the production of natural growth hormone into fish eggs. Although the foreign genes do not always function, when they do, the gene is passed on to the next generation. These transgenic fish and their offspring have shown from 20 to 46 percent faster growth than the wild stock. Another method is to transfer genes that will increase the immunity of the fish and so boost their survival. Many farm-raised halibut and Atlantic salmon die in winter when their body fluids freeze. A gene from the Arctic winter flounder has been transferred into Atlantic salmon, and preliminary results indicate that these fish survive very cold temperatures better than normal fish.

Another technique produces fish carrying an extra set of chromosomes, or **triploids.** Triploid salmon and trout are produced by exposing the eggs to temperature, pressure, or chemical shock shortly after fertilization. The shock interferes with the division of the egg nucleus, and the egg retains a third set of chromosomes. Triploid salmon and trout do not mature sexually but continue to eat and reach a larger size since they are not spending energy on egg or sperm production. This improved growth and survival has led to widespread production of farmed triploid trout in the United Kingdom.

The gender of fish can also be controlled by chromosome manipulation. Researchers expose fish sperm to ultraviolet light, denaturing its chromosomes. Eggs fertilized with the inactivated chromosome sperm are briefly chilled, causing the eggs to develop with only the female's chromosomes. The result is all female offspring. Many female food fish grow bigger and live longer than males. Female flounder grow to twice the size of males; female Coho salmon have firmer and more flavorful flesh, and all-female sturgeon populations would bring higher profits to caviar producers. It is also possible to inactivate the egg chromosomes and double the sperm chromosomes to produce an all-male population of fish.

Shellfish are also being genetically engineered to produce better commercial forms. Normal oysters enter a summer reproductive phase during which their meat is poor in quality and the oysters are unmarketable. Triploid oysters do not produce sperm and egg but continue to grow steadily, becoming significantly larger than normal oysters and ready for the summer market. Currently triploid oysters are being farmed on about 450 acres of tidelands in Washington and northern California; the 1996 harvest totaled about 500,000 gallons with a wholesale value of about $21 million. The blue mussel is also ready to market in triploid form.

> What research techniques are being used to improve farmed fish and shellfish?

Summary

Benthic algae grow attached to a solid surface. These algae have a holdfast, a stipe, and photosynthetic blades but no roots, stems, or leaves. Green algae grow nearest the surface; brown algae including kelps grow at moderate depths; and red algae are found primarily below the low tide level. Each group's pigments trap the available sunlight at these depths. Seaweeds provide food, shelter, and substrate for other organisms. There are benthic diatoms and a few flowering plants, including marine grasses and mangroves.

Benthic animals are subdivided into the epifauna, which live on or attached to the bottom, and the infauna, which live buried in the substrate. Animals that inhabit the rocky littoral region are sorted by the stresses of the area into a series of zones. Organisms that live in the supralittoral zone spend long periods of time out of water. The animals of the midlittoral zone experience nearly equal periods of exposure and submergence; they have tight shells or live close together to prevent drying out. The lower littoral zone, a less stressful environment, is home to a wide variety of animals. Tide pools provide homes for some organisms in the littoral zone.

The littoral is crowded, and competition for space is great. Organisms of the littoral zone are herbivores and carnivores, each with its specialized lifestyle and adaptations for survival.

Sand, and gravel areas are less stable than rocky areas; these substrates show little zonation. The size of the spaces between the substrate particles determines the water and oxygen content of the substrate. Some sediments have a higher organic content than others. Few algae can attach to soft sediments, so few grazers are found here. Eelgrass and surf grass provide food and shelter for specialized communities. Most organisms that live associated with soft sediments are detritus feeders or deposit feeders.

The environment of the deep sea floor is very uniform. Most animals are infaunal deposit feeders. Burrowers like sea cucumbers continually rework the sediments.

Fouling organisms settle on pilings, docks, and boat hulls; borers drill holes into other organisms. Gribbles and shipworms bore into wood and cause great damage. Symbiotic relationships are intimate relationships between organisms. Mutualism, commensalism, and parasitism are types of symbiosis found in the marine environment.

Tropical coral reefs are specialized, self-contained systems. The coral animals require warm, clear, clean, shallow water and a firm substrate. Photosynthetic dinoflagellates, called zooxanthellae, live in the cells of the corals and the giant clams. The reef exists in a complex but delicate biological balance, which can be easily upset. Reefs have a typical zonation and structure associated with depth and wave exposure. The reefs are presently threatened by warming ocean temperatures and human activities.

Shellfish are valuable world food resources. Aquaculture can be used to increase the shellfish and shrimp harvests. Algae are gathered in many countries and are cultivated in Japan. Some algae are used directly as food; others yield substances that are used as stabilizers and emulsifiers in foods and other products. Benthic organisms produce biologically active compounds that are being screened for use as therapeutic drugs.

Gene transfers are used to improve fish growth and increase immunity of stocks for fish farming. Triploid fish and shellfish, because they do not mature sexually, continue to grow to larger than normal size.

Key Terms

agar, 315
algin, 315
bioturbation, 307
blade, 297
carrageenan, 315
commensalisms, 309
epifauna, 298
holdfast, 297
infauna, 298
intertidal zonation, 300
kelp, 297

motile, 298
mutualism, 309
parasitism, 309
polyp, 309
sessile, 298
stipe, 297
substrate, 297
symbiosis, 309
triploid, 316
zooxanthellae, 309

Critical Thinking

1. Design an original organism to inhabit the supralittoral, the littoral, or the sublittoral zone. Consider its requirements for food, shelter, and protection from predators, its adaptations to its environment, and its life history.
2. Why are bacteria important to benthic organisms?
3. How are coral reefs able to support rich and varied populations when the water surrounding the reef is clear and has no planktonic primary producers?
4. Discuss the advantages and disadvantages of the genetic manipulation of fish and shellfish. Do the advantages outweigh the possible disadvantages?
5. Discuss the food-gathering strategies of motile and sessile organisms in the littoral and sublittoral zones.

Suggested Readings

Levinton, J. S. 1985. *Marine Biology: Function, Biodiversity, Ecology.* Oxford University Press, N.Y.

The Plants and Animals

Birkeland, C. 1989. The Faustian Traits of the Crown-of-Thorns Starfish. *American Scientist* 77(2):154–63.
Childress, J., H. Felbeck, and G. Somero. 1987. Symbiosis in the Deep Sea. *Scientific American* 256(5):114–20.
Genthe, H. 1998. The Incredible Sponge. *Smithsonian* 29(5):50–58.
Grall, G. 1992. Life on a Wharf Piling. *National Geographic* 182(1):95–115.

Peterson, C. H. 1991. Intertidal Zonation of Marine Invertebrates in Sand and Mud. *American Scientist* 79(3):236–49.
Tennesen, M. 1992. Kelp: Keeping a Forest Afloat. *National Wildlife* 30(4):4–11.

Coral Reefs

Ariyoshi, R. 1997. Halting a Coral Catastrophe. *Nature Conservancy* 47(1):20–25. (Cyanide harvesting)
Brown, B. E., and J. C. Ogden. 1993. Coral Bleaching. *Scientific American* 268(1):64–70.
Chadwick, D. 1999. Corals in Peril. *National Geographic* 195(1):31–37.
Golden, F. 1991. Reef Raiders. *Sea Frontiers* 37(1):18–25.
Hughes, T. P. 1994. Catastrophes, Phase Shifts, and Large-Scale Degradation of a Caribbean Coral Reef. *Science* 264(5178):1547–51.
Leal, J. H. 1991. Australia's Vast Offshore Maze. *Sea Frontiers* 37(4):48–51. (The Great Barrier Reef.)
Levine, J. 1993. The Changing of the Guard on the Coral Reef. *Smithsonian* 24(7):104–115.
Littler, M., and D. Littler. 1995. Impact of CLOD Pathogen on Pacific Coral Reefs. *Science* 267(5202):1356–60.
Ross, J. 1998. The Miracle of the Reef. *Smithsonian* 28(11):88–96.
Torrance, D. C. 1991. Deep Ecology: Rescuing Florida's Reefs. *Nature Conservancy* 41(4):8–17.
Ward, F. 1990. Florida's Coral Reefs Are Imperiled. *National Geographic* 178(1):115–32.

Harvesting the Benthos

Chamberlain, G., and H. Rosenthal. 1995. Aquaculture in the Next Century. *World Aquaculture* 26(1):21–25.
Colwell, R. 1995. Marine Biotechnology — A Potential Being Realized. *Sea Technology* 36(1):27–28.
Faulkner, D. J. 1992. Biomedical Uses for Natural Marine Chemicals. *Oceanus* 35(10):29–35.
Fischetti, M. 1991. A Feast of Gene-Splicing Down on the Fish Farm. *Science* 153(5019):512–13.
Flam, F. 1994. Chemical Prospectors Scour the Seas for Promising Drugs. *Science* 266(5189):1324–25.
Harvey, D. 1995. Aquaculture Outlook. *Aquaculture Magazine* 21(3):38–43.
Standish, J. A., Jr. 1988. Triploid Oysters Ensure Year-Round Supply. *Oceanus* 31(3):58–63.
Wright, A., and P. McCarthy. 1994. Drugs from the Sea at Harbor Branch. *Sea Technology* 35(8):1014, 17–18.

internet references

worldwide websites

The Benthos

Web Lift to Information on Taxa
www.ucmp.berkeley.edu/help/taxaform.html
Go to Animals and Seaweed

Cephalopod Page: Octopus, Squid, Cuttlefish, and Nautilus
is.dal.ca/~ceph/TCP/index.html

Life in a Massachusetts Tide Pool
www.umassd.edu/Public/People/Kamaral/thesis/tidepools.html

Monterey Bay Aquarium: Kelp Forest—Tide Pool Cam
www.mbayaq.org/efc/efc_hp/hp_rcky_cam.asp

Low Tide Zone
web.mit.edu/corrina/tpool/zonel.html

Seaweed Information Server—Images
seaweed.ucg.ie/seaweed.html

Marine Algae
ceres.ca.gov/ceres/calweb/coastal/plants/algae.html

Uses of Seaweeds
seaweed.ucg.ie/seaweedusesgeneral/seaweeduses.html

Corals and Coral Reefs

Reef Check 2000—Home Page
www.ust.hk/~webrc/ReefCheck/reef.html

Coral Health and Monitoring Program
coral.aoml.noaa.gov/

Global Relief Effort for Coral Reefs
www.nos.noaa.gov/aa/ia/cri.html

NOAA's Coral Reef
www.coralreef.noaa.gov/

Coral Reef and Marine Conservation
www.reefnet.org/

Coral Reef Photo Gallery
octopus.gma.org/onlocation/belize/gallery2.html

Fisheries and Aquaculture

Ocean Planet Over Fishing
seawifs.gsfc.nasa.gov/OCEAN_PLANET/HTML/peril_fishing.html
Go to Crab and Shrimp

The Fishing Industry
www.gov.nf.ca/publicat/tags/text/page 23.htm
See Shellfish and Crabs

Aquaculture Trends
www.fao.org/WAICENT/FAOINFO/FISHERY/trends/aqtrends/aqtrend.asp
Go to Shellfish

Biodiversity or biological diversity is the variety of life forms (number of species) found in a defined area. An area is made up of living and non-living parts that form a stable, self-sustaining system known as an **ecosystem** (ecological system). In an ecosystem organic molecules are manufactured and nutrients are recycled between the organisms and their environment.

mated 5–100 million species, less than 2 million identified). It is estimated that the ocean's coral reef ecosystems support at least one million species. The deep sea floor was once thought to be barren and inhospitable to life, but scientists now know this is not true. For example, scientists who surveyed 21 square meters of sea floor at depths between 1500 and 2500 m (5000–8000 ft) found 798 species, 46

Biodiversity in the Oceans

How much diversity is there? To answer requires that we know what species are present in the oceans, but the total number of species in the oceans is unknown. Only about 275,000 marine species have been described (compared to the earth with an esti-

new to science. Some biologists think the great area of the deep sea floor could be home to another 10 million species.

The oceans are made up of many distinctly different ecosystems, such as coral reefs, coastal areas, estuaries, open ocean waters,

Figure 1

Tropical coral reefs provide a large number of habitats; they are estimated to support at least one million species. (a) Corals and a sea fan at Mana Island, Fiji. (b) A butterfly fish (left) and a moorish idol (right) at Soma Soma Straits, Tareuni Island, Fiji.

deep sea floor, seamounts, hydrothermal vents, and cold seeps. Nearly 90% of the ocean's water lies below 100 m (330 ft), a homogeneous environment—cold and dark, but this deep-water environment is also extremely patchy in space and time. Nutrient patches caused by sinking phytoplankton blooms, fish, and the carcasses of marine mammals provide environments for many hundreds of other species. Small differences in such an environment allow for greater diversity than might be expected. There is competition for food and space, predation by other species and natural disturbances; all act to control an area's biodiversity.

The arrangement of the continents and oceans combined with latitude and related climate zones provides a series of areas with different water properties and patterns of circulation. Boundaries include currents and also changes in temperature, light, and density. As latitude decreases from the poles to the equator, species diversity tends to increase; the same is true on land. Because of the coral reefs and the diverse habitats they provide, the Pacific has a greater number of species than the Atlantic.

How much diversity is there? How fast is biodiversity being degraded or lost? What can slow or prevent an increasing rate of loss? There is much we do not know, but we do know the main cause and the most direct cause of biodiversity loss in the oceans is habitat destruction or deterioration due to human interference. For centuries people thought that humans could not drive any ocean-living species to extinction, for the sea was too big, too deep, its inhabitants too numerous, prolific, and widespread. In the last 200 years only one species of marine mammal (Stellar's sea cow) and four species of marine shellfish have become extinct, but the estimate of extinctions among small organisms in coastal areas and on coral reefs range from 100 to more than 1000. Consider the diversity of the ocean's food chains and remember that if you lose one component, you upset the entire ecosystem and other components suffer also. Losing a strand in the web of life can have unexpected and serious consequences for other organisms and humans as well.

For More Information:

Malakoff, D. 1997. Extinction on the High Seas. *Science* 277 (5325):486–88.

Mills, C. E., and J. T. Carlton. 1998. Rationale for a System of International Reserves for the Open Ocean. *Conservation Biology* 12 (1):244–47.

Solow, A. R. 1997. Biological Diversity in the Oceans. *Sea Technology* 38 (1):50–52.

Internet References

What is Biological Diversity and Why Is It Critical to Human Well-Being?
www.brrc.unr.edu/data/docs/biodiv.html

Centre for Biodiversity Research
www.bcu.ubc.ca/~otto/Biodiversity.html

Convention on Biological Diversity—Explanatory Leaflet
www.biodiv.org/conv/leaflet.html

Convention on Biological Diversity
www.biodiv.org/

Latitude and Longitude

To determine a location on the surface of the Earth, a grid of surface lines that cross each other at right angles is used. These lines are latitude and longitude. Lines of latitude begin at the equator, which is created by passing a plane through the Earth halfway between the poles and at right angles to the Earth's axis. The equator is marked as 0° latitude, and other latitude lines are drawn around the Earth parallel to the equator, north to 90°N, the North Pole, and south to 90°S, the South Pole (fig. A.1a). Lines of latitude are also termed parallels because they are parallel to the equator and to each other. Lines of latitude describe smaller and smaller circles as the poles are approached. All parallels of latitude must be designated as an angle either north or south of the equator. The latitude value is determined by the internal angle (ϕ or phi) between the latitude line, the Earth's center, and the equatorial plane (fig. A.1b).

Lines of longitude, also called meridians, are formed at right angles to the latitude grid (fig. A.2). Longitude begins at 0°, a line on the Earth's surface extending from the North Pole to the South Pole that passes directly through the Royal Naval Observatory in Greenwich, England. The 0° longitude line is known as the prime meridian. Longitude lines are identified by the angular displacement (θ or theta) to the east and west of 0° longitude (fig. A.2b). Directly opposite the prime meridian, on the other side of the Earth, 180° longitude approximates the international date line. All meridians are the same size; they mark the intersection of the Earth's surface with planes passing through the Earth's center at right angles to the parallels of latitude. Any circle that passes through the Earth's center is called a "great circle," and all longitude lines form great circles; only the equator is a great circle of latitude. A great circle connecting any two points on the Earth's surface defines the shortest distance between them.

To identify any location on the Earth's surface, use the crossing of the latitude and longitude lines; for example, 158°W, 21°N is the location of the Hawaiian Islands, and 18°E, 34°S identifies the Cape of Good Hope at the southern tip of Africa. For greater accuracy, one degree of latitude or longitude is divided into 60 minutes of arc; each minute is divided into 60 seconds of arc. One minute of arc length of latitude or longitude at the equator is equal to one nautical mile (1.852 km or 1.15 land miles).

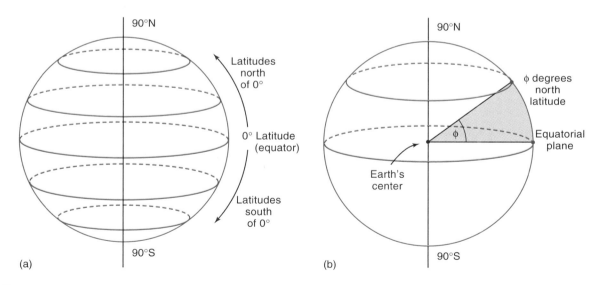

Figure A.1

(a) Latitude lines are drawn parallel to the equatorial plane. (b) The value of a latitude line is expressed in angular degrees determined by the angle formed between the equatorial plane and the latitude line to the Earth's center. This is the angle ϕ (phi). The degree of ϕ must be noted as north or south of the equator.

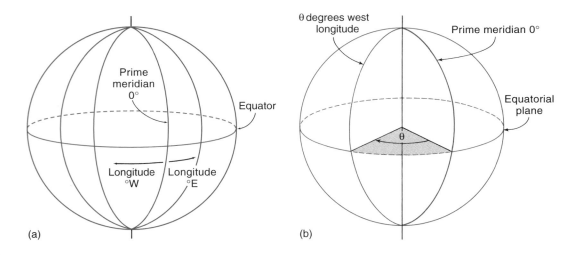

Figure A.2

(a) Longitude lines are drawn with reference to the prime meridian. (b) The value of a longitude line is expressed in angular degrees determined by the angle formed between the prime meridian and the longitude line to the Earth's center. This is the angle θ (theta). The value of θ is given in degrees east or west of the prime meridian.

Appendix B

Classification Summaries

Classification systems devised by biologists for identification and study of living organisms are works in progress. More than two hundred years ago Carolus Linnaeus introduced a two-kingdom system that divided all living organisms into plants and animals. The most widely used system today is the five kingdom system, placing living organisms in the kingdoms: Monera, Protista, Fungi, Plantae, and Animalia (see below). A new system based on genetic and biochemical research pro-poses three domains above the kingdom level; the Monera are placed in the Bacteria and Archaea domains and the Eukarya domain includes all the other kingdoms (see figure B.1). Consensus favoring the three-domain system is building among biologists and will be used more commonly during the next few years. Keep in mind that any classification system is a human construct that remains useful only until additional information and agreement among scientists leads to the adoption of a new system.

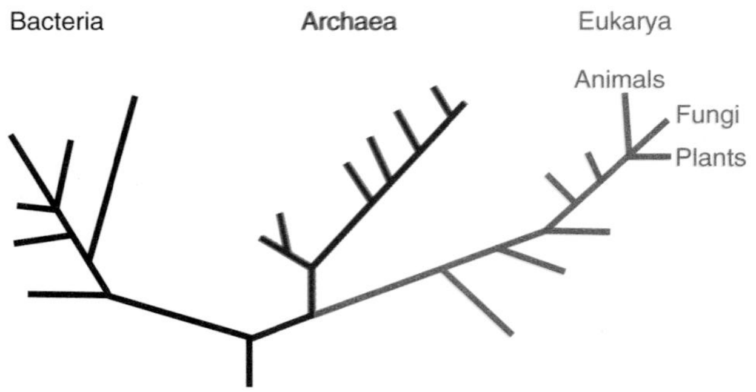

Bacteria   Archaea   Eukarya

Animals
Fungi
Plants

Figure B.1

The new family tree of life is divided into three domains: Bacteria, Archaea, and Eukarya. The Bacteria and Archaea are single cells without nuclei. The Eukarya domain includes many single-celled organisms with nuclei as well as the animals, plants, and fungi.

Classification Summary of the Plankton

The types of plankton can be categorized in kingdoms, divisions, phyla, and classes, as in the following list.

 I. **Kingdom Monera:** cells, simple and unspecialized; single cells, some in groups or chains.
   A. *Bacteria:* single cells, in chains or groups; autotrophic and heterotrophic, aerobic and anaerobic; important as food source and in decomposition.
   B. *Cyanobacteria:* blue-green algae; autotrophic single cells, in chains or groups; produce some red blooms in sea; phytoplankton.
 II. **Kingdom Protista:** grouping of microscopic and mostly single-celled organisms; autotrophs (algae) and heterotrophs (protozoa).

A. *Phylum Chrysophyta:* golden-brown algae; yellow to golden autotrophic single cells, in groups or chains; contribute to deep-sea sediments; phytoplankton.
  1. *Class Bacillariophyceae:* diatoms.
  2. *Class Chrysophyceae:* coccolithophores, silicoflagellates, and other flagellates.
B. *Phylum Dinoflagellata:* fire algae; single cells with flagella; produce most red tides; bioluminescence common; usually considered phytoplankton.
C. *Phylum Sarcodina:* microscopic heterotrophs, moving with pseudopodia; radiolarians and foraminiferans.
D. *Phylum Ciliophora:* microscopic heterotrophs, moving with cilia; ciliates, zooplankton.

III. **Kingdom Plantae:** plants; primarily nonmotile, multicellular, photosynthetic autotrophs.
A. *Division Phaeophyta:* brown algae; *Sargassum* maintains a planktonic habitat in the Sargasso Sea.

IV. **Kingdom Animalia:** animals; multicellular heterotrophs with specialized cells, tissues, and organ systems; zooplankton.
A. *Phylum Cnidaria:* coelenterates; radially symmetrical with tentacles and stinging cells.
  1. *Class Hydrozoa:* jellyfish or seajellies as one stage in the life cycle, including such colonial forms as the Portuguese man-of-war.
  2. *Class Scyphozoa:* jellyfish or seajellies.
B. *Phylum Ctenophore:* comb jellies; translucent; move with cilia; often bioluminescent.
C. *Phylum Chaetognatha:* arrowworms; free-swimming, carnivorous worms.
D. *Phylum Mollusca:* mollusks; the snail-like pteropod is planktonic.
E. *Phylum Arthropoda:* animals with paired, jointed appendages and hard outer skeletons.
  1. *Class Crustacea:* copepods and euphausiids.
F. *Phylum Chordata:* animals, including vertebrates, with dorsal nerve cord and gill slits at some stage in development.
  1. *Subphylum Urochordata:* saclike adults with "tadpole" larvae; salps.

Meroplankton: larval forms from the phyla *Annelida* (segmented worms), *Mollusca* (shellfish and snails), *Arthropoda* (crabs and barnacles), *Echinodermata* (starfish and sea urchins), and *Chordata* (fish).

Classification Summary of the Nekton

The nekton, all members of the kingdom Animalia, can be classified in the following phyla and classes.

I. **Kingdom Animalia:** animals; multicellular heterotrophs with specialized cells, tissues, and organ systems.
A. *Phylum Mollusca:* mollusks.
  1. *Class Cephalopoda:* squid; the octopus is a member of the benthos.

B. *Phylum Chordata:* animals with a dorsal nerve cord and gill slits at some stage in development.
  1. *Subphylum Vertebrata:* animals with a backbone of bone or cartilage.
    a. *Class Agnatha:* jawless fish; lampreys and hagfish.
    b. *Class Chondrichthyes:* jawed fish with cartilaginous skeletons; sharks and rays.
    c. *Class Osteichthyes:* bony fish; all other fish, including commercial species such as salmon, tuna, herring, and anchovy.
    d. *Class Reptilia:* air breathers with dry skin and scales; young develop in self-contained eggs; turtles, sea snakes, iguanas, and crocodiles.
    e. *Class Aves:* warm-blooded; skin covered with feathers, forelimbs are wings; embryo enclosed in shell; birds.
    f. *Class Mammalia:* warm-blooded; body covering of hair; produce milk; young born live.
      (1) *Order Cetacea:* whales.
        (a) *Suborder Mysticeti:* baleen whales, including blue, gray, right, sei, finback, and humpback whales.
        (b) *Suborder Odonticeti:* toothed whales, including the killer and sperm whale, porpoise, and dolphin.
      (2) *Order Carnivora:* sea otters, seals, sea lions, and walruses.
      (3) *Order Sirenia:* dugongs and manatees.

Classification Summary of the Benthos

Benthic organisms include members of the following kingdoms, divisions, and phyla.

I. **Kingdom Monera:** includes the bacteria, an important food source in mud and sand environments; also cyanobacteria or blue-green algae, especially abundant on coral reefs.

II. **Kingdom Protista:** convenience grouping of microscopic, usually unicellular plants and animals.
A. *Phylum Chrysophyta:* golden-brown algae; mainly planktonic, but benthic diatoms are abundant.
B. *Phylum Protozoa:* infauna include many members among sand and mud particles, including amoeboid and foraminiferan species.

III. **Kingdom Plantae:** plants; primarily nonmotile, multicellular, photosynthetic autotrophs.
A. *Phylum Chlorophyta:* green algae; seaweeds such as *Ulva* and *Codium.*
B. *Phylum Phaeophyta:* brown algae; coastal seaweeds, including the large kelps *Nereocystis, Macrocystis,* and *Postelsia. Sargassum* maintains a planktonic habit in the Sargasso Sea.
C. *Phylum Rhodophyta:* red algae; littoral and sublittoral seaweeds of varied form.

D. *Phylum Tracheophyta:* plants with vascular tissue; true roots, stems, and leaves present.
   1. *Class Angiospermae:* flowering plants; seeds enclosed in fruit; eelgrass, surf grass, and mangrove trees.

IV. **Kingdom Animalia:** animals; multicellular heterotrophs with specialized cells, tissues, and organ systems.
  A. *Phylum Porifera:* sponges; simple, nonmotile filter feeders.
  B. *Phylum Cnidaria:* coelenterates; radially symmetric, with tentacles and stinging cells.
   1. *Class Anthozoa:* sea anemones, corals, and sea fans.
  C. *Phylum Platyhelminthes:* flatworms; free-living and parasitic forms.
  D. *Phylum Nemertina:* ribbon worms.
  E. *Phylum Nematoda (Aschelminthes):* round worms, or nematodes.
  F. *Phylum Brachiopoda:* lampshells.
  G. *Phylum Annelida:* segmented worms.
   1. *Class Polychaeta:* free-living marine carnivores, including tube worms, *Nereis,* and clam worms.
  H. *Phylum Pogonophora:* deep-sea tube-dwelling worms; giant members found around hot vents on the sea floor.

I. *Phylum Mollusca:* mollusks.
   1. *Class Polyplacophora:* chitons.
   2. *Class Scaphopoda:* tusk shells.
   3. *Class Gastropoda:* single-shelled mollusks; snails, limpets, and sea slugs.
   4. *Class Pelecypoda* or *Bivalvia:* bivalved mollusks; clams, mussels, scallops, and oysters.
   5. *Class Cephalopoda:* octopus. The squid is a member of the nekton.
J. *Phylum Arthropoda:* animals with paired, jointed appendages and outer skeletons.
   1. *Class Crustacea:* crabs, shrimp, and barnacles.
K. *Phylum Echinodermata:* radially symmetric, spiny-skinned animals with water-vascular system, including starfish, sea urchins, sand dollars, and sea cucumbers.
L. *Phylum Hemichordata:* acorn worms.
M. *Phylum Chordata:* animals, including vertebrates, with dorsal nerve cord and gill slits at some stage in development.
   1. *Subphylum Urochordata:* filter-feeding, saclike adults with "tadpole" larvae; sea squirt.

Appendix C

Units and Notation

Scientific (or Exponential) Notation

The writing of very large and very small numbers is simplified by using exponents, or powers of 10, to indicate the number of zeroes required to the left or to the right of the decimal point. The numbers that are equal to some of the powers of 10 are as follows:

$$1,000,000,000. = 10^9 = \text{one billion}$$
$$1,000,000. = 10^6 = \text{one million}$$
$$1000. = 10^3 = \text{one thousand}$$
$$100. = 10^2 = \text{one hundred}$$
$$10. = 10^1 = \text{ten}$$
$$1. = 10^0 = \text{one}$$
$$0.1 = 10^{-1} = \text{one tenth}$$
$$0.01 = 10^{-2} = \text{one hundredth}$$
$$0.001 = 10^{-3} = \text{one thousandth}$$
$$0.000001 = 10^{-6} = \text{one millionth}$$
$$0.000000001 = 10^{-9} = \text{one billionth}$$

The number 149,000,000 is rewritten by moving the decimal point eight places to the left and multiplying by the exponential number 10^8, to form 1.49×10^8. In the same way, 605,000 becomes 6.05×10^5.

A very small number such as 0.000032 becomes 3.2×10^{-5} by moving the decimal point five places to the right and multiplying by the exponential number 10^{-5}. Similarly, 0.00000372 becomes 3.72×10^{-6}.

To add or subtract numbers written in exponential notation, convert the numbers to the same power of 10. For example,

$$
\begin{array}{ccccc}
1.49 \times 10^3 & & 1.490 \times 10^3 & & 14.90 \times 10^2 \\
+6.05 \times 10^2 & = & 0.605 \times 10^3 & = & 6.05 \times 10^2 \\
\hline
& & 2.095 \times 10^3 & & 20.95 \times 10^2
\end{array}
$$

$$
\begin{array}{ccccc}
2.36 \times 10^3 & & 23.60 \times 10^2 & & 2.360 \times 10^3 \\
-1.05 \times 10^2 & = & -1.05 \times 10^2 & = & -0.105 \times 10^3 \\
\hline
& & 22.55 \times 10^2 & & 2.255 \times 10^3
\end{array}
$$

To multiply, the exponents are added and the numbers are multiplied:

$$
\begin{array}{c}
4.6 \times 10^3 \\
\times 2.2 \times 10^2 \\
\hline
10.12 \times 10^5 = 1.012 \times 10^6
\end{array}
$$

To divide, subtract the exponents and divide the numbers:

$$\frac{6.0 \times 10^8}{2.5 \times 10^3} = 2.4 \times 10^5$$

The following prefixes correspond to the powers of 10 and are used in combination with metric units:

SI Units

The *Système International d'Unités*, or International System of Units, is a simplified system of metric units (known as SI units) adopted by international convention for scientific use.

Exponential Notation

Exponential Value	Prefix	Symbol
10^{18}	exa	E
10^{15}	peta	P
10^{12}	tera	T
10^{9}	giga	G
10^{6}	mega	M
10^{3}	kilo	k
10^{2}	hecto	h
10^{1}	deka	da
10^{-1}	deci	d
10^{-2}	centi	c
10^{-3}	milli	m
10^{-6}	micro	µ
10^{-9}	nano	n
10^{-12}	pico	p
10^{-15}	femto	f
10^{-18}	atto	a

Basic SI Units

Quantity	Unit	Symbol
length	meter	m
mass	kilogram	kg
time	second	s
temperature	kelvin	K

Derived SI Units

Quantity	Unit	Symbol	Expression in SI Base Units
area	meter squared	m^2	m^2
volume	meter cubed	m^3	m^3
density	kilogram per cubic meter	kg/m^3	kg/m^3
speed	meter per second	m/s	m/s
acceleration	meter per second per second	m/s^2	m/s^2
force	newton	N	$(kg)(m)/s^2$
pressure	pascal	Pa	N/m^2
energy	joule	J	$(N)(m)$
power	watt	W	$J/s; (N)(m)/s$

Length: The Basic SI Unit Is the Meter

Unit	Metric Equivalents	Traditional Equivalents	Other
meter (m)	100 centimeters 1000 millimeters	39.37 inches 3.281 feet	0.546 fathom
kilometer (km)	1000 meters	0.621 land mile	0.540 nautical mile
centimeter (cm)	10 millimeters 0.01 meter	0.394 inch	
millimeter (mm)	0.1 centimeter 0.001 meter	0.0394 inch	
land mile (mi)	1609 meters	5280 feet	0.869 nautical mile
nautical mile (nm)	1852 meters	1.151 land mile 6076 feet	1 minute of latitude
fathom (fm)	1.829 meters	6 feet	

Area: Derived from Length

Unit	Metric Equivalents	Traditional Equivalents	Other
square meter (m^2)	10,000 square centimeters	10.76 square feet	
square kilometer (km^2)	1,000,000 square meters	0.386 square land mile	0.292 square nautical mile
square centimeter (cm^2)	100 square millimeters	0.151 square inch	

Volume: Derived from Length

Unit	Metric Equivalents	Traditional Equivalents	Other
cubic meter (m^3)	1,000,000 cubic centimeters 1000 liters	35.32 cubic feet 264 U.S. gallons	
cubic kilometer (km^3)	1,000,000,000 cubic meters	0.239 cubic land mile	0.157 cubic nautical mile
liter (L) or (l)	1000 cubic centimeters	1.06 quarts, 0.264 U.S. gallon	
milliliter (mL)	1.0 cubic centimeter		

Mass: The Basic SI Unit Is the Kilogram

Unit	Metric Equivalents	Traditional Equivalents
kilogram (kg)	1000 grams	2.205 pounds
gram (g)		0.035 ounce
metric ton, or tonne (t)	1000 kilograms 1,000,000 grams	2205 pounds
U.S. ton	907 kilograms	2000 pounds

Time: The Basic SI Unit Is the Second

Unit	Metric and Traditional Equivalents
minute	60 seconds
hour	60 minutes; 3600 seconds
day	86,400 seconds ⎫ 24 hours ⎭ mean solar day
year	31,556,880 seconds
	8765.8 hours ⎫ 365.25 solar days ⎭ mean solar year

Temperature: The Basic SI Unit Is the Kelvin

Reference Points	Kelvin (K)	Celsius (°C)	Fahrenheit (°F)
absolute zero	0	−273.2	−459.7
seawater freezes	271.2	−2.0	28.4
fresh water freezes	273.2	0.0	32.0
human body	310.2	37.0	98.6
fresh water boils	373.2	100.0	212.0
conversions	K = °C + 273.2	$°C = \dfrac{(°F - 32)}{1.8}$	°F = (1.8 × C) + 32

Speed (Velocity): The Derived SI Unit Is the Meter per Second

Unit	Metric Equivalents	Traditional Equivalents	Other
meter per second (m/s)	100 centimeters per second 3.60 kilometers per hour	3.281 feet per second 2.237 land miles per hour	1.944 knots
kilometer per hour (km/hr)	0.277 meter per second	0.909 foot per second	0.55 knot
knot (kt)	0.51 meter per second	1.151 land miles per hour	1 nautical mile per hour

Acceleration: The Derived SI Unit Is the Meter per Second per Second

Unit	Metric Equivalents	Traditional Equivalents
meter per second per second (m/s^2)	12,960 kilometers per hour per hour 100 centimeters per second per second	8048 miles per hour per hour 3.281 feet per second per second

Force: The Derived SI Unit Is the Newton

Unit	Metric Equivalents	Traditional Equivalents
newton (N)	100,000 dynes	0.2248 pound force
dyne (dyn)	0.00001 newton	0.000002248 pound force

Pressure: The Derived SI Unit Is the Pascal

Unit	Metric Equivalents	Traditional Equivalents	Other
pascal (Pa)	1 newton per square meter 10 dynes per square centimeter		
bar	100,000 pascals 1000 millibars	14.5 pounds per square inch	0.927 atmosphere 29.54 inches of mercury
standard atmosphere (atm)	1.013 bars 101,300 pascals	14.7 pounds per square inch	29.92 inches of mercury or 76 cm of mercury

Energy: The Derived SI Unit Is the Joule

Unit	Metric Equivalents	Traditional Equivalents
joule (J)	1 newton-meter 0.2389 calorie	0.0009481 British thermal unit
calorie (cal)	4.186 joules	0.003968 British thermal unit

Power: The Derived SI Unit Is the Watt

Unit	Metric Equivalents	Traditional Equivalents
watt (W)	1 joule per second 0.2389 calorie per second 0.001 kilowatt	0.0569 British thermal unit per minute 0.001341 horsepower

Density: The Derived SI Unit Is the Kilogram per Cubic Meter

Unit	Metric Equivalents	Traditional Equivalents
kilogram per cubic meter (kg/m^3)	0.001 gram per cubic centimeter (g/cm^3)	0.0624 pound per cubic foot

To Learn More About Metrics:
U.S. Metric Association (USMA) Inc. at
lamar.colostate.edu/~hillger/

glossary

a

absorption taking in of a substance by chemical or molecular means; change of sound or light energy into some other form, usually heat, in passing through a medium or striking a surface.

abyssal pertaining to the great depths of the ocean below approximately 4000 meters.

abyssal hill low, rounded submarine hill less than 1000 m high.

abyssal plain flat, ocean basin floor extending seaward from the base of the continental slope and continental rise.

active margin *See* leading margin.

adsorption attraction of ions to a solid surface.

advection horizontal or vertical transport of seawater, as by a current.

advective fog fog formed when warm air saturated with water vapor moves over cold water.

agar substance produced by red algae; the gelatin-like product of these algae.

algae marine and freshwater plantlike organisms (including seaweeds) that are single-celled, colonial, or multicelled, with chlorophyll but no true roots, stems, or leaves and with no flowers or seeds.

algin complex organic substance found in or obtained from brown algae.

amplitude for a wave, the vertical distance from sea level to crest or from sea level to trough, or one-half the wave height.

anion negatively charged ion.

anoxic deficient in oxygen.

Antarctic bottom water densest oceanic water type, formed at the surface below sea ice, flows northward along the floor of the Atlantic Ocean.

Antarctic Circle *See* Arctic and Antarctic Circles.

Antarctic intermediate water water with a salinity of 34.4% and a temperature of 5°C, produced at the 40°S surface convergence in the Atlantic.

antinode portion of a standing wave with maximum vertical motion.

aphotic zone that part of the ocean in which light is insufficient to carry on photosynthesis.

aquaculture cultivation of aquatic organisms under controlled conditions. *See also* mariculture.

Arctic and Antarctic Circles latitudes $66^1/_2°$N and $66^1/_2°$S, respectively, marking the boundaries of light and darkness during the summer and winter solstices.

asthenosphere upper, deformable portion of the Earth's mantle, the layer below the lithosphere; probably partially molten; may be site of convection cells.

atmospheric pressure pressure exerted by the atmosphere due to the weight of the column of air lying directly above any point on Earth.

atoll ring-shaped coral reef that encloses a lagoon in which there is no exposed preexisting land and which is surrounded by the open sea.

attenuation decrease in the energy of a wave or beam of particles occurring as the distance from the source increases; caused by absorption, scattering, and divergence from a point source.

autumnal equinox *See* equinoxes.

b

backshore beach zone lying between the foreshore and the coast, acted upon by waves only during severe storms and exceptionally high water.

baleen whalebone; horny material growing down from the upper jaw of plankton-feeding whales; forms a strainer, or filtering organ, consisting of numerous plates with fringed edges.

bar offshore ridge or mound of sand, gravel, or other loose material, which is submerged, at least at high tide; located especially at the mouth of a river or estuary, or lying a short distance from and parallel to the beach.

barrier island deposit of sand, parallel to shore and raised above sea level; may support vegetation and animal life.

barrier reef coral reef that parallels land but is some distance offshore, with water between reef and land.

basalt fine-grained, dark igneous rock, rich in iron, magnesium, and calcium; characteristic of oceanic crust.

basin large depression of the sea floor having about equal dimensions of length and width.

bathymetric pertaining to the study and mapping of seafloor elevations and the variations of water depth; pertaining to the topography of the sea floor.

beach zone of unconsolidated material between the mean low water line and the line of permanent vegetation, which is also the effective limit of storm waves; sometimes includes the material moving in offshore, onshore, and longshore transport.

beach face section of the foreshore normally exposed to the action of waves.

benthic of the sea floor, or pertaining to organisms living on or in the sea floor.

benthos organisms living on or in the ocean bottom.

berm nearly horizontal portion of a beach (backshore) with an abrupt face; formed from the deposition of material by wave action at high tide.

berm crest ridge marking the seaward limit of a berm.

biodiversity species richness, the number of species in an area compared with the number of individuals.

biogenous sediment sediment having more than 30% material derived from organisms.

bioluminescence production of light by living organisms as a result of a chemical reaction either within certain cells or organs or outside the cells in some form of excretion.

biomass the total mass of all or specific living organisms, usually expressed as dry weight or grams of carbon per unit area or unit volume. *See also* standing crop.

bioturbation reworking of sediments by organisms that burrow and ingest them.

blade flat, photosynthetic, "leafy" portion of an alga or seaweed.

bloom high concentration of phytoplankton in an area, caused by increased reproduction; often produces discoloration of the water. *See also* red tide.

bore *See* tidal bore.

breaker sea surface-water wave that has become too steep to be stable and collapses.

breakwater structure protecting a shore area, harbor, anchorage, or basin from waves; a type of jetty.

buffer substance able to neutralize acids and bases, therefore able to maintain a stable pH.

buoyancy ability of an object to float due to the support of the fluid the body is in or on.

by-catch *See* incidental catch.

C

calcareous containing or composed of calcium carbonate.

calorie amount of heat required to raise the temperature of 1 gram of water 1°C at 15° to 16°C.

carbonate an ion composed of one carbon and three oxygen atoms, CO_3^{2-}.

carbonate compensation depth (CCD) the depth at which the amount of calcareous material preserved falls below 20% of the total sediment.

carnivore flesh-eating organism.

carrageenan substance produced by certain algae, used as a thickening agent.

cation positively charged ion.

centrifugal force outward-directed force acting on a body moving along a curved path or rotating about an axis.

centripetal force inward-directed force necessary to keep an object moving in a curved path or rotating around an axis.

cetacean any member of the order Cetacea, mostly marine mammals, includes whales, dolphins, and porpoises.

chemosynthesis formation of organic compounds with energy derived from inorganic substances such as ammonia, sulfur, and hydrogen.

chlorinity measure of the chloride content of seawater in grams per kilogram.

chlorophyll group of green pigments that are active in photosynthesis.

cilia microscopic, hairlike processes of living cells, which beat in coordinated fashion and produce movement.

Cnidaria phylum of radially symmetrical marine organisms with tentacles and stinging cells; includes jellyfish, sea anemones, and corals.

coast strip of land of indefinite width that extends from the shore inland to the first major change in terrain that is unaffected by marine processes.

coastal circulation cell longshore transport cell pattern of sediment moving from a source to a place of deposition.

coastal zone land and water areas including cliffs, dunes, beaches, bays, and estuaries.

coccolithophore microscopic, planktonic alga surrounded by a cell wall with embedded calcareous plates (coccoliths).

coelenterate *See* Cnidaria.

colonial (organism) organism consisting of semi-independent parts that do not exist as separate units; groups of organisms with specialized functions that form a co-ordinated unit.

commensalism an intimate association between different organisms in which one is benefited and the other is neither harmed nor benefited.

conduction transfer of heat energy through matter by internal molecular motion; also heat transfer by turbulence in fluids.

conservative ions seawater ions whose concentration changes only as a result of physical processes and not as a result of biological or chemical processes; for example, salinity.

continental drift motion of the continents due to plate tectonics.

continental margin zone separating the continents from the deep-sea bottom, usually subdivided into shelf, slope, and rise.

continental rise gentle slope formed by the deposition of sediments at the base of a continental slope.

continental shelf zone bordering a continent, extending from the line of permanent immersion to the depth at which there is a marked or rather steep descent to the great depths.

continental shelf break zone along which there is a marked increase of slope at the outer margin of a continental shelf.

continental slope relatively steep downward slope from the continental shelf break to depth.

contour line on a chart or graph connecting the points of equal value for elevation, temperature, salinity, and so on.

convection transmission of heat by the movement of a heated gas or liquid; vertical circulation resulting from changes in density of a fluid.

convection cell density-driven transfer of heat by circulation of liquid or gas in which warm, low-density material rises and cold, high-density material falls.

convergence situation in which fluids of different origins come together, usually resulting in the sinking, or downwelling, of surface water and the rising of air.

copepod small, shrimplike members of the zooplankton.

coral colonial animal that secretes a hard, outer, calcareous skeleton; the skeletons of coral animals form in part the framework for warm-water reefs.

core vertical, cylindrical sample of bottom sediments, from which the nature of the bottom can be determined; also the central zone of the Earth, thought to be liquid or molten on the outside and solid on the inside.

corer device that plunges a hollow tube into bottom sediments to extract a vertical sample.

Coriolis effect apparent force acting on a body in motion, due to the rotation of the Earth, causing deflection to the right in the Northern Hemisphere and to the left in the Southern Hemisphere; the force is proportional to the speed and latitude of the moving body.

cosmogenous sediment sediment particles with an origin in outer space; for example, meteor fragments and cosmic dust.

crest *See* berm crest, reef crest, wave crest.

crust outermost shell of the Earth, separated from the mantle by the Moho; continental crust is comprised of granite-type rock and oceanic crust is comprised of basalt-type rock.

crustacean member of a class of primarily aquatic organisms with paired jointed appendages and a hard outer skeleton; includes lobsters, crabs, shrimp, and copepods.

CTD commonly used abbreviation for conductivity-temperature-depth sensor, electronic instruments used to measure salinity, temperature, and depth of seawater.

ctenophores transparent, planktonic animals, spherical or cylindrical in shape with rows of cilia; comb jellies.

Curie temperature the temperature at which the magnetic signature of a rock is frozen into the rock as it cools from the molten state.

current horizontal movement of water.

current meter instrument for measuring the speed and direction of a current.

cusp evenly spaced, crescent-shaped depressions along a sand or gravel beach.

cyclone *See* typhoon, hurricane.

d

decomposer microorganisms (usually bacteria and fungi) that break down nonliving organic matter and release nutrients.

deep scattering layer (DSL) layer of organisms that move away from the surface during the day and toward the surface at night; the layer scatters or returns vertically directed sound pulses.

deep-sea reversing thermometer (DSRT) mercury-in-glass thermometer that records seawater temperature upon being inverted and retains its reading until returned to its upright position.

deep-water wave wave in water where depth is greater than one-half its wavelength.

delta area of unconsolidated sediment deposit, usually triangular in outline, formed at the mouth of a river.

density property of a substance defined as mass per unit volume and usually expressed in grams per cubic centimeter or kilograms per cubic meter.

depth recorder *See* echo sounder.

desalination process of obtaining fresh water from seawater.

detritus any loose material, especially decomposed, broken, and dead organic materials.

diatom microscopic unicellular alga with an external skeleton of silica.

diffraction process that transmits energy laterally along a wave crest.

dinoflagellate one of a class of planktonic organisms, often bioluminescent; may form red tides.

dipole magnetic field with two magnetic poles of opposite polarity.

dispersion (sorting) sorting of waves as they move out from a storm center; occurs because long waves travel faster in deep water than short waves.

diurnal tide tide with one high water and one low water each tidal day.

divergence horizontal flow of fluids away from a common center, associated with upwelling in water and descending motions in air.

doldrums nautical term for the belt of light, variable winds near the equator. *See also* intertropical convergence.

downwelling sinking of water toward the bottom, usually the result of a surface convergence or an increase in density of water at the sea surface.

dredge cylindrical or boxlike sampling device made of metal, net, or both, which is dragged across the bottom to obtain biological or geological samples.

dugong *See* sea cow.

dune wind-formed hill or ridge of sand.

dynamic tide an actual tide as it occurs in the ocean basin.

e

earth sphere depth uniform depth of the Earth below the present mean sea level, if the solid Earth surface were smoothed off evenly (2440m).

ebb tide falling tide; the period of the tide between high water and the next low water.

echo sounder (depth recorder) instrument used to measure the depth of water by measuring the time interval between the release of a sound pulse and the return of its echo from the bottom.

ecology the scientific study of the interactions among and between organisms and between the organisms and all living and non-living aspects of their environment.

eddies circular movements of water.

Ekman spiral in a theoretical ocean of infinite depth, unlimited extent, and uniform viscosity, with a steady wind blowing over the surface, the surface water moves 45° to the right of the wind in the Northern Hemisphere. At greater depths the water moves farther to the right with decreased speed, until at some depth (approximately 100 m) the water moves opposite to the wind direction. Net water transport is 90° to the right of the wind in the Northern Hemisphere. Movement is to the left in the Southern Hemisphere.

Ekman transport net water transport in a theoretical ocean of infinite depth with a steady wind blowing over the surface, moves 90° to the right of the wind in the Northern Hemisphere and 90° to the left of the wind in the Southern Hemisphere.

electromagnetic radiation waves of energy formed by simultaneous electric and magnetic oscillations; the electromagnetic spectrum is the continuum of all electromagnetic radiation from low-energy radiowaves to high-energy gamma rays, including visible light.

El Niño wind-driven reversal of the Pacific equatorial currents resulting in the movement of warm water toward the coasts of the Americas, so called because it generally develops just after Christmas. *See also* La Niña.

El Viejo *See* La Niña.

emergent coast a coast that is rising relative to sea level.

epicenter the point on the surface of the Earth directly above the focus of an earthquake. It is specified by two coordinates: latitude and longitude.

epifauna animals living attached to the sea bottom or moving freely over it.

episodic wave abnormally high wave unrelated to local storm conditions.

equator 0° latitude, determined by a plane that is perpendicular to the Earth's axis and is everywhere equidistant from the North and South Poles.

equilibrium tide theoretical tide formed by the tide-producing forces of the Moon and Sun on a water-covered Earth.

equinoxes times of the year when the Sun stands directly above the equator, so that day and night are of equal length around the world. The vernal equinox occurs about March 21, and the autumnal equinox occurs about September 22.

estuary semi-isolated portion of the ocean, which is diluted by freshwater drainage from land.

eukaryote cell characterized by an organized nucleus and other membrane-bound subcellular structures.

euphausiid planktonic, shrimplike crustacean. *See also* krill.

evaporation process by which a liquid becomes a vapor. *See also* hydrologic cycle.

evapotranspiration the combined effect of evaporation and transpiration. *See also* evaporation, transpiration.

extremophile microorganism that thrives under extreme conditions of temperature, lack of oxygen, or high acid or salt levels; conditions that kill other organisms.

f

fathom a traditional unit of length equal to 1.83 m or 6 ft; used to measure water depth.

fault break or fracture in the Earth's crust, in which one side has been displaced relative to the other.

fetch continuous area of water over which the wind blows in essentially a constant direction.

fjord narrow, deep, steep-walled inlet of the ocean formed by the submergence of a mountainous coast or by the entrance of the ocean into a deeply excavated glacial trough after the melting of the glacier.

flagellum (pl. flagella) long, whiplike extension from a living cell's surface that by its motion moves the cell.

flood tide rising tide; the period of the tide between low water and the next high water.

flushing time length of time required for an estuary to exchange its water with the open ocean.

focus (of an earthquake) the point of origin of an earthquake; specified by three coordinates: latitude, longitude, and depth.

fog visible assemblage of tiny droplets of water formed by condensation of water vapor in the air; a cloud with its base at the surface of the Earth.

food chain sequence of organisms in which each is food for the next member of the sequence. *See also* food web.

food web complex of interacting food chains; all the feeding relations of a community taken together; includes production, consumption, decomposition, and the flow of energy.

foraminifera minute, one-celled animals that usually secrete calcareous shells.

foreshore portion of the shore that includes the low tide terrace and the beach face.

fringing reef reef attached directly to the shore of an island or a continent and not separated from it by a lagoon.

g

Gondwanaland the southern portion of Pangaea consisting of Africa, South America, India, Australia, and Antarctica.

grab sampler instrument used to remove a piece of the ocean floor for study.

granite crystalline, coarse-grained, igneous rock composed mainly of quartz and feldspar.

gravitational force mutual force of attraction between particles of matter (bodies).

gravity Earth's gravity; acceleration due to the Earth's mass is 981 cm/s²; g is used in equations.

greenhouse effect warming effect caused by the atmosphere's transparency to incoming sunlight and its opacity to outgoing radiation.

Greenwich Mean Time (GMT) *See* Universal Time.

group speed speed at which a group of waves travels (in deep water, group speed equals one-half the speed of an individual wave); the speed at which the wave energy is propagated.

guyot submerged, flat-topped seamount.

gyre circular movement of water, larger than an eddy; usually applied to oceanic systems.

h

half-life time required for half of an initial quantity of a radioactive isotope to decay.

heat budget accounting for the total amount of the Sun's heat received on Earth during one year as being exactly equal to the total amount lost from the Earth due to radiation and reflection.

heat capacity quantity of heat needed to produce a unit change of temperature in a unit mass.

herbivore animal that feeds only on plants.

higher high water higher of the two high waters of any tidal day in a region of mixed tides.

higher low water higher of the two low waters of any tidal day in a region of mixed tides.

high-pressure zone region of air with greater than average density and atmospheric pressure.

high water maximum height reached by a rising tide.

holdfast organ of a benthic alga that attaches the alga to the sea floor.

holoplankton organisms living their entire life cycle in the floating (planktonic) state.

horse latitudes regions of calms and variable winds that coincide with latitudes at approximately 30° to 35°N and S.

hot spot surface expression of a persistent rising jet of molten mantle material.

hurricane severe, cyclonic, tropic storm at sea, with winds of 120 km/hr (73 mph) or more; generally applied to Atlantic Ocean storms. *See also* typhoon.

hydrogenous sediment sediment formed from substances dissolved in seawater.

hydrologic cycle movement of water among the land, oceans, and atmosphere due to changes of state, vertical and horizontal transport, evaporation, and precipitation.

hydrothermal vent seafloor outlet for high-temperature groundwater and associated minerals; a hot spring.

hypsographic curve graph of land elevation and ocean depth versus area.

i

iceberg mass of land ice that has broken away from a glacier and floats in the sea.

incidental catch (by-catch) animals caught by those fishing for other species.

infauna animals that live buried in the sediment.

inner core inner solid portion of the Earth's core.

international date line a theoretical line approximating the 180° meridian. Regions on the east side of the line are one day earlier in calendar date than regions on the west side.

intertidal *See* littoral zone.

intertidal zonation vertical distribution of plants and animals over tidal and subtidal areas.

intertropical convergence meteorological term for the belt of light, variable winds near the equator. *See also* doldrums.

invertebrate any animal without a backbone or spinal column.

ion positively or negatively charged atom or group of atoms.

island arc chain of volcanic islands formed when plates converge at a subduction zone.

isotope atoms of the same element having different numbers of neutrons.

i

jet stream a high-speed, generally westerly wind of the upper atmosphere between 30°N and 50°N.

jetty structure located to influence currents or protect the entrance to a harbor or river from waves (U.S. terminology). *See also* breakwater.

k

kelp any of several large, brown algae, including the largest known algae.

knot a traditional unit of speed equal to 0.51 m/s or 1 nautical mile per hour.

krill term used by whalers for the small, shrimp-like crustaceans found in huge masses in polar waters and eaten by baleen whales.

l

lagoon shallow body of water, which usually has a shallow, restricted outlet to the sea.

La Niña (El Viejo) condition of colder-than-normal surface water in the eastern tropical Pacific. *See also* El Niño.

larva (pl. larvae) immature juvenile form of an animal.

latitude line on the Earth's surface paralleling the equator (0° latitude), value is expressed in degrees north and south of the equator.

Laurasia the northern portion of Pangaea composed of North America and Eurasia.

leading margin on a lithospheric plate moving toward a subduction zone, the margin of a continent closest to the subduction zone.

lithogenous sediment sediment composed of rock particles eroded mainly from the continents by water, wind, and waves.

lithosphere outer, rigid portion of the Earth; includes the continental and oceanic crust and the upper part of the mantle.

littoral (intertidal) area of the shore between mean high water and low water; the intertidal zone.

longitude line on the Earth's surface drawn between the poles whose value is expressed in degrees east and west of the prime meridian (0° longitude).

longshore current current produced in the surf zone by waves breaking at an angle with the shore; runs roughly parallel to the shoreline.

longshore transport movement of sediment by the longshore current.

loran navigational system in which position is determined by measuring the difference in the time of reception of synchronized radio signals; derived from the phrase "long-range navigation."

lower high water lower of the two high waters of any tidal day in a region of mixed tides.

lower low water lower of the two low waters of any tidal day in a region of mixed tides.

low-pressure zone region of air with less than average density and atmospheric pressure.

low tide terrace flat section of the foreshore seaward of the sloping beach face.

low water minimum height reached by a falling tide.

lysocline the depth at which calcareous skeletal material first begins to dissolve.

m

magma molten rock material that forms igneous rocks upon cooling; magma that reaches the Earth's surface is referred to as lava.

magnetic pole either of the two points on the Earth's surface where the magnetic field is vertical.

major constituents the most abundant salts in seawater, present in concentrations expressed in parts per thousand or grams per kilogram.

manatee *See* sea cow.

manganese nodules rounded, layered lumps found on the deep-ocean floor that contain as much as 20% manganese and smaller amounts of iron, nickel, and copper; a hydrogenous sediment.

mantle main volume of the Earth between the crust and the outer core; temperature gradients produce density variations leading to slow convection of the material.

mariculture cultivation of marine organisms under controlled conditions. *See also* aquaculture.

mean average, a value intermediate between other values.

mean sea level average height of the sea surface, based on observations of all stages of the tide over a nineteen-year period in the United States.

meridian circle of longitude passing through the poles and any given point on the Earth's surface.

meroplankton floating developmental stages (eggs and larvae) of organisms that as adults belong to the nekton and benthos.

mesosphere lower portion of the mantle below the asthenosphere; temperature gradients and density variations drive slow convection.

minus tide tide level that falls below the zero tide depth reference level.

mixed tide type of tide in which large inequalities between the two high waters and the two low waters occur in a tidal day.

Moho (Mohorovičić discontinuity) chemical boundary between crust and mantle, marked by a rapid increase in seismic wave speed.

mollusks marine animals, usually with shells; includes mussels, oysters, clams, snails, and slugs.

monoculture cultivation of only one species of organism in an aquaculture system.

monsoon name for seasonal winds; first applied to the winds over the Arabian Sea, which blow for six months from the northeast and for six months from the southwest; now extended to similar winds in other parts of the world; in India, the term is popularly applied to the southwest monsoon and also to the rains that it brings.

motile moving or capable of moving.

mutualism an intimate association between different organisms in which both organisms benefit.

n

nautical mile unit of length equal to 1852 m or 1.15 land miles, or one minute of latitude.

neap tides tides occurring near the times of the first and last quarters of the Moon, when the range of the tide is least.

nebula large dense cloud of gas and dust in space.

nekton pelagic animals that are active swimmers; for example, adult squid, fish, and marine mammals.

neritic shallow water marine environment extending from low water to the edge of the continental shelf. *See also* pelagic.

node point of least or zero vertical motion in a standing wave.

nonconservative ions seawater ions whose concentration changes as a result of biological or chemical processes as well as physical processes; for example, nutrients and oxygen in seawater.

North Atlantic deep water Atlantic Ocean water with a salinity of about 34.9% and a temperature of 2 to 4°C, formed at the surface at approximately 50 to 60°N.

nudibranch soft-bodied, slug-like mollusk; sea slug.

nutrient in the ocean, any one of a number of inorganic or organic compounds or ions used primarily in the nutrition of primary producers; nitrogen and phosphorus compounds are examples.

o

oceanic pertaining to the ocean water seaward of the continental shelf; the "open ocean." *See also* pelagic.

ocean ranching raising of salmon to a juvenile stage, releasing of the juveniles to sea, and harvesting of adults on their return.

ocean thermal energy conversion (OTEC) method of extracting energy from the oceans that depends on the temperature difference between surface water and water at depth.

offshore direction seaward from the shore.

onshore direction toward the shore.

onshore current any current flowing toward the shore.

ooze fine-grained deep-ocean sediment composed of at least 30% sand or silt-sized calcareous or siliceous remains of small marine organisms, the remainder being clay-sized material.

orbit in water waves, the path followed by the water particles affected by the wave motion; also, the path of a body subjected to the gravitational force of another body, such as the Earth's orbit around the Sun.

osmosis tendency of water to diffuse through a semipermeable membrane to make the concentration of water on one side of the membrane equal to that on the other side.

outer core outer liquid portion of the Earth's core, located between the inner core and the mantle.

overturn sinking of more dense water and its replacement by less dense water from below.

oxygen minimum zone in which respiration and decay reduce dissolved oxygen to a minimum, usually between 800 and 1000 meters.

ozone a form of oxygen that absorbs ultraviolet radiation from the Sun.

p

paleomagnetism study of ancient magnetism as recorded in rocks; includes the study of changes in the location of the Earth's magnetic poles through time and reversals in the Earth's magnetic field.

Pangaea single supercontinent of 200 million years ago.

parallel circle on the surface of the Earth parallel to the plane of the equator and connecting all points of equal latitude; a line of latitude.

parasitism an intimate association between different organisms in which one is benefited and the other is harmed.

passive margin *See* trailing margin.

pelagic primary division of the sea, which includes the whole mass of water subdivided into neritic and oceanic zones; also pertaining to the open sea.

period *See* tidal period; wave period.

pH measure of the concentration of hydrogen ions in a solution; the concentration of hydrogen ions determines the acidity of the solution; $pH = -\log_{10}(H^+)$, where H^+ is the concentration of hydrogen ions in gram atoms per liter.

phosphorite form of phosphate precipitated from seawater, found in muds, sands, and nodules.

photic zone layer of a body of water that receives ample sunlight for photosynthesis; usually less than 100 meters.

photosynthesis manufacture by plants of organic substances and release of oxygen from carbon dioxide and water in the presence of sunlight and the green pigment chlorophyll.

physiographic portrayal of the Earth's features by perspective drawing.

phytoplankton microscopic plant forms of plankton.

pinniped member of the marine mammal group, characterized by four swimming flippers; for example, seals and sea lions.

plankton passively drifting or weakly swimming organisms.

plate tectonics theory and study of the Earth's lithospheric plates, their form-ation, movement, interaction, and destruction; the attempts to explain the Earth's crustal changes in terms of plate movements.

polar easterlies winds blowing from the poles toward approximately 60°N and S; winds are northeasterly in the Northern Hemisphere and southeasterly in the Southern Hemisphere.

polar reversal a change in the Earth's magnetic field between normal and reversed polarities.

polar wandering curve a curve mapping the apparent past movement of the Earth's magnetic poles.

polyculture cultivation of more than one species of organism in an aquaculture system.

polyp sessile stage in the life history of certain coelenterates (Cnidaria); sea anemones and corals.

precipitation falling products of condensation in the atmosphere, such as rain, snow, or hail; also the falling out of a substance from solution.

prevailing westerlies *See* westerlies.

primary coast a coast formed by processes that occur at the land-air boundary.

prime meridian meridian of 0° longitude, used as the origin for measurements of longitude; internationally accepted as the meridian of the Royal Naval Observatory, Greenwich, England.

production (primary) the amount of living matter synthesized by photosynthetic or chemosynthetic organisms, usually expressed in grams of carbon per m^3. In practice the terms "production" and "productivity" are often used interchangeably.

productivity (primary) the rate at which biomass is produced by photosynthesis and chemosynthesis, usually expressed as grams of carbon produced per day. In practice the terms "production" and "productivity" are often used interchangeably.

projection in mapmaking, the systematic construction of lines on a flat surface that corresponds to the latitude and longitude lines from the curved surface of the Earth or the celestial sphere.

prokaryote cells that lack structural complexity and defined nucleus of eukaryotes. Bacteria are prokaryotes.

pteropod pelagic snail whose foot is modified for swimming.

P-wave primary or compressional seismic wave with a higher velocity than other seismic waves; energy propagated by alternate compressions and dilations; passes through solids, liquids, and gases. *See also* S-wave.

r

radar system of determining and displaying the distance of an object by measuring the time interval between transmission of a radio signal and reception of the echo return; derived from the phrase "Radio detecting and ranging."

radiation energy transmitted as rays or waves without the need of a substance to conduct the energy.

radiative fog fog formed as a result of warm moist air cooling at night when winds are calm, often in bays, inlets, and land valleys.

radiolarians minute, single-celled animals with siliceous skeletons.

radiometric dating determining ages of geological samples by measuring the relative abundance of radioactive isotopes and comparing isotope systems.

rain shadow low-precipitation zone on the sheltered side of islands and mountains.

red clay red to brown fine-grained lithogenous deposit, of predominantly clay size, which is derived from land, transported by winds and currents, and deposited far from land and at great depth; also known as brown mud and brown clay.

red tide red coloration, usually of coastal waters, caused by large quantities of microscopic organisms (generally dinoflagellates); some red tides result in mass fish kills, others contaminate shellfish, and still others produce no toxic effects.

reef crest highest portion of a coral reef on the exposed seaward edge of the reef.

reef flat portion of a coral reef landward of the reef crest and seaward of the lagoon.

reflection rebounding of light, heat, sound, waves, and so on, after striking a surface.

refraction change in direction or bending of a wave due to change in speed.

reservoir natural or artificial collection or accumulation of water.

residence time mean time that a substance remains in a given area before replacement, calculated by dividing the amount of a substance by its rate of addition or subtraction.

respiration metabolic process by which food or food-storage molecules yield the energy on which all living cells depend.

reverse osmosis process of obtaining fresh water from seawater by applying pressure to the salt water and forcing the water molecules through a membrane that does not pass salt molecules.

ridge long, narrow elevation of the sea floor, with steep sides and irregular topography.

rift valley trough formed by faulting along a zone in which plates move apart and new crust is created, such as along the crest of a ridge system.

rift zone *See* spreading center.

rip current strong surface current flowing seaward from shore; the return movement of water piled up on the shore by incoming waves and wind.

rise long, broad elevation that rises gently and generally smoothly from the sea floor.

s

salinity measure of the quantity of dissolved salts in seawater. It is formally defined as the total amount of dissolved solids in seawater in parts per thousand (‰) by weight when all the carbonate has been converted to oxide, all the bromide and io-dide have been converted to chloride, and all organic matter is completely oxidized.

salinometer instrument for determining the salinity of water by measuring the electrical conductivity of a water sample of a known temperature.

salt wedge intrusion of salt water along the bottom; in an estuary, the wedge moves upstream on high tide and seaward on low tide.

satellite body that revolves around a planet; a moon; a device launched from Earth into orbit.

satellite navigation system series of signal-emitting, Earth-orbiting satellites that allow ships to determine their position relative to the satellites.

saturation value amount of material that can be held in solution without gain or loss.

scarp elongated and comparatively steep slope change separating flat or gently sloping areas on the sea floor or on a beach.

scattering random redirection of light or sound energy by reflection from an uneven sea bottom or sea surface, from water molecules, or from particles suspended in the water.

sea same as the ocean; subdivision of the ocean; surface waves generated or sustained by the wind within their fetch, as opposed to swell.

sea cow (dugong, manatee) large herbivorous marine mammal of tropic waters.

seafloor spreading movement of crustal plates away from the mid-ocean ridges; process that creates new crustal material at the mid-ocean ridges.

sea ice ice formed from freezing of seawater.

sea level height of the sea surface above or below some reference level. *See also* mean sea level.

seamount isolated volcanic peak that rises at least 1000 m from the sea floor.

sea ranching *See* ocean ranching.

sea smoke type of fog caused by dry, cold air moving over warm water.

sea stack isolated mass of rock rising from the sea near a headland from which it has been separated by erosion.

sea state numerical or written description of the roughness of the ocean surface relative to wave height and wind speed.

Secchi disk white disk used to measure the transparency of the water by observing the depth at which the disk disappears from view.

secondary coast a coast formed by ocean processes.

sediment particulate organic and inorganic matter that accumulates in loose, unconsolidated form.

seiche standing wave oscillations of an enclosed or semienclosed body of water that continue, pendulum fashion, after the generating force ceases.

seismic pertaining to or caused by earthquakes or Earth movements.

seismic sea wave *See* tsunami.

seismic wave shock wave or vibration produced by an earthquake or Earth movement.

semidiurnal tide tide with two high waters and two low waters each tidal day.

sessile permanently fixed or sedentary; not free-moving.

set direction in which the current flows.

shallow-water wave wave in water where depth is less than one-twentieth the average wavelength.

siliceous containing silica.

slack water state of a tidal current when its velocity is near zero; occurs when the tidal current changes direction.

sofar channel natural sound channel in the oceans, in which sound can be transmitted for very long distances; the depth of minimum sound velocity; derived from the phrase "sound fixing and ranging."

solstices times of the year when the Sun stands directly above $23^1/_2°$N or S latitude. The winter solstice occurs about December 22, and the summer solstice occurs about June 22.

sonar method or equipment for determining, by underwater sound, the presence, location, or nature of objects in the sea; derived from the phrase "sound navigation and ranging."

sorting, of waves *See* dispersion.

sound shadow zone area of the ocean into which sound does not penetrate because the density structure of the water refracts the sound waves.

South Atlantic surface water lens of surface water in the center of the South Atlantic current gyre.

spit (sand spit) low tongue of land, or a relatively long, narrow shoal extending from the shore.

spreading center region along which new crustal material is produced.

spring tides tides occurring near the times of the new and full Moon, when the range of the tide is greatest.

standing crop biomass present at any given time.

standing wave type of wave in which the surface of the water oscillates vertically between fixed points called nodes, without progression; the points of maximum vertical rise and fall are called antinodes.

steepness, wave *See* wave steepness.

stipe portion of an alga between the holdfast and the blade.

storm center area of origin for surface waves generated by the wind; an intense atmospheric low-pressure system.

storm tide (storm surge) along a coast, the exceptionally high water accompanying a storm, owing to wind stress and low atmospheric pressure, made even higher when associated with a high tide and shallow depths.

stratosphere the layer of the atmosphere above the troposphere where temperature is constant or increases with altitude.

subduction zone plane descending away from a trench and defined by its seismic activity, interpreted as the convergence zone between a sinking plate and an overriding plate.

sublimation direct change in state of water from solid (ice) to gaseous state (water vapor). To sublimate 1 gram of water requires 600 calories of heat.

sublittoral (subtidal) benthic zone from the low tide line to the seaward edge of the continental shelf.

submarine canyon relatively narrow, V-shaped, deep depression with steep slopes, the bottom of which grades continuously downward across the continental slope.

submergent coast a coast that is sinking relative to sea level.

submersible a research submarine, designed for manned or remote operation at great depths.

subsidence sinking of a broad area of the crust without appreciable deformation.

substrate material making up the base on which an organism lives or to which it is attached.

subtidal *See* sublittoral zone.

subtropical convergence downwelling zone produced by converging currents at approximately 30° to 40°N and S (centers of oceanic gyres).

summer solstice *See* solstices.

supersaturated solution holding dissolved material in excess of the saturation value.

supralittoral benthic zone above the high tide level that is moistened by waves, spray, and extremely high tides.

surface tension tendency of a liquid surface to contract owing to bonding forces between molecules.

surf zone area of wave activity between the shoreline and the outermost limit of the breakers.

surimi refined fish protein used to form artificial crab, shrimp, and scallop meat.

S-wave secondary or shear seismic wave, travels more slowly than P-wave; energy propagated as vibrations perpendicular to the direction of travel; passes only through solids—not liquids or gases. *See also* P-wave.

swell long and relatively uniform wind-generated ocean waves that have traveled out of their generating area.

symbiosis living together in intimate association of two dissimilar organisms.

tablemount *See* guyot.

tectonic pertaining to processes that cause large-scale deformation and movement of the Earth's crust.

temperate pertaining to intermediate latitudes, usually considered to be between $23^1/_2°$N and S and $66^1/_2°$N and S.

terranes fragments of the Earth's crust bounded by faults, each fragment with a history distinct from each other fragment.

terrigenous of the land; sediments composed predominantly of material derived from the land.

thermohaline circulation vertical circulation caused by changes in density; driven by variations in temperature and salinity.

tidal bore high tide crest that advances rapidly up an estuary or river as a breaking wave.

tidal current alternating horizontal movement of water associated with the rise and fall of the tide.

tidal day time interval between two successive passes of the Moon over a meridian, approximately 24 hours and 50 minutes.

tidal period elapsed time between successive high waters or successive low waters.

tidal range difference in height between consecutive high and low waters.

tombolo deposit of unconsolidated material that connects an island to another island or to the mainland.

trace elements elements dissolved in seawater in concentrations of less than one part per million.

trade winds wind systems occupying most of the tropics, which blow from approximately 30°N and S toward the equator; winds are northeasterly in the Northern Hemisphere and southeasterly in the Southern Hemisphere.

trailing margin on a lithospheric plate moving away from a spreading center, the margin of a continent closest to the spreading center.

transform fault fault with horizontal displacement connecting the ends of an offset in a mid-ocean ridge. Some plates slide past each other along a transform fault.

transpiration loss of water from a plant to the outside atmosphere; takes place mainly through "pores" of leaves and stems. Actively growing plants transpire 5–10 times their water content daily.

trench long, deep, narrow depression of the sea floor with relatively steep sides, associated with a subduction zone.

triploid condition in which cells have three sets of chromosomes.

trophic level relating to nutrition; a trophic level is the position of an organism in a food chain or food (trophic) pyramid.

Tropics of Cancer and Capricorn latitudes $23^1/_2°$N and $23^1/_2°$S, respectively, marking the maximum angular distance of the Sun from the equator during the summer and winter solstices.

troposphere the lowest layer of the atmosphere, where the temperature decreases with altitude.

trough long depression of the sea floor, having relatively gentle sides; normally wider and shallower than a trench. *See also* wave trough.

tsunami (seismic sea wave) long-period sea wave produced by a submarine earthquake or volcanic eruption. It may travel across the ocean unnoticed from its point of origin and build up to great heights over shallow water at the shore.

tube worm any worm or wormlike organism that builds a tube or sheath attached to a submerged substrate.

turbidite sediment deposited by a turbidity current, showing a pattern of coarse particles at the bottom, grading gradually upward to fine silt.

turbidity loss of water clarity or transparency owing to the presence of suspended material.

turbidity current dense, sediment-laden current flowing downward along an underwater slope.

typhoon severe, cyclonic, tropic storm originating in the western Pacific Ocean, particularly in the vicinity of the South China Sea. *See also* hurricane.

U

Universal Time clock time set to 12 noon when the Sun is directly above the prime meridian.

upwelling rising of water rich in nutrients toward the surface, usually the result of diverging surface currents.

V

vernal equinox *See* equinoxes.

vertebrate animal with a skull surrounding a well-developed brain and a skeleton of cartilage or bone including a vertebral column or backbone extending through the main body axis: fish, amphibians, reptiles, birds, and mammals.

W

water budget balance between the rates of water added and lost in an area.

wave periodic disturbance that moves through or over the surface of a medium with a speed determined by the properties of the medium.

wave crest highest part of a wave.

wave height vertical distance between a wave crest and the adjacent trough.

wavelength horizontal distance between two successive wave crests or two successive wave troughs.

wave period time required for two successive wave crests or troughs to pass a fixed point.

wave steepness ratio of wave height to wavelength.

wave train series of similar waves from the same direction.

wave trough lowest part of a wave.

westerlies (prevailing westerlies) wind systems blowing from the west between latitudes of approximately 30° and 60°N and 30° and 60°S; they are southwesterly in the Northern Hemisphere and northwesterly in the Southern Hemisphere.

western intensification tendency of currents flowing from low to high latitudes on the western sides of the Pacific and Atlantic Oceans to be strong, swift, and narrow.

wetlands land on which water significantly dominates development of soil and types of organisms present.

wind wave wave created by the action of the wind on the sea surface.

winter solstice *See* solstices.

Z

zonation parallel bands of distinctive plant and animal associations found within the littoral zones and distributed to take advantage of optimal conditions for survival.

zooplankton animal forms of plankton.

zooxanthellae symbiotic microscopic organisms (dinoflagellates) found in corals and other marine organisms.

credits

Chapter 1—**Opener:** © Sean Ellis/Stone; **1.1:** © Stephen D. Thomas; **1.2:** Courtesy Tom Rice, Sea and Shore Museum, Port Gamble, WA; **1.3:** from Ptolemy's *Geographia*, 15th century Italy; **1.5:** from Johannes van Keulen's *Sea-Atlas;* **1.8:** National Maritime Museum, London; **1.10:** Franklin-Folger, 1769; **1.11** This data image was obtained from the National Aeronautics & Space Administration (NASA) Physical Oceanography Distributed Active Archive Center (PO.DAAC) at the Jet Propulsion Laboratory, California Institute of Technology. The data were retrieved from the Advanced Very High Resolution Radiometer (AVHRR) on board the National Oceanic and Atmospheric Administration (NOAA) Polar Orbiting series of satellites. Processing was done at the Rosenstiel School of Marine and Atmospheric Sciences (RSMAS), University of Miami; **1.12:** from Matthew F. Maury's *Physical Geography of the Sea*, 1855; **1.13:** Vol. 25, *Fauna and Flora dus Golfes von Neaple*, 1899; **1.14a-d:** Challenger Expedition: Dec 21, 1872-May 24, 1876. Engravings from *Challenger* Reports, vol. 1, 1885; **1.15a & b, 1.16a & b:** Courtesy The Norwegian Maritime Museum; **1.17a:** © Scripps Institution of Oceanography; **1.17b:** © D. Weisman/Woods Hole Oceanographic Institution; **1.18a:** Courtesy Joe Creager, University of Washington; **1.18b:** Courtesy Ocean Drilling Program, Texas A&M University; **1.19:** CZCS image provided courtesy of the CZCS Project, Goddard Space Flight Center, and the Goddard Distributed Active Archive Center; **1.20b:** Courtesy of U.S. Argo; **Figure 1:** Mary Rose Trust; **Figure 2:** Hans Hammerskiold for Vasa Museum, Stockholm, Sweden; **Figure 3a & b:** © 1989 Quest Group, Ltd..

Chapter 2—**Opener:** © Keith A. Sverdrup; **2.1:** NASA; **2.5:** © Alyn & Alison Duxbury; **2.11:** Courtesy Stephen P. Miller, UC-Santa Barbara; **2.18:** Department of Defense.

Chapter 3—**Figure 3.4:** Paleogeographic Maps by Christopher R. Scotese, PALEOMAP Project, University of Texas at Arlington (www.scotese.com) **3.5:** B. Hazeen & M. Tharp: *World Ocean Floor,* © Marie Tharp, 1977, South Nyack, NY 10960. Reproduced by permission; **3.7:** U.S. Geological Survey/National Earthquake Information Center **3.8:** © Dr. Robert Keickhefer; **3.21:** OAR/National Underseas Research Program/NOAA; **3.25:** P. W. Lipman/U.S. Geological Survey; **3.29d:** Paleogeographic Maps by Christopher R. Scotese, PALEOMAP Project, University of Texas at Arlington (www.scotese.com) **3.30:** Courtesy Paul Johnson/School of Oceanography, University of Washington; **3.31:** Dr. Verena Tunnicliffe/university of Victoria, British Columbia; **3.32:** Courtesy John Delaney/School of Oceanography, University of Washington; **Figure 2, 3 & 5:** Courtesy Deborah Kelley, University of Washington.

Chapter 4—**Opener:** NASA/The Image Works; **4.5:** © Alyn & Alison Duxbury; **4.12:** Courtesy Walter H. F. Smith/ NOAA; **4.14a & b:** Courtesy Williamson & Associates, Seattle, WA; **4.16a & b:** Courtesy Ken Adkins/School of Fisheries, University of Washington; **4.16c & d:** Courtesy of Steve Nathan & R. Mark Leckie, University of Massachusetts; **4.17:** Courtesy Joe Creager/University of Washington; **4.18:** Courtesy Dean McManus/School of Oceanography, University of Washington; **4.19:** Courtesy Kathy Newell/School of Oceanography, University of Washington; **4.20b:** Courtesy Sara Barnes/School of Oceanography, University of Washington; **4.20c:** © Alyn & Alison Duxbury; **4.20d:** Courtesy Dr. Richard Sternberg/School of Oceanography, University of Washington; **4.21:** Courtesy Chevron Corporation; **Box 4.1 & 4.3:** © Peter J. Auster; **Box 4.5:** Courtesy James Morrison, Polar Science Center, HM-10, Applied Physics laboratory, Seattle, WA.

Chapter 5—**Opener:** © Ed Simpson/Stone; **5.4:** Courtesy Sandra Hines/News & Information, University of Washington; **5.8:** © Alyn & Alison Duxbury; **5.10:** Courtesy EPC Laboratories, Inc., Danvers, MA; **5.19:** John Conomas/U. S. Geological Survey.

Chapter 6—**Opener:** © Jack Swenson/Tom Stack & Assocs; **6.6a & b:** M. Chahine/JPL & J. Sussking/Goddard Space Flight Center; **6.7a & b:** J. Zwally/Goddard Space Flight Center; **6.7c & d:** D. J. Cavaneri & P. Gloersen/Goddard Space Flight Center; **6.8a:** Edward Joshberger/U. S. Geological Survey; **6.8b:** Courtesy Kathy Newell/School of Oceanography, University of Washington; 6.20: JPL, University of CA, & Los Angeles Atmospheric Environment Service; **6.24a:** © William Rosenthal/Jeroboam, Inc.; **6.24b** © Bill Silliker,Jr.; **6.25a–c:** Images processed by Dr. Xiao-Hai Yan/College of Marine Studies, University of Delaware; **6.27:** National Hurricane Service, NOAA; **6.28:** National Weather Service; **Figure 2-4:** © Alyn & Alison Duxbury; **Feature Figure 1 & 2:** © Keith A. Sverdrup.

Chapter 7—**Opener:** © NASA/CORBIS; **7.7:** © Alyn & Alison Duxbury; **7.8:** Courtesy James R. Postel/School of Oceanography, University of Washington; **7.9:** Courtesy of Chelsea Instruments, Ltd; **7.10:** © News Office/ Woods Hole Oceanographic Institution; **7.11** Courtesy Dana Swift/School of Oceanography, University of Washington; **7.21:** Courtesy Otis Brown; **7.22:** Courtesy Will Patterson/School of Oceanography, University of Washington; **7.23a:** Courtesy Kathy Newell/School of Oceanography, University of Washington; **7.23b** Courtesy of Aanderaa Instruments; **Figure 1:** Bauer, Moynihan & Johnson, Seattle, WA; **Figure 2:** Courtesy Rowland Studio, Seattle, WA.

Chapter 8—**Opener:** © Darrell Gulin/CORBIS; **8.1:** © Eda Rogers; **8.5:** US Army Corps of Engineers; **8.9:** Courtesy Exxon Corporation, photo by Leonard Wolhuter; **8.17a & b:** © Alyn & Alison Duxbury; **8.18:** © SuperStock; **8.19a, b & c:** National Geophysical Data Center, NOAA; **8.20, 8.21:** Courtesy Dr. Yoshinobu Tsuji; **8.31a & b, 8.32:** Tourism New Brunswick; **8.34:** Courtesy Nova Scotia Power Corporation; **Box 8.2:** Courtesy Steve Niland/Datasonics, Cataumet, MA.

Chapter 9—**Opener:** © Vince Streano/Stone; **9.1a:** © Cheasapeak Bay Foundation, Inc.; **9.1b:** Courtesy Joe Creager/School of Oceanography, University of Washington; **9.1c:** NASA; **9.1d:** © Alyn & Alison Duxbury; **9.1e:** Courtesy Dr. Sherwood Maynard/University of Hawaii; **9.1f:** © Terra-Mar Resource Information Services, Inc.; **9.2a:** Courtesy Dr. Richard Sternberg/School of Oceanography, University of Washington; **9.2b:** © Alyn & Alison Duxbury; **9.2c:** © Alex S. MacLean/Landslides; **9.2d, e, f & g, 9.4, 9.5:** © Alyn & Alison Duxbury; **9.6:** © Carr Clifton; **9.7:** © Alyn & Alison Duxbury; **9.8:** Courtesy Dr. Richard Sternberg/School of Oceanography, University of Washington; **9.9:** U.S. Army Corps of Engineers; **9.10, 9.11, 9.19:** © Alyn & Alison Duxbury; **9.20:** © Bob Evans/La Mer Bleu Productions; **9.21:** Courtesy Port of Seattle; **9.22:** © Alex S. MacLean/Landslides; **9.23a, b & c:** Courtesy Joe Lucas/Marine Entanglement Research Program/National Marine Fisheries Service, NOAA; **9.23d:** Courtesy R. Herron/Marine Entanglement Research Program/National Marine Fisheries Service, NOAA; **9.25a–c:** Courtesy Hazardous Materials Response Branch/Jeremy Gault/NOAA; **9.25d:** NOAA; **9.25e:** Courtesy Hazardous Materials Response Branch/Jeremy Gault/NOAA; **9.25f:** © Alyn & Alison Duxbury; **Box 9.2:** © Francois Gohier/Photo Researchers, Inc.; **Box 9.3:** © Larry Lipsky/Tom Stack & Associates.

Chapter 10—**Opener:** © Brandon D. Cole/CORBIS; **10.5:** Courtesy Field Museum of Natural History, Chicago, neg. #Z82860; **10.12a–d:** Courtesy David English/School of Oceanography, University of Washington. Images from NASA Global Ocean Color data set, Gene Feldman/Goddard Space Flight Center, NASA; CZCS data processed using University of Miami DSP system developed by Otis Brown and Robert Evans with maintenance support from NASA; **10.18:** © Richard A. Lutz **10.19:** Courtesy Dr. James M. Brooks/Geochemical and Environmental Research Group, College Station, TX; **10.20a & b:** Courtesy Dr. John Baross/School of Oceanography, University of Washington; **Figure 1:** Dr. Fred Nicholas/U.S. Geological Survey, Menlo Park, CA; **Figure 2:** © Caroline Kopp; **Figure 3:** © Alyn & Alison Duxbury.

Chapter 11—**Opener:** © Fred McConnaughey/Photo Resaerchers, Inc.; **11.4:** Tiffany Vance/NOAA; **11.6:** © Alyn & Alison Duxbury; **11.8:** Courtesy Kathy Newell/School of Oceanography, University of Washington; **11.15:** Courtesy Mark Holmes, USGS/School of Oceanography, University of Washington; **11.21a:** Courtesy Dr. Albert Erickson/Fisheries Research Institute, University of washington; **11.21b:** Courtesy David Withrow/National Marine Fisheries Service, NOAA; **11.26, 11.27:** © Alyn & Alison Duxbury; **11.28:** H. Wes Pratt/Apex Predator Investigation, NOAA, NMFS Narragansett Laboratory, RI; **11.30:** Bruce W. Buls, Seattle; **11.31:** Courtesy Sea Farm Washington, Inc.; **Box 1:** © Dr. Craig R. Smith/Dept. of Oceanography, University of Hawaii; **Box 2:** © H. Kukert/Courtesy Craig R. Smith.

Chapter 12—**Opener:** © Brandon D. Cole/CORBIS; **12.10a, b & c:** Courtesy Verena Tunnicliffe/University of Victoria, B.C.; **12.11a & b:** © Alyn & Alison Duxbury; **12.13:** © David Hall/Photo Researchers, Inc.; **12.15:** Courtesy Brad Matsen/*National Fisherman* Magazine, Seattle, WA; **12.16:** Courtesy Ken Chew/School of Fisheries, University of Washington; **Box 12.1a & b:** Courtesy Kathy Newell, School of Oceanography, University of Washington.

Credits **337**